Under the Radar

W.M. Goss · Richard X. McGee

Under the Radar

The First Woman in Radio Astronomy: Ruby Payne-Scott

Springer

Prof. W. M. Goss
National Radio Astronomy
Observatory
1003 Lopezville Road P.O.Box O
Socorro NM 87801
USA
mgoss@aoc.nrao.edu

Dr. Richard X. McGee
Commonwealth Scientific &
Industrial
Research Organisation
(CSIRO)
Radiophysics Laboratory
Epping NSW 1710
Australia

ISSN 0067-0057
ISBN 978-3-642-26073-5 e-ISBN 978-3-642-03141-0
DOI 10.1007/978-3-642-03141-0
Springer Heidelberg Dordrecht London New York

Cover illustration: Starry Night, after Van Gogh′ by Fiona Hall the daughter of Ruby Payne-Scott and Bill Hall. This 'Reconstructed painting' from 1981 shows electrical power cords as they swirl in the night sky. From page 43, *Fiona Hall* by Julie Ewington (2005). 'The Reconstructed paintings are also notable for the obvious pleasure Hall took in multiple slippages of meaning between historical templates and contemporary life.' Used by permission of Fiona Hall.

Cover design: eStudio Calamar, Spain

Printed on acid-free paper

Springer is part of Springer Science+Business Media (www.springer.com)

To
Libby, Lyn

Foreword

It is rare for a complete biography of an Australian scientist, particularly of an Australian woman scientist, to be published. It is rarer for such a book to be co-authored by an American.

Although scientists have written discourses on the history of their discipline, it is most unusual for a scientist to write a full length biography of a colleague in his field. It is also uncommon for a man to write about an Australian woman scientist; most of the work on Australian women scientists has been done by other women. However, these authors, both distinguished researchers in the field of radio astronomy, became so interested in the history of their discipline and in the career of the pioneer radio astronomer Ruby Payne-Scott that they spent some years bringing this book to fruition.

Until relatively recently, Ruby Payne-Scott had been the only woman scientist mentioned briefly in histories of Australian science or of Australian radio astronomy. This book will be an invaluable resource for anyone interested in these disciplines. Being scientists themselves, the authors explain Payne-Scott's scientific work in detail; therefore, the value and importance of her contributions can, for the first time, be recognised, not only by historians but also by scientists.

After a brilliant academic career, with an M.Sc. in physics (the highest qualification then available at any Australian university), Ruby Payne-Scott worked as a science teacher, one of the few professional positions available to Australian women in the 1930s, and especially in depression. However, the Second World War opened up opportunities for women science graduates. She was one of the first of the scientific staff members appointed to the new Radiophysics Laboratory of the Council for Scientific and Industrial Research (CSIR) from which radio astronomy developed, and notably the first woman scientist in the Laboratory.

She was part of a pioneering group of radar scientists during the Second World War, led by J.L. Pawsey, whose scientific distinction and leadership qualities have been referred to by all writers in the field. Although it was mainly due to him that radio astronomy developed in Australia from 1944, she was one of the key people contributing to Australia's pre-eminence in the world in radio astronomy for many decades. Pawsey valued her judgment and experience so highly that when she was absent from a meeting, he would often not make a final decision until she had been consulted. She became the overall advisor to the group on scientific issues, engineering planning, and mathematics, and she also made major contributions to the development of radio astronomical techniques.

By 1951, when she left the discipline, she had been promoted to the highest research category short of the leader and was being paid the second highest salary on the scientific staff. Her standing was confirmed later by a member of this distinguished group who himself became an important radio astronomer but who disliked Payne-Scott; nevertheless, he considered her, as the authors record, to have been "one of the best physicists at Radiophysics – no, one of the best physicists in Australia".

In telling Payne-Scott's story, the authors highlight the inferior position of women in the work force at that time. Married women could not become permanent employees in the public service. The practice of requiring women to choose between marriage and their careers inevitably deprived Australia of unknown talent.

Payne-Scott had to suffer the indignity of keeping her marriage secret from CSIR for some five years. When the marriage was discovered, she fought vehemently against the injustice of this regulation but was forced to become a temporary employee, losing all her superannuation entitlements in the process. She finally resigned in 1951 when pregnant with her first child, as there was no maternity leave at that time.

The War years provided some measure of equal pay for women. After the War, the old discriminatory practices returned. Payne-Scott, together with other colleagues, campaigned unsuccessfully for the recognition of the principle of equal pay.

"Women's rights" was not the only issue about which she felt strongly and for which she argued publicly and vigorously. During the War, it was natural that the type of work which the group was engaged in was classified; but after the War, she was bitterly opposed to secret research in the CSIR. She believed that it was impossible to do good research in the atmosphere of limitations imposed by a sponsoring body, particularly when that body was the military. She wrote to CSIR: "Frightened men do not produce great research".

The Australian Security Intelligence Organisation kept files on her that have only recently been made available to researchers. A subsequent media release by the National Archives of Australia, headed "The Secret Life of Ruby Payne-Scott", states that she was "passionate about both the independence of scientific research and human rights. These sentiments were deemed a security risk."

The work of pioneering Australian women scientists is gradually being recognised. CSIRO offers OCE Science Team Career Awards. One of these is the OCE Payne-Scott Career Award for researchers returning from family-related career breaks. The life and work of a feisty, brilliant woman is finally being recognised.

Nessy Allen

Preface

Ruby Payne-Scott (1912–1981): Remarkable Scientist and Champion for Women's Rights

Almost 60 years after her retirement in 1951, why is the life of Ruby Payne-Scott of significance to us? She was a unique scientist working in one of the first major solar radio astronomy groups after the end of World War II. This fortunate circumstance was due to the experience she gained working on radar at the major Australian laboratory during World War II. Payne-Scott was a pioneer Australian scientist leading the charge for equality of women in the work place.

In 1997, Dick McGee and I began a discussion about the possibility of writing a short biography of Ruby Payne-Scott. Dick had known her for a brief period in the early 1950s when he joined the radio astronomy group in Sydney, Australia. We first envisaged an article for the *Publications of the Astronomical Society of Australia*, similar in scope to our previous collaboration published in a conference proceedings in 1996, *The Discovery of the Radio Source Sagittarius A (Sgr A)*. As we collected material and carried out initial interviews in Australia, the complex nature of this study of Payne-Scott's life became evident. A year after a trip of several weeks' duration to Australia that I made in early 1999,[1] McGee began to write the first draft. He reported to me at that time that he had been increasingly impressed with the scope of Payne-Scott's contributions to solar radio astronomy. Gradually over the next few years, we both became convinced that a longer version of the Payne-Scott story was required.

New themes developed as Dick and I learned more about her remarkable life: her battles over discrimination against women, her success as a scientist and educator, her membership in the Communist Party of Australia (CPA), her remarkable family and her passion for bush-walking. In particular, her pioneering work in the new field

[1] I gave the SAFA (Sydney Association for Astrophysics) lecture on 9 February 1999, on the topic of Ruby Payne-Scott; a number of her former colleagues were in the audience, who provided helpful comments after the lecture.

of solar radio astronomy (including a major role in the invention of radio aperture synthesis) had not been fully appreciated. Also we met many of her friends and colleagues, who opened up new facets of her life. Payne-Scott's children, Peter and Fiona Hall, were especially helpful in revealing the character of their mother in numerous interviews and visits in both the US and Australia.

We also discovered that Payne-Scott had largely been neglected in treatments of the early years of radio astronomy. Numerous accounts of the history of post World War II science in Australia did provide a cursory glance at her career, often with some distortion. An example is the mistaken assertion by Collis in his history of CSIRO (*Fields of Discovery,* 2002) that: "... in 1951 [Payne-Scott] was forced by public service rules to leave her job when it was discovered she had been secretly married since 1944. Married women were not allowed to hold permanent staff position." Dick and I correct this common misconception in Chap. 4.

An example of neglect can be found in the comprehensive and influential, popular book *The Changing Universe* published in 1956 by the *Scientific American* author John Pfeiffer. He did an admirable job of visiting radio astronomers in the US, the UK, France, and the Netherlands. He corresponded with a number of prominent radio astronomers in Australia. Many of the pioneering results from the Australian group were summarised in the chapter "The Sun in Action", including a description of the remarkable Type II outburst of 8 March 1947, with a whimsical cartoon of the effects of solar outbursts on terrestrial communication. This result was published by Payne-Scott, Yabsley, and Bolton (see Chap. 7). No mention of Payne-Scott appears in the Pfeiffer volume, even though most of her Australian colleagues are named.

Finally, there is an example of neglect in the modern era, the plaque shown in Fig. P.1; this comprehensive display at Dover Heights in Sydney (Rodney Reserve) is the site of the major astronomical discoveries made in the era 1946–1948 by the Radiophysics Laboratory (RPL) group. This plaque and a replica of the 100 MHz Yagii antenna were unveiled on 20 July 2003 by Her Excellency, the Governor of New South Wales, Professor Marie Bashir.[2] Although there is a brief description of the solar radio work at this site on the plaque, the text describes the work done by the men at Dover Heights: Bolton, Pawsey,[3] Stanley, Slee (at the ceremony), and McGee, with an emphasis on the radio sources discovered at this site (Taurus A, Virgo A, and Centaurus A). The absence of any mention of Payne-Scott on the plaque is surprising. A number of visitors to the Dover Heights monument since

---

[2]I was the master of ceremonies, and talks were given by Paul Pierce, Mayor of Waverley Council, Woody Sullivan of the University of Washington in Seattle, and Ron Ekers of the CSIRO Australia National Telescope Facility.

[3]Pawsey was appointed to the CSIR RPL in London in October 1939 and started work in Sydney on 2 February 1940. During the War, he was a major leader in the development of radar in Australia. After the War, he was the Deputy Chief of RPL after E.G. ("Taffy") Bowen became Chief in May 1946. When Pawsey began radio astronomy in 1944 at RPL, Payne-Scott became the leader of the first scientific efforts, while Lindsay McCready was the leader of the engineering efforts. Pawsey was one of the inventors of the new phrase "radio astronomy" in early 1948.

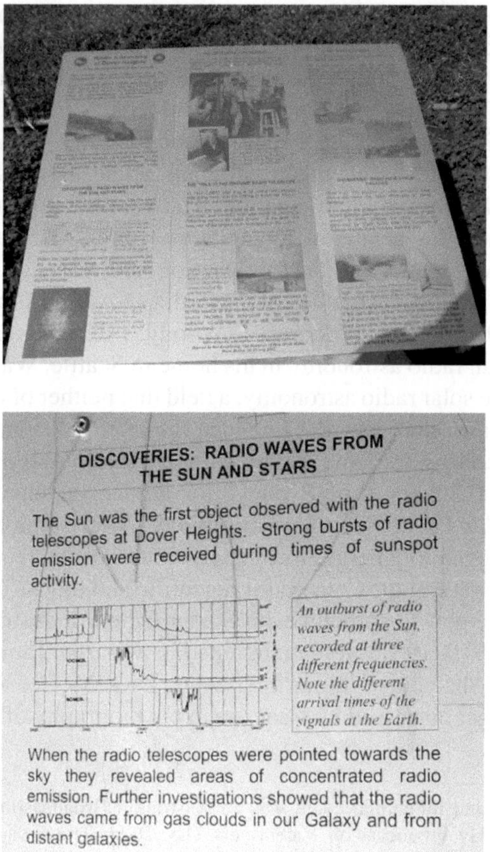

Fig. P.1 Radio astronomy display at Dover Heights (Rodney Reserve), Sydney, New South Wales, Australia. The plaque was constructed by the Australia Telescope National Facility, CSIRO to commemorate the pioneering radio astronomy done at this site from 1946 to 1954. The opening of the scientific memorial was by Her Excellency, the Governor of New South Wales, Prof. Marie Bashir, 20 July 2003. The 25th General Assembly of the International Astronomical Union was being held in Sydney during this period (13–26 July, 2003). Photo by Goss. (**a**) shows the entire plaque; the solar radio astronomy is described in a single panel shown in (**b**). Photo by Goss

2003 have pointed this absence out to me. The printed program for the unveiling ceremony did include a mention of Payne-Scott, although not a photograph.[4]

Only two recent publications deal with Payne-Scott in detail: Claire Hooker in her 2004 book, *Irresistible Forces: Australian Women in Science*, has a thorough

[4]A photograph of the two element 100 MHz Yagi at Dover Heights was included in the published program for the event of Sunday 20 July 2003: "...this was first used by Ruby Payne-Scott, Don Yabsley and John Bolton to study solar radio emission." There is no mention that Payne-Scott was a major player in the first ever radio astronomy interferometer observation as the sun rose on 26 January 1946. The location of this ground breaking event is not certain but was at either Dover Heights or Collaroy.

treatment in Chap. 10, "The Sun, Ruby Payne-Scott and the Birth of Radio Astronomy." Woody Sullivan, in his monumental study, *Cosmic Noise: A History of Early Radio Astronomy* (2009), has described her work in detail in a sub-section of his Chap. 13, "The Radio Sun: Payne-Scott's work."

During the initial years of the twenty-first century, a number of newspaper articles and a radio and a television program about Payne-Scott appeared in Australia with Dick's and my participation. The positive response encouraged us to continue our quest for details of Payne-Scott's life.[5] Dick and I followed the trail using interviews, letters and the Division of Radiophysics Archives in Marsfield (Sydney), the National Archives of Australia (NAA) in Canberra and Sydney (Chester Hill), and Professor Woodruff ("Woody") T. Sullivan III's remarkable archive of Australian radio astronomy in his home in Seattle, Washington. We were forced to learn some solar radio astronomy, a field that neither of us had worked in as professional radio astronomers (McGee since the late 1940s and I since the early 1960s). I relied on colleagues Tim Bastian (National Radio Astronomy Observatory) and Don Melrose (University of Sydney) for advice on many aspects of solar physics, in particular Payne-Scott's contributions. After some time, the nature of her scientific life began to emerge; fortunately, much of the records from the time that Payne-Scott worked at the Council for Scientific and Industrial Research (CSIR, 1941–1949) and the new Commonwealth Scientific and Industrial Research Organisation (CSIRO, 1949–1951) were maintained in a *written* form after this interval of 50–60 years.[6] In the course of our interviews, Dick McGee and I re-discovered the well-known effect discussed by Isaacson in his biography of Einstein published

---

[5]The most successful and influential event was an Australian Broadcasting Corporation (ABC) Radio National Saturday broadcast on Valentine's Day 2004. The broadcast was in the long running series, "The Science Show", by Robyn Williams, directed in a thorough fashion by Pauline Newman Davies. Elizabeth (Betty) Hall, McGee, Claire Hooker, Fiona Hall, Carolyn Little and I were interviewed. The world wide web distribution of the transcript has led to numerous helpful comments to the authors. By contrast, the television program in the *Rewind* series on 7 February 2005 by the ABC was a disappointment. In spite of a heroic effort by the director, Laurie Critchley, the original concept for the "History Detectives" series fell victim to internal infighting within the ABC. The final version is a watered-down presentation that does not capture the essence of Critchley's original production. In particular the fascinating interview with Bruce Slee at Dover Heights (Sydney) was cut as well as a humorous interview with McGee at his home in Eastwood, Sydney. I have film copies of these un-broadcast interviews on DVD.

[6]The NAA has an extensive collection of RPL material in the series C3830 (727 entries, Sydney). The series C4659 (12 entries) contains correspondence of J.L. Pawsey. An especially important series is NAA: C3830, F1/4/PAW/1 Part 1 (US) and Part 2 (Europe), containing correspondence during the trip that Pawsey took to the US, Canada, and Europe in 1947–1948. From the NAA in Canberra, Payne-Scott's personnel record was obtained (NAA: A8520, PH/PAY/002) as well as a redacted Australian Security Intelligence Organisation (ASIO) file (NAA: A6119). In addition, the service record in World War II of her brother, Henry Payne-Scott (Service Number 20769, NAA: A9301), was obtained from the NAA in Canberra. Woody Sullivan provided a number of copies of RPL files that he found during his visits to Australia. A few key items were found in his files that were not located in Australia in the NAA. Eventually, we will deposit the Payne-Scott archive that we have collected with the National Radio Astronomy Observatory Archives in Charlottesville, Virginia, under the direction of Ellen Bouton.

in 2007: "... remind us of the caution needed when writing history based on dimming recollections." Dick McGee and I discovered numerous inconsistent statements about Payne-Scott; wherever possible, we rely on contemporary records in place of recalled memories of her colleagues.

Of all researchers working on the history of Australian radio astronomy, we owe a huge debt to Sally (Sallie) Atkinson BEM,[7] who had been at RPL since the early days of World War II (she and Payne-Scott appear on the staff list at RPL in June 1942 for the first time). Atkinson had been Secretary to E.G. Bowen from 1946 to 1971, the year of the retirement of Bowen. She had been Honorary Archivist of RPL from 1971 until about 1992. McGee and I visited and corresponded with Sally several times in the 1990s (up to 1999). She had carried out the monumental task of organizing the RPL archives before they were transferred to the NAA (files from 27 November 1940 to April 1988). During W.T. Sullivan III's two visits to RP in March 1978 and April 1981, Atkinson provided hundreds of files for his inspection. In fact, many of the documents in the NAA have in Atkinson's handwriting "Hold for WTS" or "W.T. Sullivan".[8]

Ideally, Dick or I would have interviewed Payne-Scott in the 1960s–1970s. Since this did not occur, we only have Sullivan's attempted telephone interview with her on 3 March 1978, 3 years before her death. Sullivan was in Australia, beginning the interviews for his book on the history of radio astronomy. On this date, she stated to Sullivan that her memory was failing and that she was not very coherent; after 30 min of conversation, Sullivan reluctantly agreed. She told Sullivan that she had been trained as a physicist and had worked on engineering projects and in addition carried out scientific research. She credited a large component of her success to the excellent workshop at RPL and to Alec Little for his major contributions to their joint project at Potts Hill (swept-lobe interferometer, Chap. 9). She especially praised Joe Pawsey: "he seemed to have a vision of what to do next. He was terribly enthusiastic – you were lucky to get home for dinner when he showed up at the end of the day." Payne-Scott did describe the marriage crisis of 1950 (both professional and non-professional women had to resign their permanent employment status on marriage, Chap. 4); she stated that there was considerable surprise among some colleagues and a few showed indignation because they thought that she had been disloyal and dishonest. On a subsequent visit to Australia 3 years later, Sullivan rang the Hall home on 30 March 1981. A man, likely Bill Hall, answered. Sullivan was told that Payne-Scott was in a nursing home and had no memory of the past. She died 2 months later.

Since 1999, I have given lectures about the life of Payne-Scott at the Raman Research Institute, Bangalore, India; at the National Centre for Radio Astrophysics (Tata Institute) in Pune, India; at Agnes Scott College, Atlanta, Georgia; at the Adler Planetarium, Chicago, at the University of Sydney (twice), at the National

---

[7]British Empire Medal (Civil) awarded 1978 for "Recognition of service to the public service with CSIRO".

[8]Woodruff T. Sullivan, III. In 2008, Atkinson celebrated her 95th birthday.

Radio Astronomy Observatory, Socorro, New Mexico and Charlottesville, Virginia, at the University of Adelaide, at the University of Wisconsin, Madison, and at James Cook University, Townsville, Australia. My research and thoughts about this topic have benefited greatly from the interactions with the audience at each lecture.

Throughout this book, we use much of the radio astronomy terminology of the late 1940s and early 1950s. However, we have decided to use the modern MHz (mega Hertz, one million) instead of the previous terms used in 1950, Mc/s (Mega Cycles per second). In addition, in most cases, we use the modern terms for the types of solar radio bursts and outbursts. As Wild (1985) has written: "Before the 'origin of the species' could be identified there had to be an exercise in taxonomy." This exercise was quite important in clarifying the perception of these new phenomena. We introduce the newer terms for solar bursts in the relevant Chaps. 7–9. Also we use the unit of radio astronomy intensity-Jansky: or 10^{-26}W $M^{-2}Hz^{-1}$. We do not use the often-quoted solar flux unit, favoured by solar radio astronomers – s.f.u. or 10^4 Jansky (Jy). For the CSIRO and the CSIR Division of Radiophysics, we use the term RPL (Radiophysics Laboratory), which was common during World War II and the immediate post War period; RP became the common designation later in the 1950s.

Dick McGee and I are aware of the advantages and disadvantages of *scientists*' attempting to write history, in contrast to historians of science. The geologist William Glen[9] (1988) has discussed this issue in detail:

> [Scientists] suggest that history [of science] written by historians which is based on critical examination of the scientific ideas themselves, will be wanting. The historians, on the other hand, through the methods and techniques of history, social studies and philosophy – tools not often possessed by scientists – can show scientists how their discipline has come to be, how it is bedded in society, and how it derives its esteem and support for being. Practicing scientists can seldom understand their intellectual past and predecessors as historians do. Historians only rarely come to know all that scientists understand about the ideas, methods, instruments, and practice of science. Scientists and historians have much to offer each other; fully furnished history requires contributions from both.

Dick and I hope that we have brought the understanding of scientists to the study of this remarkable physicist, Ruby Payne-Scott.

Now, after more than half a century has passed, we can attempt to re-interpret the fascinating and complex life of Ruby Payne-Scott from a new point of view. As J.D. North has suggested (*The Measure of the Universe: A History of Modern Cosmology*, 1965):

> The past, as elicited by the historian, is not something which was simply "there" and now awaits description. It is a product of the minds of both author and reader, and hence of the circumstances under which it was written and read. Different individuals may, of course, ask the same questions and answer them differently.

[9] I am grateful to Ron Bracewell (1921–2007) for introducing me to Glen in Palo Alto, California, in January 2007.

Dick McGee and I hope that the readers will ask additional questions about the life of Ruby Payne-Scott.

A word to our readers: non-scientists may possibly want to skip Chaps. 5–9; these deal with the details of Payne-Scott's scientific achievements concerning War time radar and the initial years of solar radio astronomy in Australia in the era 1944–1951. A short summary of Payne-Scott's astronomical work is presented in Chap. 1. Some of the appendices are also technical in nature; Appendices C–G, L and M are intended for readers with a background in physics or astronomy.

Numerous individuals have contributed to this study of the life of Ruby Payne-Scott. Her children, Peter Hall and Fiona Hall (Chap. 12), were especially helpful; Betty Hall (Dr. Elizabeth Hall), a lifelong friend of Ruby and Bill Hall (Ruby's husband), was crucial to our understanding of Payne-Scott; she is not related to Bill Hall! Harry Wendt of Vaucluse (Sydney) played an important role in helping me to find my way through the labyrinth of the National Archives in Sydney; he provided me with numerous copies of relevant files. Harry has recently received a Ph.D. degree from James Cook University with a thesis on a history of Potts Hill and Murray Bank field stations of RPL. Alison Muir of the School of Physics, Sydney University, provided the title of the book during a conversation in early 2007. Claire Hooker has provided assistance and insight for many years and gave us her Payne-Scott archive. Woody Sullivan has provided inspiration and assistance for years. Barnaby Norris at the Australia Telescope National Facility has helped considerably with the photo-archive from the Division of Radiophysics. My son, Andrew M. Goss, historian at the University of New Orleans, has provided advice and comments on the writing of history. The detailed editing of this volume has been carried out by my daughter-in-law, Pax Bobrow. Finally, I thank Nessy Allen (retired, School of Science and Technology Studies, University of New South Wales) for writing the foreword to this book. She is an expert on "women in science", and in particular on "women scientists in Australia".[10] In addition, she has provided copious suggestions for improvements to the text of our book. Numerous chapters have been improved thanks to her meticulous editing.

The following individuals have provided assistance and advice:

In Adelaide, Glenys Edwards, Dawn Geyer, Mary Carver, Jill Colley, the late Jocelyn Pedler, Elizabeth Hallett, Susan Bennett, Wendy Davis and Alan and Libby Brissenden (thanks for the wonderful hospitality) and Jill Fitch.

In Sydney: Rachel Makinson, Anne Manefield, John Brooks, Mal Sinclair, Peter Murphy, Ron and Jay Ekers, Vicki Fraser, Sue Little, Christine Vanderleeuw, Laurie Critchley, Martin Smith, Rita Nash, John Freeland, Sue Brian, Jan Christensen, Lyn and Fred Brown, Philip Edwards, Jessica Chapman, the late Brian Robinson, John O'Byrne, Bob (Robert) and Margaret Frater, Bruce McAdam, Jim (J.A.) Roberts, Bernard and Crys Mills, the late Harry Minnett, John and Joan Murray,

[10] Her important comparative study of Rachel Makinson (Chap. 11) and Joan Freeman (Chap. 11) was published in 1990: "Australian Women in Science – A Comparative Study of Two Physicists", *Metascience*, Vol. 8, No. 2, pp. 75–85.

Tim Bedding, Vicki Drazenovic, Lewis Ball, Tony Crawshaw, Harvey Cohen, Harvey Green, Carolyn Little, Adele Little, Renata Watkinson, the late Brian Cooper, Fred Blackman, Marjorie Thompson (*née* McKechnie), Bruce Slee, the late Don Yabsley, Paul Wood, Melanie Grogan, Kerrie Jarvis, Katherine Newton-McGee, Grace Noble, the late Marie Alewood, Bob and Christa Younger, Alex Colley, the late Dot Butler, Hastings and Liz Pawsey, Brian Bolton, the late Merle Watman, Margaret Giannesca, Bill Holland, Patrick James, Shirley Hokin, Shami Chatterjee, Lyn Frost, Renata Mancini, the late Arthur Gilroy, Letty Bolton, Renata Wakinson, Robert Makinson, Peter Davidson, Bruce Yabsley, Lindsay Baudinet, Marjorie Lobban, Eric Aarons, Doug Milne, Roy MacLeod, Margaret Holmes, Elaine Pacey, and John O'Byrne.

In Grafton, NSW: Frank Mack.

In Coonabarabran, NSW: Judith Hadfield, Jean Dow, Ann Savage, and Robina Otrupcek.

In Coomba Park, NSW: John B. Whiteoak

In Newcastle, NSW: Jennifer Wilson and Brian English.

In Bateau Bay, NSW: Betty Siegman

In Mullabula, NSW: R.N. (Ray) Smartt.

In Queensland, Australia: Wayne Orchiston, Hugh Hamersley, Ron Green, Ann Whittingham, Malcolm and Cheryl McKellar, and Graeme White.

In Canberra: the late J.P. Wild, the late W.N. Christiansen, John Curran, Rob Birtles, William Edwards, David Bell, Caroline Webber, Lawrence Cram, Ben Gascoigne, and David and Dorothy Jauncey.

In Melbourne: R.W. Home.

In Hobart: Jim Lovell

In the United Kingdom: the late Joan Freeman, Professor Sir Bernard Lovell, Professor Sir Francis Graham-Smith, Professor Rod Davies, Katherine Blundell, Robert Coulson, Richard Schilizzi, Biddy Schilizzi, Alan Pedlar, Pauline Pedlar, and Cedric Ashton.

In the US: Kate Goss, Robert Hayward, Lori Appel, Betty Trujillo, the late Terry Romero, Barry Clark, A. Richard ("Dick") Thompson, Jim Moran, the late Ron Bracewell, Helen Bracewell, Mark Bracewell, Marsha Bishop, Ellen Bouton, Megan Nunemaker, Douglas Bock, Marshall Cohen, John A. Graham, Stew Gillmor, Holli Glassel, Doug Roberts, Martha Stahr Carpenter, Lance Utley, Sarah Streb, Marie McCabe, John Jefferies, Stuart Pawsey, and Ken Kellermann.

In Belgium: Inge Heleu and Inge Lievens.

In Auckland, New Zealand: Gary Tee and Sergei Gulyaev.

In Argentina: Gloria Duber.

In Switzerland: Dame Joan Sutherland.

In France: J.F. Denisse, Monique Pick, and Francois Viallefond.

In Spain: Elizabeth Hallet.

I especially thank Professor Anne Green of the University of Sydney for her sabbatical support in early 2007 (Dennison Distinguished Visitor) at the School of Physics. Also, I thank Professor Eric Wilcots and Professor Ed Churchwell for sabbatical support at the University of Wisconsin in mid-2007.

I extend a special word of appreciation to my former advisor, colleague, and friend, Dr. Nan Dieter Conklin. Nan has been a steadfast supporter for 44 years; she is the first US woman radio astronomer.[11]

The extensive archive of material for this book will be deposited in the National Radio Astronomy Observatory Archive under the direction of Ellen Bouton; this will include interviews, archival material, and video material. The National Radio Astronomy Observatory is a facility of the National Science Foundation operated under cooperative agreement by Associated Universities, Inc.

W.M. Goss

[11]See her autobiography, *Two Paths to Heaven's Gate*, published by the National Radio Astronomy Observatory, 2006 (available from NRAO).

Contents

Chapter 1
Introduction: The Life of Ruby Violet Payne-Scott – 28 May 1912 (Grafton NSW, Australia) to 25 May 1981 (Sydney, Australia)

In *Irresistible Forces: Australian Women in Science* (2004), Claire Hooker has written, "Yet if [Payne-Scott] was a feminine flare in radio astronomy, she was a bright one, both as a physicist and as a woman." In Fig. 1.1, a photograph of Payne-Scott as a young student in the 1930s is displayed.

The field of radio astronomy in Australia grew out of the radar research carried out during World War II at the Council for Scientific and Industrial Research (CSIR) Radiophysics Laboratory (RPL; see Appendix A). Payne-Scott joined the new institute in 1941 (Chap. 4), as one of the first scientific staff as well as one of the first woman scientists (Joan Freeman had joined a few months earlier). RPL played a key role in the War effort, producing numerous copies of the aircraft warning radars that were used so successfully in the Southwest Pacific Area by both US and Australian military personnel from 1942 to 1945 in the war against Japan. Payne-Scott made major contributions to this top secret radar research; she became the Australian expert on the theory of the detection of enemy aircraft using the display system that had been invented in the UK, named the PPI or Plan Position Indicator (Chap. 5). She was also an experienced radio engineer; her experience with B.Y. Mills in the development of an experimental high frequency (25 cm) aircraft warning radar contributed to her success as an experimental radio astronomer starting in mid-1945.

World War II provided an opportunity for women to join the work force in Australia. With many young men overseas in the armed forces, there was a personnel shortage and thus a need for any able-bodied workers, including the hitherto under-utilized female population. In the later stages of World War II, women in the civil service in the Federal government were paid wages equal to that of their male counterparts, which was in great contrast to the previous convention of paying women only two-thirds of the male wage. In 1949, Payne-Scott was involved in a public controversy when the CSIR began to withdraw wage parity. It was only in 1969 and 1972 that Australian women were given wage equality based on rulings of the Australian Conciliation and Arbitration Commission (Appendix H).

Payne-Scott became a radio astronomer by accident. Her first observation was made from the RPL building on the campus of Sydney University in March 1944,

W.M. Goss and R. McGee, *Under the Radar*, Astrophysics and Space Science Library 363,
DOI 10.1007/978-3-642-03141-0_1,

Fig. 1.1 Photograph of Ruby Payne-Scott as a student in the 1930s, possibly while she was studying at the University of Sydney 1929–1932, working on a B.Sc. degree in physics. Bill Hall family collection, used by permission of Peter Hall

during a test of radar equipment at 10 cm (Chap. 6). Thus Payne-Scott became one of the first radio astronomers, as well as the first woman radio astronomer. This observation also represented the first astronomical project with Joseph L. Pawsey, an association that became very profitable in the years 1945–1951.

An explosive growth of radio astronomy occurred in Australia starting in late 1945. RPL became one of the pre-eminent radio astronomy institutes in the world under the direction of Pawsey and Edward ("Taffy") G. Bowen. Within a few years, Australia established its international leadership in radio astronomy. Payne-Scott participated in the first publication of the budding Australian radio astronomers; she wrote one of the first radio astronomy summary papers in December 1945 (Chap. 6). From 1945 to 1952, the RPL radio astronomers published 62 papers in radio astronomy; Payne-Scott was a participating author in nine of these ground-breaking investigations.

In the short period from 1945 to her resignation in July 1951, she became a driving force in the early radio astronomy efforts in Australia; she was the first scientific leader in the solar radio group, directed by Pawsey (Chaps. 7–9). Her leadership role was diminished in the post World War II era as new recruits (J.P. Wild, J.G. Bolton, W.N. ("Chris") Christiansen, B.Y. Mills and F.J. Kerr) became prominent in the late 1940s to the early 1950s; she continued, however, to play a significant role. In 1946 she discovered Type III solar radio bursts (Chaps. 7 and 8) which originated at long radio wavelengths in the solar corona and played an important part in the discovery of Type I bursts (1946) and Type II solar radio outbursts (1947).[1] Payne-Scott, together with Alec Little (1925–1985), even detected Type IV solar radio outbursts with the Potts Hill swept-lobe interferometer in 1949–1951, several years before they were recognized as distinct physical entities by the French group of Boishot and Denisse in the mid-1950s (Chap. 9). In Fig. 1.2, we show the well-known photograph of Ruby Payne-Scott, Chris Christiansen and Alec G. Little, taken at Potts Hill sometime between 1949 and 1951. This is the only photo of Payne-Scot that has traditionally been displayed in a number of publications in Australia in the last 20 years.

Payne-Scott also made major contributions to the development of radio astronomical techniques. Two prominent examples were: (1) the mathematical development of "aperture synthesis" (Chap. 7), the technique utilized by many of the advanced radio astronomy instruments of the modern era (e.g. the Very Large Array, the Atacama Large Millimetre Array, the Multi Element Radio Linked Interferometer, the Australia Telescope Compact Array, etc.); and (2) the swept-lobe interferometer at Potts Hill, developed by her and Little. With this instrument the scientists could make a crude movie (25 frames a second), showing the motions of the solar radio bursts as the emitting gas moved outwards in the corona at high velocities. In addition, there is strong evidence that she was the first person in Australia to recognize the importance of confusion in radio astronomy (the necessity to achieve high angular resolution in addition to sensitivity in order to recognize distinct radio sources).

In these first years of growth in radio astronomy after World War II, Payne-Scott began to come into conflict with one prominent colleague and with the bureaucracy of CSIR.

Payne-Scott was subjected to discrimination against women, prevalent in all aspects of Australian society in the 1940s and 1950s.[2] These controversies have been mentioned in a number of popular articles in the Australian press and also in books providing summaries of Australian astronomy. The latter have been correct

[1]Curiously, Payne-Scott soon lost confidence in the reality of the important Type II outbursts, associated with large optical flare events on the sun and delayed (by about a day) aurora and magnetic storms on the earth.

[2]There is a temptation to evaluate these issues with the viewpoint of the more egalitarian society of the early twenty-first century; as a number of correspondents have pointed out to the authors, the draconian treatment of Payne-Scott in the mid-twentieth century was consistent with practices in many walks of life. The characteristic that distinguished Payne Scott was her resistance to these inequalities.

Fig. 1.2 The most commonly published photograph of Ruby Payne-Scott. This was taken at the Potts Hill Reservoir, likely in late 1948. "Chris" Christiansen is to the right with Alec Little in the middle. Payne-Scott and Little were working on observations of the sun at 97 MHz using the newly constructed swept-lobe interferometer (Chap. 9). (ATNF Historical Photographic Archive: B14315; permission granted by Jessica Chapman.)

in attributing her 1951 resignation to the birth of her first child in late 1951. The nature and the consequences of the discovery in 1950 of her "secret marriage" of 1944 have been described with considerable distortion in popular articles.[3]

[3]One source claims erroneously that "She was obligated to resign when her marriage was exposed". (See also the Preface.) Another source suggests that her resignation when pregnant in 1951 was a protest against the marriage bar; we have found no evidence for this. An ASIO report (Chap. 13) of 2 March 1959 "... identified a Ruby Payne-Scott, a Research Officer at Radio Physics [sic], NSW, whom [BLANK-redacted] had referred to as 'a Red' and who was dismissed from CSIRO for failing to give notification of her marriage". At a subsequent location in the file, the correct reason for her resignation "because of her child birth" was stated.

The major conflict that Payne-Scott experienced in her career at CSIR and CSIRO[4] occurred in the period February–May 1950, probably starting when she met in person with Sir Ian Clunies Ross, the Chairman of the new CSIRO. Ruby Payne-Scott and William "Bill" H. Hall married in September 1944; this was known by most of her colleagues at RPL. The rule against married women at CSIRO maintaining their permanent employment status was challenged head on by Payne-Scott; Clunies Ross wrote her a series of forceful letters with equally strong replies from her side (see Appendix I for Clunies Ross's scathing reply of 3 March 1950). As expected, Payne-Scott lost this battle and became a temporary employee of the CSIRO. The major losses were her superannuation (pension) rights, the non-return of the CSIR/ CSIRO contributions (1946–1950), and the loss of accrued interest when her own contributions were returned. Details of this conflict are presented in Chap. 4.

A chronic conflict with John G. Bolton was a part of Payne-Scott's life at the RPL after he joined the CSIR in September 1946; Bolton was demobilized from the British Navy in late 1945, having served as a radar officer on the Royal Navy *Unicorn* (aircraft carrier) for about a year in the East Indies and the Pacific.[5] By the time of Pawsey's overseas trip in September 1947 to October of the following year, the conflict reached a boiling point. The sharing of the Dover Heights site produced continual conflict and Payne-Scott was "exiled" to the Hornsby field station (Chaps. 7 and 8).

During World War II there was an environment of complete secrecy at the RPL, due to the military nature of the research on radar. Each of the scientists had signed an oath of secrecy. Most documents were classified with different levels of secrecy. Thus the 1944 report on the first radio astronomy experiments at 10 cm (Chap. 6) carried the designation "confidential"; in 1945 or 1946 this was changed to "unclassified". This aspect of Payne-Scott's employment as a scientist and government employee is discussed further in Chap. 4.

After World War II Payne-Scott was concerned that CSIR would continue carrying out secret research, even though Australia was not then at war. In 1948, Payne-Scott took a strong stand against continued secret research within the CSIR. She wrote an un-equivocal letter to *The Sydney Morning Herald* on 29 July 1948 (Chap. 13), signed by many colleagues, pointing out that classified research in the

[4]The transition from the CSIR to the CSIRO (Commonwealth Scientific and Industrial Research Organisation) occurred in the period March–May 1949 with the passing of the Science and Industry Research Act 1949 by the Australian Parliament. The change from "Council" to "Commonwealth" was chosen to emphasise the national character of the new organisation and the word "Organisation" was used to highlight the changed character of the administration by the new CSIRO Executive of five members (including three scientists) (Schedvin 1987).

[5]D.B. Melrose and H.C. Minnett (1917–2003) have written a *Bibliographical Memoir* (1998) of Jack H. Piddington (1910–1997), who worked at RPL and then the Division of Physics for much of his career. He had played an important role in the Darwin anti-aircraft radar events of February 1942. Melrose and Minnett have quoted a colleague at RPL as mentioning that there was "a triangle of antagonism between John Bolton, Ruby Payne-Scott and Jack Piddington" at RPL in the late 1940s. Minnett also acknowledged these antipathies and described them as "creative tensions between very different personalities". RPL was blessed with some strong personalities!

CSIR would inhibit creativity. This issue would play a role in the transition to the CSIRO in May 1949; a major issue of contention was whether the CSIR would continue secret military research in the peace time environment.[6] Most classified research was moved out of the new CSIRO.

Payne-Scott had been known as a "left winger" at the RPL in the 1940s (Chap. 13). The Australian Security Intelligence Organisation (ASIO) maintained a large dossier concerning her and suspected that she was, in fact, a member of the Communist Party of Australia (CPA). ASIO had no proof of this affiliation at the time of a 1950 report. In 1999, however, the authors discovered that she had been a member of the CPA, possibly breaking with the Party later in the 1950s.[7] Behind her back, she was often referred to as "Red Ruby", even by her closest friends.[8] ASIO was aware of Payne-Scott's participation in the letter to *The Sydney Morning Herald*, along with numerous other colleagues not considered left-wing, including J. Paul Wild, future Chairman of CSIRO from 1978 to 1985.

In July 1951, Payne-Scott resigned from the RPL, with an advance notice of only 2 days. She was pregnant; her son Peter G. Hall – future Professor of Mathematics at the University of Melbourne and Fellow of the Royal Society of London – was born on 20 November 1951. A daughter, the famous Australian artist Fiona Hall, was born 2 years later (Chap. 12).[9]

After the birth of her children, Payne-Scott remained at home in Oatley (a suburb of Sydney), while she and Bill Hall raised the two young children (Chap. 12). After the children were about 10 and 12 years old, Payne-Scott became a mathematics and science teacher at Danebank Anglican School for Girls in nearby Hurstville; she was in this position from 1963 to 1974 (Chap. 14). It is likely that she developed Alzheimer's disease at an early age; she died in Sydney a few days before her 69th birthday (25 May 1981). Bill Hall died 21 July 1999.

An important aspect of Payne-Scott's career at RPL was the interaction and support of other prominent women colleagues during World War II. These colleagues were Joan Freeman Jelly and K. Rachel Makinson (Chap. 11). Joan Freeman's autobiography (*A Passion for Physics*, 1991) preserves a number of famous anecdotes about Payne-Scott. Rachel Makinson has also described a number of

---

[6]Collis, in a history of CSIRO, *Fields of Discovery: Australia's CSIRO* (2002), has described this controversy. As an example, the CSIR Division of Aeronautics was moved in 1949 from the CSIR to the Department of Supply and Development of the Australian Government (then Aeronautical Research Laboratories). The Chairman of the CSIR, Sir David Rivett, resigned on 18 May 1949; Rivett had been the founding CEO of CSIR (from 1927 to the end of 1945, then Chairman until May 1949)

[7]Peter Murphy, the Sydney Secretary of the Social Education and Research Concerning Humanity (SEARCH) Foundation, has been of considerable assistance (Chap. 13) in sorting out Payne-Scott's connections with the CPA. Her redacted ASIO file is in NAA: A6119/83, 1679, with the title "Payne Scott, Ruby Violet a.k.a. Hall".

[8]Rachel Makinson, February 2007.

[9]Payne-Scott experienced a miscarriage earlier. The date is uncertain; based on attendance at meetings of the radio astronomy group we surmise that this event may have occurred in late 1946 (see Chap. 7).

relevant events in the life of Payne-Scott in interviews with Goss in 2003 and 2007.[10]

Payne-Scott's legacy consists of two major components. First, she was an unconscious crusader for the rights of women in the scientific work place in Australia. Other women had experienced discrimination; Payne-Scott complained loudly about the treatment. She helped pave the way for future generations of women. Secondly, Payne-Scott was one of the first three pioneers in the new field of radio astronomy, which burst into prominence at RPL in Sydney in 1944–1945. Within a few years, Australia and the United Kingdom became the leaders in this new astronomy. Pawsey and Payne-Scott initially provided the key leadership for the rapid growth in solar physics that solar radio astronomy created in the first decades after World War II. When she retired in mid-1951, the leadership roles in solar physics were maintained by Pawsey and Paul Wild.

In 2008, the CSIRO recognized Payne-Scott with the establishment in June 2008 of the "Payne-Scott Award". This award is "for researchers returning from family related career breaks". The description states:

> The grant [$A35,000 for 1 year] will provide support to researchers to re-establish themselves and re-connect with the research underway in their field and related fields of research.

The grant targets women in research taking extended leave for family care; men involved in primary family care are also eligible. In the first year of the new grant program, ten individuals applied and six awards were granted. Given the level of conflict that Payne-Scott had over her marriage in 1950 (Chap. 4) and the career-ending nature of her pregnancy in 1951, this grant in her name is appropriate. (The "Payne-Scott" Award was announced at the same time as the "Newton Turner" Award for senior scientists at CSIRO, named after the prominent agricultural statistician at the Division of Animal Genetics from 1946 to 1976. In Appendix H a possible connection between Payne-Scott and Newton Turner is described.)

Most of the 16 chapters of this book have been summarised throughout this introduction. In Chap. 2 the ancestors of Payne-Scott, her early childhood, her move to Sydney to attend high school and her university years at the University of Sydney, are described. In Chap. 3 we describe her period as a science teacher at Woodlands School in Glenelg (Adelaide), as well as her short period working at

[10]The most famous colleague of the RPL during the War was certainly Dame Joan Sutherland, the famous opera singer. Although she may have heard of Payne-Scott while she was a clerical assistant at the RPL from April 1944 to January 1945, she never met Payne-Scott (Letter from Sutherland to Goss, 23 April 2007). In 1944, Sutherland was the typist of a report written by E.G. Bowen concerning the meteorological effects on radar reception. The most famous story about Joan Sutherland concerns the Musical Revue, *Hush-Hush*, in late 1944, staged by Robert Coulson. This was a spoof of the *Mikado* with a skit involving the "Lord High Clerical Officer" draped in red tape. Joan Freeman was in the chorus together with Sally Atkinson. The 18 year old Sutherland auditioned for a part and was rejected (confirmed by a letter from Dame Joan to Goss, April 2007). It appears that this rejection had no adverse effect on her career; already in 1947 she had made her concert debut in Sydney as Dido in Purcell's *Dido and Aeneas*.

Fig. 1.3 Bushwalking, a passion of Ruby Payne-Scott. Here she is probably in the Blue Mountains in the 1940s. Many photos of Payne-Scott show her eating or drinking at the time of the photographic session. Her daughter Fiona suggested that her mother likely thought that posing for photos was a waste of time; thus she could be more efficient when combining posing with eating or drinking! A companion photograph is shown in Appendix J. (Fig. J.3a). Bill Hall family collection, used by permission of Peter Hall

Amalgamated Wireless, Australasia (AWA)[11] in Sydney. Chapter 4 contains the details of Payne-Scott's personnel interactions with CSIR and CSIRO from 1941 to 1951; the details of the marriage controversy of 1950 and the equal pay conflicts of the post War era are summarised. Chapter 5 provides a detailed summary of her research in the area of War time radar. Chapter 6 contains the description of the first radio astronomy observations of 1944 and the transition to peace time in 1945. Chapter 7 contains the formative period of the first sea-cliff interferometer at Dover

[11]Amalgamated Wireless (Australasia) Ltd., formed in 1913 from the merger of Marconi's Wireless Telegraph Co. Ltd. and the Australian Wireless Company.

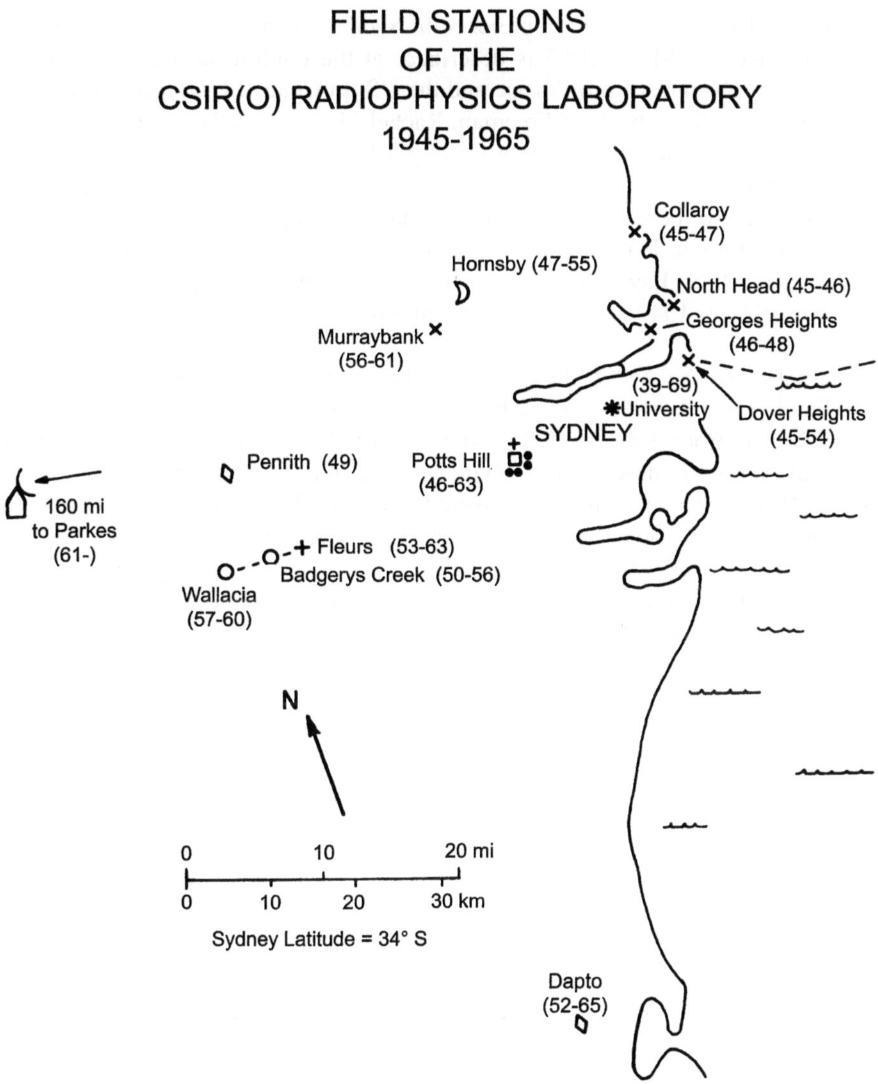

Fig. 1.4 A schematic map of the various RPL sites in New South Wales. The sites where Payne-Scott worked in 1945–1951 were Dover Heights, Hornsby and Potts Hill. (CSIRO, ATNF Archive)

Heights in Sydney (1946–1947), while Chap. 8 details the important year 1948 at Hornsby and the elucidation of the properties of Type III bursts; the nature of the growing conflict with John Bolton is also described in Chap. 8. Chapter 9 provides the details of the final observational campaign of Payne-Scott at Potts Hill; radio "movies" of the motions of solar bursts were created providing, for the first time, direct determinations of the ejection speeds of the bursts in the solar corona.

In Chap. 10, the important conference in Sydney of the International Union of Radio Science (URSI) in 1952 is described; at the conference Payne-Scott met many of her colleagues from Europe and the US. In Chap. 11 we recount reminiscences of Payne-Scott by Joan Freeman, Rachel Makinson, Harry Minnett, Bernie Mills, Chris Christiansen, Elizabeth ("Betty") Hall, Bruce Slee, Gordon Stanley, and Lyn Brown.

In Chap. 12 we describe Ruby and Bill Hall's family, with details of their two remarkable children, Peter and Fiona. An important part of Payne-Scott's life was bushwalking. Ruby Payne-Scott and Bill Hall met in the Sydney Bush Walkers (SBW) in 1941 or 1942; a well known picture of her during the period is shown in Fig. 1.3.

The nature of the ASIO file concerning Payne-Scott is presented in Chap. 13, including her membership of the CPA.

The period from 1963 to 1974, when Payne-Scott was a science teacher at Danebank School is described in Chap. 14. The final years of Payne-Scott's life up to 1981 and the conclusions are covered in Chap. 15. Chapter 16 is an Epilogue which includes the ironical motivation for our research on the life of Ruby Payne-Scott, which started in 1997.

To aid in the location of the various field stations of RPL in New South Wales, Australia, a schematic map of sites is shown in Fig. 1.4.

Chapter 2
Ruby Payne-Scott: the Early Years: 1912–1938, Ancestors, Childhood, Secondary School and Sydney University

Ruby Payne-Scott's Paternal Ancestors in the UK

Thanks to census data, birth, death and marriage certificates, and the occasional public notice in a local newspaper, we can in fact trace Ruby Payne-Scott's paternal ancestors back three generations to the United Kingdom.[1] What we find is a history of men trying their hand at a number of professions with perhaps limited success, and women giving birth to children well into their thirties, with extended families who tended to stay nearby and remain connected as younger generations were born and raised.

We start in the early nineteenth century with Ruby's paternal great-grandparents in the town of Tiverton, in the County of Devon, England. Tiverton is an old town on the River Exe, in the southwest of England, once a flourishing centre of the wool industry. Due to industrialization and competition from abroad, however, the wool trade declined in the late eighteenth century. Things picked up again in 1815 with the opening of the John Heathcoat and Co. lace-making factory, which utilized new machinery and the recently popularized synthetic dyes. Here we find Ruby's great-grandfather, John Scott, who made a living as a dyer in the lace factory.

John Scott was born in 1801, and in late 1830 he married Margaret Payne, who was born in 1808. John and Margaret had four children; Martha, Henry, Mary and Hubert. The eldest son, Henry Thomas Scott[2] was born on 18 May 1835. Ruby's own grandfather, Hubert Payne Scott was born 8 years later in 1843, in the family home at 41 Peter Street in Tiverton.

---

[1] Details of Payne-Scott's ancestors researched by Dr. Elizabeth Hall. Much of the Tiverton information was contributed by Cedric Ashton, Research Assistant, Tiverton Museum.

[2] Henry Scott had a most remarkable life. Like his younger brother, he became a physician. He was widowed in 1868 when his wife Sarah and daughter Margaret (born c. 1860) died in the Chincha Islands, off the coast of Peru. There is no information as to why the family was in South America. Henry then apparently remarried in the mid-1870s and became a clergyman in Swettenham in Cheshire. His name was still registered in the medical register. His second wife was Annie with a son William, born c. 1878. In the 1881 and 1891 census records the family was living in Cheshire. He last appeared in a census record of 1901, visiting Mannington Bruce in Wiltshire.

W.M. Goss and R. McGee, *Under the Radar*, Astrophysics and Space Science Library 363,
DOI 10.1007/978-3-642-03141-0_2,

At the time of the 1851 census Martha, aged 18, was single and still living at home, Henry, aged 16, was a pupil-teacher,[3] and Mary and Hubert, aged 12 and 8 respectively, were still at school. By 1861, we can see from the census data that John Scott must have passed away, but that Ruby's great-grandmother, Margaret, was still living at the same address, along with Hubert, who by then, aged 19, was listed as the head of the household, and a grain merchant by trade. From that same census we see that another "Payne" family was living next door, most likely Margaret's brother, Frederick.

The next we see of Hubert in public records is in the 1871 census which shows him living with a Sarah Payne, probably his maternal aunt, in Marylebone, a neighbourhood in central London. He may have been studying medicine at the time, and at some point courting his future wife, Agnes Duppuy. Agnes, born in 1847, was living with her widowed mother, Maria, at their home on George Street, in Croydon, a neighbourhood in south London. Though we have no idea how they met, the couple married in early 1872, in Hanover Square, which is in Mayfair, the neighbourhood adjacent and to the south of Marylebone. 1872 was an eventful year for Ruby's grandparents; apart from getting married at the start of the year, Hubert became licensed to practise as a homoeopath by the London Society of Apothecaries in December.

By 1881 Hubert was a general practitioner by trade. There in their home at 8 Amherst Road, in New Cross, part of the borough of Lewisham in the southeast of London, Hubert and Agnes had three children: Margarita, born in 1875; Valerie Violet, born in 1877; and Ruby's father, Cyril Herman, born in 1880. They also shared their home with Agnes's mother, Maria, whose age, health and role in the family is unknown, though, if she were still spry enough, she would have been ideally placed to help with the three children.

Emigration to Australia by Hubert and Agnes Scott

What happened next is not clear, but by 1886, when we next catch a trace of Hubert in public records,[4] he, his wife Agnes, and only two of their children, Valerie Violet and Cyril Herman, are in Australia. Agnes's mother, Maria, is not listed, and more tellingly, neither is their eldest daughter, Margarita. No records are found for Margarita either in Australia or the UK, and one can only assume that the child died before reaching the age of ten. Between 1886 and 1903, Hubert practised at various addresses in Sydney; beginning with 181 Macquarie Street, which was and

---

[3] A student-teacher would still be a student but assisting in the education of younger children. Henry was fortunate in attending the well-known Blundell's School, which had links to colleges at Cambridge and Oxford, and at the age of 28 he became a licentiate of the College of Physicians.

[4] In 1886 an entry in the *Australian Medical Directory* has been located: "Scott, Hubert Payne, 181 Macquarie Street, Sydney, NSW, LSA (Licenciate of the Society of Apothecaries), London."

Fig. 2.1 The house of Hubert Scott (the grandfather of Ruby Payne-Scott) and his family from 1897. The house at 76 Edgeware Road, Newtown, Sydney, in late 2006 is shown. Photo by and permission given by Jan Christensen

is a fashionable location for the medical profession. This address on Macquarie Street appeared in medical registries up to 1897, in addition to other Sydney locations on King Street (1896, 1899) and George Street (1900, 1903).[5]

It is hard to know if the family would have lived in the same location as the address listed for Hubert's medical practice. We see a great moving about from one address to another in Sydney until 1902. In 1897, Hubert Scott was listed in residence at 76 Edgeware Road in Newtown, near The University of Sydney. This house as it appeared in 2007 is shown in Fig. 2.1. In 1898, a listing at 3 Trafalgar Street, Newtown (Fig. 2.2, as it was in 2007) appeared, while the next year a new address at 63 London Street was published. In 1900 and 1902, a listing on Marrickville Road in Marrickville was published.[6]

It is then, in 1903, that we find the mysterious disappearance of Hubert Scott from all public records. He neither appears in medical directories nor the electoral rolls. Very strangely, however, there is no record of his death. Agnes is listed in the postal directory as a music teacher, implying that she now needed to support

[5]The following is a quotation from Dr. Elizabeth Hall: "1886–1892: the Scotts have only a single address. According to Alyson Daley, the Librarian at the Royal Australian College of Physicians, it was not unusual at the time for medical men to live over their surgeries. In 1893 two addresses appear and after 1896 Hubert seems to have given up his Macquarie-Street address. As Marrickville Council Rate Books do not indicate whether a building was used for private or professional purposes, it is not clear whether or not Hubert was practising from home."

[6]All the address information referenced in this section was gathered by Dr. Elizabeth Hall from the Sands Directory, Directory of New South Wales, Medical Registers, Council Rates Books and Electoral Rolls.

Fig. 2.2 The house of Hubert Scott and family from 1898. The house in Newtown, Sydney at 3 Trafalgar Street in late 2006 is shown. Photo by and permission given by Jan Christensen

herself, which corroborates the notion that Hubert had left the family. Also in 1903, Agnes and the two children, Valerie and Cyril, suddenly adopt the hyphenated name "Payne-Scott". By this time, the children were adults; Valerie would have been about 26 and Cyril 23. Valerie may have moved into her own home, as we have a listing for only Agnes and Cyril living in "Pretoria", a house at 75 London Street in the neighborhood of Newton.[7] Cyril and Agnes would remain physically close over the next 3 years, as Cyril remained a householder at 75 London Street, while Agnes took up residence with a housemate, Mrs. Sarah King, in the adjoining house around the corner, at 49 Liberty Street (Fig. 2.3). Agnes would remain at that

[7]The following is a quotation from Dr. Elizabeth Hall: "76 Edgeware Road, 3 Trafalgar Street, 63 London Street and probably Arundel Terrace, all rented by Hubert, are modest three-bedroom residences. 'Pretoria' at 75 London Street, is larger and more imposing. This was rented by Cyril. It is mentioned in a history of the area but there is no indication of whether or not it was used commercially. Agnes is shown as living there in 1903 but Hubert's name does not appear in connection with it or at any future address occupied by the family."

Fig. 2.3 The house at 49 Liberty Street, Newtown in 2006. This house was the home Agnes Scott for some years into the twentieth century. After 1903 she was Agnes Payne-Scott, her new name. Her son (the father of Ruby Payne-Scott) lived nearby until moving to Grafton, New South Wales, a few years later (perhaps circa 1908). Photo by and permission given by Jan Christensen, late 2006

address, even after Cyril moved on until 1911, when she would move to 12 William Street in Ashfield, a neighbourhood to the west of "Pretoria" in Newtown.[8]

Some time later, Cyril moved some 600 km to the north to an address in South Grafton, New South Wales, an area near the coast in northern NSW, 340 km south of Brisbane.[9] His mother, and most likely his sister, remained in Sydney. He met his wife, Amy Neale[10] in Grafton and they were married on 15 November 1910 at the Church of England in South Grafton. The next day, an article in the local newspaper, *Grafton Argus,* under the heading "Wedding Bells", mentions that it "...was the scene of an interesting wedding. ... Mr. and Mrs. Payne-Scott are spending the honeymoon in the north". It lists Amy's older sister, Ruby Pearl Neale, as the bridesmaid, and Cyril's occupation as "accountant".

[8]From 1919 to 1921 Agnes lived at 8 Holborow Street, Croydon, Sydney. On 6 February 1921, her death was registered in Croydon by her daughter Valerie (1877–1948). Hubert P. Scott's whereabouts remained a mystery; the marital status was left blank on Agnes's death certificate. In addition, her burial location remains a puzzle; the death certificate stated that burial was at Gore Hill, Sydney; however, the cemetery has no record of the grave. We know little of Ruby's aunt Valerie Violet either, aside from her name appearing on a copy of Robert Louis Stevenson's *The Child's Garden of Verses* given to a young Ruby as a birthday present.

[9]He appeared on the electoral roll in 1909 as: "Cyric [sic] Payne-Scott, Thorough [sic, in fact 'Through Street' in South Grafton] Street, Clerk."

[10]Amy was a school teacher born 1875 in Picton, NSW. Her parents were William Neale, an auctioneer born in 1840, who most likely died before 1910, and Ada Mary Moffitt, born 1846.

Ruby's Early Childhood

Two years after the nuptials, our protagonist, Ruby Violet Payne-Scott was born, on 28 May 1912. She was followed a year later by her younger brother Henry, on 8 June 1913. Ruby was named after both her maternal and paternal aunts, Ruby Pearl Neale and Valerie Violet Payne-Scott. Ruby was born at Runnymede Private Hospital, a maternity clinic on Fitzroy Street in South Grafton owned by Dr. Henry. The birth certificate was signed by Dr. Earle Page, Australia's shortest serving Prime-Minister[11] and Nurse Riordan. The hospital from circa 1910 is shown in Fig. 2.4. It is likely that the young Payne-Scott family lived with Amy's mother in "Uloom", the elegant Neale family house on Bent Street in Grafton, shown in Fig. 2.5.[12]

By 1915, although several members of the Neale family continued to live in Grafton, the Payne-Scott family seems to have moved again. The details of this

Fig. 2.4 The maternity hospital where Ruby Payne-Scott was born on 28 May 1912, Runnymede Private Hospital, South Grafton, New South Wales. Dr. Earle Page was the attending physician. Photo, circa 1910, courtesy of Frank Mack, President of the Clarence River Historical Society in Grafton. Permission given by Mack

---

[11] Probably Page was a visiting physician at the Runnymede Private Hospital as he owned his own hospital, Clarence House on Through Street, South Grafton. Sir Earle Page (1880–1961) was born in Grafton; after finishing medical school in Sydney he opened the hospital in 1903. He was a member of the Australian Federal Parliament from 1919 to 1961. When Lyons died in 1939, Page was appointed Prime Minister and served for 20 days (7–26 April 1939). Page was a minister of Health from 1949 to 1956.

[12] Details about South Grafton and the Neale family provided in 2007 by Frank Mack, President of the Clarence River Historical Society, founded in 1931 by Dr. Earle Page. Mack wrote an article about Payne-Scott in the 18 November 2006 Clarence River Historical Society Newsletter (No. 94).

Fig. 2.5 A modern day view (circa 2007) of the house "Uloom" in Grafton where the Payne-Scott family lived for a few years with Amy Payne-Scott's family (Neale) after 1911. Photo courtesy of and permission given by Frank Mack

portion of their life remain quite uncertain. Based on family memories,[13] it is likely that the family lived for some years in the rural town of Coonabarabran, NSW (the future home of the Anglo-Australian Telescope towards the end of the twentieth century at a distance of 450 km from Sydney). However, no archival evidence of the family can be found in Coonabarabran. There is no trace of them in the electoral rolls, school records or copies of local newspapers.[14] At the time of the move, Ruby was about three and Henry two. Thus valuable information about Payne-Scott's early school years is missing. A possible reason may be that the children were home-schooled. Her brother, Henry, told a Royal Australian Air Force psychiatrist[15] in late 1942 that he was schooled by his mother, a trained school teacher, at home until he was about ten, i.e., about 1923 or even 1924. Henry's troubled war record is described in Appendix N; from a time starting in the mid- to late-1930s until his death in 1970, he and his sister were estranged with little or no contact. A possible reason for this is described in Appendix N.

---

[13]Interview with Peter Hall, February 2007. His mother had vivid memories of coming home to Coonabarabran from Sydney in the late 1920s on school holidays. She described the animals on their property (cats, dogs and rabbits) and visits from relatives from Grafton who complained about the summer heat in this inland community.

[14]With the assistance of Dr. Ann Savage of Coonabarabran, contacts were made with the Local Family History Group of Coonabarabran and a letter to the *Coonabarabran Times* was placed in early March 2007; no information about the Payne-Scott family could be elicited. Additional contacts were made with several individuals in the area (Judith Hadfield and Jean Dow); again no trace of the family could be established in Coonabarabran in 2007.

[15]For Sgt. Henry Payne-Scott's war record (NAA: A9301, 20769), see Appendix N.

Secondary School Education: Sydney

Ruby Payne-Scott must have been a noticeably bright child, and her parents must have planned to give her every opportunity they could for a good education, even if they could not afford to pay for her schooling at an elite private school. We can see this by the path her schooling took once she reached her pre-teens. Generally, a young person would attend school in Australia until the age of 14, at which point she/he would sit the examination for an Intermediate Certificate. One could not, however, enter University with an Intermediate Certificate. To do that, one would have to stay in school for another 2 years and pass the examination for a Leaving Certificate. Provided the requisite matriculation subjects had been taken, entry into university was then possible. When Ruby's parents moved to the remote country town of Coonabarabran, there would have been fairly limited educational institutions, and even her mother Amy's dedication and training as the children's at home teacher would have taken them only so far.

While her parents and brother remained in Coonabarabran, Ruby went off to attend the Cleveland Street School[16] in Sydney. From her Intermediate Certificate[17] of 1925, we can deduce that she would have been living in Sydney to attend the school from 1923 to 1925, making her eleven at the beginning and just reaching 13 by graduation, a younger age than average. It is possible that she lived with her maternal Aunt, Eva Mary Neale[18] while living in Sydney. This would emphasize both a level of trust in extended family to care for children, as well as a fostering of independent thought and behaviour for the sake of education and reaching one's highest potential. Ruby certainly did reach for her highest potential. She achieved excellent marks with As in five subjects: English, Maths I, Maths II, French and Botany; and Bs in History, Geography and Latin. There was a report on 7 May 1924 in *The Sydney Morning Herald*, the major Sydney newspaper, that "Among girls who have gained noteworthy examination results [from Cleveland Street High School] were ... Ruby Payne-Scott".[19]

It would appear the family moved back to Sydney in 1925, when her father, Cyril Payne-Scott, reappeared on the electoral rolls. We can assume she moved back into her parents' home once they returned to Sydney. In 1925, Cyril was recorded as

[16]In 1912 or 1913, Cleveland Street High School became an Intermediate High School for both boys and girls. In 1929 (after Payne-Scott had left), the school became an Intermediate High School for boys only with the girls transferred to schools at Crown Street and Marrickville; the Cleveland Street name was maintained at the Crown Street site (letter from Margaret Giannasca of Cleveland Street High School to Peter Hall, 8 March 1999).

[17]Susan Brian obtained a replacement Intermediate Certificate in August 2008 from the New South Wales Board of Studies.

[18]Eva Mary Neale was a younger sister of Amy Neale, born in 1881 in Grafton, died in 1953 in Sydney. She was a witness at Ruby's very private wedding in 1944.

[19]Article received from Margaret Giannesca, School Promotion, Cleveland Street High School, Sydney, February, 1999.

running a mixed business at 118 Liverpool Street, Enfield.[20] The next year he was reported as having an additional address at 6a Burwood Road, Burwood.[21]

With an Intermediate Certificate completed, Ruby was able to aim higher in the hierarchy of Sydney public schools. The Cleveland Street Intermediate High School would have acted as a feeder school for the prestigious Sydney Girls High School (SGHS), especially for girls from the country.[22] Many of the public high public schools in Sydney at the time were sex segregated; SGHS was one of the outstanding public high schools for girls in Australia. With a strong background from SGHS, Ruby was well positioned to begin a university course in science.

The SGHS records indicate that Ruby entered the school on 9 February 1926 and finished her Leaving Certificate with honours in late 1928, when she was only 16. She was not quite 17 when she entered the University of Sydney in early 1929 (Fig. 2.6).[23] The final school report for Ruby at SGHS was indeed impressive with a pass at almost the highest level possible: first class honours in Maths I and Maths II and Botany. Her grades in English, Latin and Mechanics (Physics) were at the A level, while her only B was in French.[24] Apparently Payne-Scott repeated the last year at SGHS, at first glance a surprise for such a bright student; it was the practice in those days for a younger student to stay at school an extra year in order to enhance the chances of an improved exam result and thus a scholarship to university.

Ruby left behind two articles in the SGHS publication, *The Chronicle*, in 1927. In the November 1927 issue she wrote an article about the "Pictures in the Library", in which we are given a piece by piece tour of the classic reproductions hanging in the school library, complete with the expectedly prosaic 15-year-old's opinions on each. Earlier, in June 1927, she had written a rather cute primer on how to succeed at SGHS, in rhyming verse:

TO A SYDNEY HIGH GIRL
(With apologies to Mr. Rudyard Kipling)
If you are always listening when you should be,
And keeping very quiet and very still;
If you can always see whate'er you should see
And tackle mathematics with a will;
If you can wait, and not be tired of waiting,

[20]This building was later demolished; in 2007 there was no indication of the structure. When Payne-Scott entered Sydney Girls' High School in early 1926 the address listed was her father's shop at 118 Liverpool Road. Ruby's Leaving Certificate of late 1928 listed the updated address, 156 Liverpool Street. Also her brother was pursuing the Intermediate School Certificate at a school in Summer Hill, only 5 km distant (NAA: A9301, 20769).

[21]The Burwood Road business was a china, glass and earthenware shop which existed for 1 year. The Liverpool Street address was also reported in 1928 and 1929. The Liverpool business continued until 1932–1933 or possibly longer.

[22]Based on an interview in 1999 with Shirley Hoskin, the archivist of SGHS.

[23]The date of this photo is uncertain.

[24]Much of the information about schools in Australia was provided by Rita Nash. Information about Sydney Girls' High School from Norman (*The Brown and the Yellow: Sydney Girls' High School 1883–1983*, 1983).

Fig. 2.6 Ruby Payne-Scott in the 1920s. Bill Hall family collection, used by permission of Peter Hall

Although you're squashed in the Assembly Hall;
If you can work, and still not work be hating,
And yet on Wednesdays chase a hockey-ball;
If you are ready for examination
And need not cram your work right at the end;
If you can face that awful French dictation
As though it were your loving, kindly friend,
If you come first or last, although you've hard tried
And yet not let a change come o'er your face;
If you can lose just on the very last stride,
And yet can cheer the winner of the race;
If you can work, and yet find time for playing;
If you can play, and yet find time for work;

Fig. 2.7 The Payne-Scott family lived in Chatswood (10 Warrane Road) after about 1936; Cyril, Amy, Ruby and Henry (born 1913) were listed as living at this address in the late 1930s; Henry, an accountant, was estranged in later years from his sister. Ruby (University Demonstrator) was listed as living here from 1936 to 1938. Photo by and permission given by Jan Christensen, late 2006

> If you are always ready for essaying
> Your task, and do not any duty shirk;
> It does not matter whether you are clever,
> It does not matter if you are a fool,
> If you are all this, yet exams not weather,
> Still will you be a credit to your school.
> R. Payne-Scott (5B)[25]

By 1936, Cyril and Amy moved to 10 Warrane Road in Chatswood (see Fig. 2.7), a Sydney suburb on the north shore. Cyril's occupation was listed as homeopathist, a return to the occupation of his father. There was no indication of a separate address for a medical practice. At this address Henry (accountant; see Appendix N for a description of the life of her estranged brother), and Ruby (demonstrator at The University of Sydney) were also listed, Henry for the years 1936, 1937 and 1939, Ruby for the years 1936–1938. Ruby left for Adelaide in 1938 and returned to Sydney in mid 1939 (see Chap. 3). Cyril died in 1942,

[25]The Leaving Certificate was taken in the fifth year of high school, thus the Fifth Form, now year 12 in NSW. "5B" was the level below "5A" of the Fifth Form classes.

followed by the death of his wife Agnes in 1943. Thus at the time of the marriage of William Hall and Ruby Payne-Scott on 8 September 1944, both parents of the bride were listed as deceased. At a rather early age in young adulthood, Ruby had no surviving parents.

Bachelor of Science in Physics at the University of Sydney

Ruby began her studies at The University of Sydney in early 1929, before reaching her 17th birthday, with a merit-based bursary award and Science Exhibition scholarship from the University. She was an outstanding student, with an impressive record[26] throughout her stay. She did the normal B.Sc. degree course in physics in 1929, 1930 and 1931 gaining outstanding marks; she had high distinctions in mathematics in all 3 years and in physics in 1930 and 1931. In 1932 she completed the Honours Physics course with first class honours. The Honours degree was awarded at the beginning of 1933.[27] In 1932, Ruby Payne-Scott started the Honours Course in physics. In 1931, she shared the Deas Thomson Scholarship (after the 19th century Vice-Chancellor and later Chancellor of the University of Sydney Sir Edward Deas Thomson) with Reginald Healy for the 'greatest proficiency in Physics III' (her course in 1931). She also was awarded the Walter Burfit scholarship for excellency in Physics III. Both awards required that the recipient be an honours student at the University of Sydney.

Ruby was only the third woman to receive a degree in physics at The University of Sydney,[28] and in fact the 90th individual to receive a physics honours degree from The University.

We know little of her experiences at The University of Sydney. From discussions with her son, Peter Hall,[29] in later years, we know that she found the mathematics lectures by H.S. (Horatio Scott) Carslaw (1870–1954) especially stimulating. Payne-Scott remembered that Carslaw read letters from his former student John C. Jaeger (1907–1979), while Jaeger was a student at Cambridge (UK) in the early 1930s. Later in her life, Payne-Scott and Jaeger became friends and colleagues, and shared a passion for cats.[30]

---

[26] Letter from Renata Mancini of The University of Sydney Archives to Peter Hall, March 1999.

[27] The Honours degree in physics required an additional year.

[28] Edna Briggs (*née* Sayce) in 1917, Phyllis Nicol in 1926, Payne-Scott in 1933 and then Joan Freeman-Jelly in 1940 (see also Hooker 2004).

[29] Interview 12 February 2007.

[30] For Jaeger's life summary see Paterson (1982). Carslaw and Jaeger collaborated on a number of well-known applied mathematics books, including the well known second edition of *The Conduction of Heat in Solids* (Carlslaw and Jaeger 1947). See Chap. 15 for more details about Payne-Scott and Jaeger.

In the final year of Payne-Scott's honours course, she published a short article in *Nature* entitled "Relative Intensity of Spectral Lines in Indium and Gallium" (Payne-Scott 1933). It is likely that this short paper represented a portion of the research for her honours physics project. There is a single reference to a paper by O.U. Vonwiller (1882–1972) (*Physics Review* of 1930 with a discussion of similar results for thallium), who had suggested the project. He was a Professor of Physics at Sydney from 1923 to 1946. This paper by the 21-year-old physicist consisted of an analysis of photographic spectra of the two elements (and alloys of indium with lead) in the range 4,033–4,511 Å. The short note presents the line ratios of the doublets of indium and gallium under various conditions; there is essentially no discussion of the results. The connection with Professor Vonwiller probably led to Payne-Scott's position at the Cancer Research Committee in the years 1932–1935.

Master of Science and the Cancer Research Committee at The University of Sydney

Ruby Payne-Scott began her association with the ill-fated Cancer Research Committee of the University of Sydney (CRC) most likely in late 1932, continuing for 3 years.[31] The M.Sc. degree was awarded in February 1936 and her thesis was published in *The British Journal of Radiology* in December 1937, having been submitted 13 months earlier.[32] It seems likely that she supported herself financially through the Cancer Research Committee in parallel with her position as a Demonstrator/Tutor in the School of Physics, while she completed her research for the M.Sc.

The history of the CRC has been presented in detail by Hamersley (1988) and summarised by Hooker (2004). The Cancer Research Committee had been formed in the early 1920s at The University of Sydney to foster research on the treatment of cancer. As Hamersley has pointed out, its history, from 1922 to the end of the enterprise in April 1938, was characterised by internal bickering and debate. Certainly the inexperience of this set of Australian academics in funding, supervising and running such a research institute played a role in the ultimate failure of the CRC.

In the late 1920s, the opportunities for physics graduates were enhanced by the need for an understanding of the physics of X and gamma rays. As Hamersley points out:

> For physicists this national ferment [new interest in the treatment of cancer by the use of X-rays] created some unique opportunities and challenges. Their participation in the

[31] Based on the in-house journal, *The Journal of the Cancer Research Committee of the University of Sydney*, 1 October 1938, article by Vonwiller.

[32] The publication lists Payne-Scott's affiliation as Demonstrator in Physics, School of Physics, University of Sydney.

> national cancer effort encouraged the perception – new for Australians – that physics, traditionally one of the abstract sciences most remote from them, had something to contribute in matters of central human and societal concern. A limited number of new career openings for physicists also appeared, as a nation-wide scheme for providing physical services for radiotherapy, based on the physics departments of the state universities, gradually evolved. Finally, for physicists in the major centres of Sydney and Melbourne, there was the prospect of obtaining from cancer funds support for basic and applied research in radiation physics ... [this] could bring significantly nearer realization of emerging ambitions to establish physics as a viable research discipline in Australia.

The CRC would have been an ideal place for a young physicist to gain professional experience and connections, earn a livelihood, and still work towards a higher university degree. Unfortunately, the CRC was almost completely devoted to researching the work of one medical doctor, Wanford Moppett. In 1924, Moppett tried out a new theory of radiation therapy on chicken eggs that were in fact cracked open at the top to allow the extremely weak radiation into the embryo. After much criticism from the larger cancer research community it was found that any effect upon the chicken eggs was probably caused by long exposure to a contaminated environment rather than the radiation Moppett used.[33]

By the time Ruby Payne-Scott joined the CRC in 1932, the rationale for the research flagship of the enterprise was collapsing. What's more, in May 1934, the Director of Research, H.G. Chapman, committed suicide amid accusations of financial malfeasance and there were also increased doubts about the reality of the Moppett Effect. In August 1934, O.U. Vonwiller and D.A. Walsh became the supervisors of research for the Cancer Research Committee. When the institute was closed in April 1938, Vonwiller wrote the apologia in the last issue of the in-house journal on 1 October 1938,[34] placing much of the blame on the misdirection of Chapman. As Hamersley has pointed out, the entire edifice established in the 1920s was flawed in contributing to this "collective folly". Even Vonwiller had played a role in committing resources to investigate the Moppett Effect.

The circumstances in which Payne-Scott joined the Cancer Research Committee are unknown.[35] We can surmise that with her previous advisor, Vonwiller, as the de facto director of the CRC for the final years 1934–1938, Payne-Scott was given a part-time appointment to support her research for her master's degree. On 10 December 1936, Payne-Scott and W.H. Love submitted a short research note, "Tissue Cultures Exposed to the Influence of a Magnetic Field", to *Nature* which was published on 15 February 1936. The goal was to investigate the influence on normal cells cultivated in vitro by a strong magnetic field. Thus 8–9 day old chick

[33] Hamersley (1988) has pointed out that a number of investigators found that the changes might have been due simply to fungi or spore-forming bacteria in the exposure in the air, passing through the holes cut through the egg shells to allow the low energy X-rays to pass into the interior of the egg.

[34] *The Journal of the Cancer Research Committee of the University of Sydney*, 1 October 1938, article by Vonwiller.

[35] In her application for a CSIR position in 1941, Payne-Scott described her association with the CRC as a two year fellowship; however, Vonwiller's article lists a 3 year period, 1932–1935.

embryos were exposed to a 5,000 gauss field. After an exposure of 3–6 h "the exposed cultures exhibited no visible abnormalities in the arrangement of chromosomes in the dividing cells". There were some connections to the methodology of the Moppett effect experiments; however, no claim of a positive effect was made. Love had been a long time employee of the Cancer Research Committee; Hamersley reports joint experiments carried out by Love and Moppett in 1927 on "further egg control experiments" related to the Moppett effect.[36]

Payne-Scott had one additional publication in 1936: "Notes on the Use of Photographic Films as a Means of Measuring Gamma Ray Dosage" (Payne-Scott 1936). This paper was not mentioned in her master's thesis and was apparently a small project performed in parallel with her major research project.

The M.Sc. paper presented by Payne-Scott was dated 28 February 1936 with a striking change to the title. The original title of "On the Amount and Distribution of the Scattered Radiation in a Medium Traversed by a Beam of X or Gamma Rays" shows that the words "On the Amount and" are crossed out and the hand-written words "The Wave-Length" have been substituted. The corrected title is thus "The Wavelength Distribution of the Scattered Radiation in a Medium Traversed by a Beam of X or Gamma Rays". The paper (with the latter title) was submitted to the *British Journal of Radiology* on 17 November 1936 and published in Volume X, New Series, No 120, December 1937 with essentially no changes from the thesis. The final sentence is "These investigations were commenced while the author was attached as physicist to the Cancer Research Committee of the University of Sydney".

The publication summarized a theoretical treatment of the variation in the amount of scattered radiation as a function of primary wavelength and an investigation of the spectral distribution of the scattered radiation. Compton scattering, in which the incoming photons lost energy to the target electrons, produced a complex mixture of primary and scattered radiation. Payne-Scott wrote, "This softer radiation [Compton scattered] may, under certain circumstances amount to a larger proportion of the total radiation absorbed at a point, and thus may be of considerable importance in the study of the biological and other effects of the radiation". Since the scattering properties of water closely matched those of a biological tissue, Payne-Scott evaluated a number of the quantities derived for the scattering process for the particular medium of water. In addition, much of the experimental work available in 1936 was based on observations of radiation scattered from water phantoms (items substituted for real tissues in order to test the efficacy of the X-ray or gamma ray imaging). The comparisons were carried out for wavelengths of 17.1 X.U. (the Siegbahn scale) (0.017 Å), 50 X.U. (0.050 Å), 200 X.U. (0.2 Å)

[36]Hamersley has described how Love attempted to replicate Moppett's work without any success. He then turned to another radio-biological topic for his thesis. Hamersley has also brought to our attention the fact that in 1933 Payne-Scott worked with The University of Sydney physicist, G.H. Biggs, in running the radon plant, extracting radon and placing this into sources to be used for medical therapy. Hamersley has also pointed out that every laboratory where this type of work was done was later on found to be seriously contaminated with radioactive radon.

and 400 X.U. (0.4 Å), corresponding to gamma rays from radium, hard X-rays, hard-medium X-rays and medium-soft X-rays. It is noteworthy that none of the references in the paper referred to work done in earlier periods at the CRC. Thus, there was no apparent connection with the earlier discredited research of this institute. The mathematical skills that were to serve Payne-Scott well in her later scientific career at CSIR and CSIRO from 1941 to 1951 were clearly evident in this publication. "She was one of the very few people who had any association with the Sydney University Cancer Research Committee who produced any good substantial work."[37]

In the October 1938 article by Vonwiller, he specially praised Payne-Scott's research. After providing a detailed summary of the thesis, he added:

> Besides her routine duties Miss Payne-Scott did some original experimental work bearing on radiation measurement, and made an important and difficult theoretical investigation on the problem of quality and intensity of scattered radiation ... Miss Payne-Scott must be credited with an important contribution, useful both for the results obtained and for the indication of methods of attack to be followed.

Payne-Scott undoubtedly saw the end of the CRC on the horizon in the course of 1935–1936. With her M.Sc. completed, and her current work situation crumbling, it would have made sense for the ever energetic Payne-Scott to go the extra mile and earn a Diploma of Education in 1937. We can imagine that the need for financial security during the lean years of the Depression may have also been responsible for her acquisition of a teaching degree. Conventional wisdom in those years was for young people to obtain a safe, and if possible, permanent position of employment. Also, we must remember that Ruby's own mother had been a school teacher, and it is doubtful that Ruby Payne-Scott was naïve about the career opportunities for women in the sciences at that time. The option of teaching was an extremely practical career choice on her part. Indeed she was to spend two periods as a secondary school teacher, 1938–1939 at Woodlands School (Chap. 3) and 1963–1974 at Danebank School (Chap. 14).

---

[37] H. Hamersley communication, 27 November 2006.

Chapter 3
Woodlands School, AWA, New Career Opportunities for Women in World War II

Woodlands Church of England Girls' grammar School, Glenelg (Adelaide)

As the Cancer Research Committee closed its doors in the late 1930s, there was a scarcity of permanent teaching or research positions at the University of Sydney; Payne-Scott must have been concerned about her future. During the course of 1937, she applied for a position as Science Mistress at Woodlands Church of England Girls' Grammar School[1] in the sea-side suburb of Glenelg in Adelaide, South Australia. Payne-Scott completed the work for a Certificate of Education in 1937, and the University of Sydney awarded her the teaching certificate in 1938. Obtaining a teaching post during those Depression years must have been an asset. Positions for physicists were indeed limited, especially for women. Employment outside universities was rare for any physicist. Payne-Scott joined the staff at Woodlands at the beginning of the first term 1938 and remained until the end of first term 1939.

Payne-Scott only stayed at Woodlands for a year and a term, resigning in May 1939. Not only did she move outside New South Wales for the first and only time in her life, but she also moved from a research environment (University of Sydney) to that of a secondary school. In the 1960s she returned to teaching at the Danebank School in Sydney (see Chap. 14). At Woodlands, Payne-Scott was a "boarding mistress", living in the school together with the student boarders. This meant that the school provided her accommodation but with the added responsibilities of the supervision of pupils after school hours. Jocelyn "Jock" Pedler (*née* Britten-Jones, deceased 11 February 2009)[2] from the class of 1939 was a boarder and remembered

[1] See endnotes for a description of this historic school, now sharing the combined histories of two institutions in South Australia.

[2] Interview with Pedler, 23 March 2007 at the Archive Museum Woodlands Old Scholars' Association Inc. in the conference room at the Law-Smith Building at St. Peter's Woodlands Grammar School, South Australia. See endnotes.

DOI 10.1007/978-3-642-03141-0_3, © Springer-Verlag Berlin Heidelberg 2010

Fig. 3.1 Photo of the Law Smith building, St. Peter's Woodlands Grammar School, Glenelg, Adelaide, South Australia in 2007. (In 1938, Woodlands Church of England Girls' Grammar School.) In 1938–1939 Payne-Scott lived and worked here as a "boarding mistress". Photo March 2007 by Goss

that Payne-Scott's room was on a mostly enclosed balcony of the Law Smith building. This building is shown in 2007 in Fig. 3.1; the balconies were removed in the late 1930s. Although Jock Pedler was not especially interested in science, she was in the science club; she remembers that Payne-Scott was a reserved, serious teacher who suffered from poor eye-sight. Elizabeth Hallett (*née* Brookman) (see Fig. 3.2), remembers that Ruby was "dedicated to the sciences and was a patient and inspirational teacher as well as having a quiet and pleasant personality ... I cannot help thinking that her spell at Woodlands could only have been a . . . low spot in her career but the fact that three of the girls in the photograph (Fig. 3.2) gained their B. Sc.'s could have given her some satisfaction".[3]

This remarkable photograph from 1939 shows Payne-Scott supervising a group of physics students, including Elizabeth Brookman (above), students performing physics laboratory experiments on the properties of inclined planes (middle) and on the density of matter with accurate weights (right). In Fig. 3.3, we see this same classroom used by a class for young children in 2007 at the renamed St. Peter's

[3] Letter from Elizabeth Brookman Hallett to Goss, 22 June 2007. Hallett has lived much of her life outside Australia, having married a well known Polish airman who had been in the Royal Air Force in the UK in World War II.

Fig. 3.2 Photograph from 1939 of a physics class. The science mistress is Ruby Payne-Scott, second from left. Elizabeth Brookman Hallett (third from the left) is the only one student living in 2008. The unidentified student to the extreme left may be Jean Gooch (class of 1939) based on information from Jock Pedler in March 2007. From "Woodlands Reflections" 1999, on the occasion of the closing of the Woodlands School after 75 years. Photo provided by Ms. Dawn Geyer of the Woodlands Old Scholars' Association Archives Museum, 2003, permission given by Geyer

Fig. 3.3 The science classroom (see Fig. 3.2) (Gillam Building, St. Peter's Woodlands Grammar School) used by Payne-Scott in 1939 (Fig. 3.2) as photographed in March 2007. In 2007 this room was a Special Education Classroom with teacher Susan Bennett. The school offers both remedial and gifted programmes for kindergarten to year 7 students. Photo by Goss

Woodlands Grammar School. In addition to the photo from 1939, a book celebrating the 50th anniversary of the Woodlands school[4] shows the entire school in 1938 just after Payne-Scott had joined the staff. Miss Monica Millington (OBE) had just become the second Headmistress of the school. Payne-Scott is clearly visible in this photograph together with approximately 23 other staff and about 220 students, including 4–5 young boys in the front row. As always, Payne-Scott can be identified by her blond hair and prominent glasses. Jocelyn "Jock" (Britten-Jones) Pedler is also visible in this photograph.

Payne-Scott was the president of the Senior and Junior Science Clubs in 1938 and the first term of 1939. In 1938 the club reported that "Miss Payne-Scott gave the Senior Club a fascinating talk on Evolution, and has helped members interested in photography to develop their own film". In 1939, the school magazine reported in the school notes section: "May 11–30. School closed for the vacation. We said good-bye to Miss Payne-Scott, who left us to take up a position in Sydney". The Science Club reported at the same time: "At the end of the first term Miss Payne-Scott left us to go to Sydney, and we would like to thank her for all she has done for us, and to welcome Dr. Gruenfeld as our new President . . . Later on in the first term [while Payne-Scott was still the President], we spent an instructive hour at the Brighton Cement Works. After becoming rather hot and dusty, we appreciated the ice creams which were so thoughtfully provided for us".

Kathleen Foy (*née* Bampton) was in Payne-Scott's botany and physiology class in 1938. This class was a requirement for the Intermediate Public Examination Certificate, which Bampton obtained. Bampton remembered a field trip to a local creek near the school during which both flora and fauna were collected.[5] Kathleen Foy crossed paths with Payne-Scott later, during the years 1963–1974, when both were teachers at Danebank School.

The year 1939, with the approaching war in Europe and Asia, saw the expansion of the electronics industry in Australia. Thus additional opportunities for physicists were available. It was at this point that Payne-Scott left her job as a girls' school boarding mistress and joined Amalgamated Wireless, Australasia (AWA) for a period of slightly over 2 years, until August 1941. The approaching world war would have a major impact on her life and professional career.

Amalgamated Wireless Australasia: 1939–1941

Payne-Scott joined AWA as a librarian in mid-1939. Her colleagues were, among others, Chris Christiansen (see Chap. 13) and A.L. Green.[6] With Marie Clark

[4] *Woodlands 1923–1973*, 1973, pp. 52–53 in Chap. 4, "Problems and Changes".

[5] Communication from Foy to Carolyn Little in 1999; communication to Dick McGee in June 2007.

[6] Green (1905–1951) was the Director of Research at AWA from 1941 to 1947. Green was a young ionospheric scientist who had been recruited by the Australian Radio Research Board by J.P.V. Madsen in the early 1930s, having finished a Ph.D. with Appleton at King's College, London in

(Hooker 2004), Payne-Scott was only one of a small number of women on the professional staff at AWA (see also Chap. 8). After a period of being in charge of the measurements laboratory, she was able to branch out into research, undoubtedly due to her expertise in physics and mathematics as well as her passion for electrical engineering. She was not pleased with the quality of the research atmosphere at AWA. Her son, Peter,[7] remembered his mother telling him that AWA had major difficulties finding qualified engineers in 1939. One striking story was about a woman with a meager background and no university qualifications who was recruited from the drawing office at AWA to join an engineering division. This individual expressed her misgivings to Payne-Scott, who replied, "Never mind. Just put on a white coat and put a slide rule in your lapel pocket. No one will find out!"

Other professional women on the AWA staff were Marie Clark and Grace Noble. Grace Noble[8] and Ruby had a brief overlap as students at Sydney Girls' High School; she left in 1926, the same year that Ruby arrived. Noble had a geology degree from the University of Sydney. She worked initially at AWA, doing mathematical calculations. When Payne-Scott left AWA in 1941, Noble was recommended by her as a replacement. Noble wrote, "I don't think I was replacing Ruby in any real sense, but I do know that radar . . . [was] being perfected." Noble worked on the problem of finding pure quartz crystals (for radar and radio receivers) in Australia, as the supply source in Brazil disappeared during the War.

Payne-Scott was involved in two publications at AWA. In 1941 she wrote an article for the *AWA Technical Review* (it is likely that she was the editor of this journal for some period). The publication, "Superheterodyne Tracking Circuits – II"[9] was the second in a series of papers (the first was by Green 1941) on the problem of obtaining solutions for the tracking between the signal and oscillator circuits of superheterodyne receivers (as in FM and TV receivers); the wide band higher frequency was "mixed" (combined) with the local oscillator to produce a difference signal that was amplified and demodulated to receive the input video or audio signal. Lookup tables were presented to facilitate computations over a wide range of conditions.

The final publication by Payne-Scott was also published in the *AWA Technical Review*, but only after she had left AWA for CSIR in August, 1941. The paper, "Note on the Design of Iron-Cored Coils at Audio Frequencies" (Payne-Scott 1943), consisted of a detailed discussion of the calculations of the "Q" factor (the magnification factor) of audio circuits for iron-cored coils at frequencies of 1–10 kHz (such as would be present in a modern stereo system). The detailed electrical engineering recipe indicated that "at any specified value of frequency and

---

1934. In 1935, he joined AWA (Schedvin 1987). From 1947 to 1951 Green was in charge of the Ionospheric Prediction Service (see Chap. 9). Also see Chap. 13. At URSI in 1952 (Chap. 10), Appleton praised Green, who had died at the early age of only 46.

[7] Interview with Goss, Socorro, New Mexico, 12 May 1998.

[8] Letter to Goss, 22 March 1999. Noble also knew Bill Hall and Ruby in the Sydney Bush Walkers organization (Chap. 12).

[9] Ruby Payne-Scott and Alfred L. Green, 1941; the paper was later reprinted in *Wireless Engineer*.

flux density (of the coil), the Q of coil wound on a given core is a function only of the mass of copper in the winding, whatever the number of turns and gauge of wire used . . ." The frequency at which the Q of the coil was maximized was found to be inversely proportional to the square root of the mass of copper which was wound on the coil. These two publications indicated that Payne-Scott was certainly adept at finding solutions to practical electrical engineering problems. These skills were to serve her well as she began the next phase of her career, working on military radars necessary for the defence of Australia in the Second World War.

New Career Opportunities for Australian Women, World War II

The beginning of World War II in September 1939 had a major impact on Payne-Scott's career. Already in May 1939 she had joined the staff at AWA. Chris Christiansen wrote in October 1997:[10]

> I knew her before anyone else in R.P. because she came to work at AWA while I was there. She was put onto rather boring work and didn't last very long before applying (Chap. 4) for a job at R.P., where she was followed by Lindsay McCready, John Downes and a couple of others.[11] I didn't go with her because I had a most interesting job in the overseas radio communication and also because I didn't want to get into "secret" work particularly as I had a record from undergraduate days of being a left-wing militant and would probably be rejected for such government work (as indeed I later found that I would have been). I had no contact with Ruby until after the end of the war and could join the group at R.P. when they were starting work on radio astronomy.

The application and appointment of Payne-Scott to the secret Division of Radiophysics of the CSIR in August 1941 is described in detail in Chap. 4. With many male scientists in the armed services of Australia (participants in the war in Europe from September 1939 and Australia and the South Pacific became a major target for the Japanese armed forces in December 1941), women scientists had, for the first time, an opportunity for employment outside the usual avenues in health services and education. In the 1930s, the University of Sydney Appointments Board had conducted surveys of job advertisements in major Australian newspapers, finding that sex was a major factor for most employers in choosing an applicant. Hence female science graduates were urged to learn to type and perform shorthand (Carey 2002). But Carey then writes of the Sydney Board in 1941:

> Never before in the history of the world has there been so great a demand for women with scientific knowledge . . . Jobs which have . . . been the prerogative of men have opened their doors wide to women . . . Employers who once refused to take women scientists are now begging for them.[12]

---

[10]Letter to Goss. Undated letter, received in Socorro, New Mexico, 7 October 1997.

[11]See Chap. 4. Payne-Scott used McCready as a reference when she applied for the position at CSIR, RPL in August 1941.

[12]"Departing from their Sphere? Australian Women in Science, 1880–1960" in Xavier Pons (ed.), *Departures: How Australia Reinvents Itself*, 2002, p. 181.

The number of women working in Australia in wartime scientific endeavours is hard to estimate. At RPL in 1945 there were two women working in a scientific staff of about 60.[13] In a publication in July 1943, *Australian Women at War*, edited by Mollie Bayne, the emphasis is on the 30,000 women in three branches of the armed forces (Army, Navy and Air Force[14]) and industry (estimated to be slightly over 200,000). At that time it was estimated that a total of 580,000 women were involved in the War effort when the population of Australia was about 7 million. Only 10% of the women in the war effort worked for government entities, including local, state and Commonwealth. Scientists did not appear as a separate category in this study by Bayne. She also has a stirring call to arms for women in the post War Australian society. She complained that there were no women in the Federal Parliament and only one woman on a Federal Government Board or Commission concerning female employment, at a time when most Australian working women were earning about half the wages of men. She compared Australia unfavorably with the achievements of the UK, the USA and the USSR in creating a more equitable work environment for women. She asked:

> Are they [the women] less capable? Or is the community [Australia] mediaeval-minded in the matter of a "woman's place?"

What about other prominent women in war related research? Florence Violet McKenzie had become an electrical engineer based on course work at the University of Sydney and later Sydney Technical College in 1923.[15] In 1940, she formed the Women's Emergency Signalling Corps and succeeded in getting 14 girls, trained as wireless telegraphists, into the WRANS (Women's Royal Australian Navy) on Anzac Day (April 25) 1941. As a volunteer, she was in the end responsible for training 12,000 servicemen and hundreds of women who all joined the women's armed services. She was made an honorary Flight Officer in the WAAAF (Women's Auxiliary Australian Air Force). Patsy Adam-Smith has written:

> After Japan entered the war, Mrs. Mac's [McKenzie] girls were already working as the nucleus of a service which, at the time of its greatest expansion, amounted to approximately 10% of the Royal Australian Navy.[16]

---

[13]Payne-Scott and Freeman. Rachel Makinson worked at Sydney University, partially supported by RPL The total staff at this time was about 300 (Evans 1970).

[14]In the UK, women played major roles in the operation of the Chain Home radar stations (Rowe 1948). Rowe wrote: "All honour to the women who shared with the men the often primitive and isolated conditions at the radar stations and who carried on with their tasks when the stations were attacked by the enemy" (p. 27). In the Australian sphere, it is likely that all radar operators were men since many of the radar stations were in the war zone in the SWPA. Many of these men were trained in the "Bailey Boys" classes at the University of Sydney; most were in the Air Force although a few were from the Navy and Army (Fielder-Gill et al. 1999). By contrast, in New Zealand a number of the radar station personnel in the Royal New Zealand Air Force were women.

[15]As was Payne-Scott, Florence McKenzie (*née* Wallace) was a graduate of Sydney Girls' High School. She was the first woman radio amateur (ham) in Australia.

[16]In *Australian Women at War*, 1984, p. 210 – another book with the same title as the Bayne volume.

There is little doubt that McKenzie, Payne-Scott, Rachel Makinson and Joan Freeman (Chap. 11) all played significant roles in the Australian War effort. Undoubtedly, there are many more.

A very rough comparison may be made between the role of women in radar research in the UK and Australia in World War II. R.V. Jones (1978, *Most Secret War*) and Louis Brown (1999, *Technical and Military Imperatives: A Radar History of World War II*) in their comprehensive publications about World War II have described the efforts of Joan Curran (*neé* Strothers, later Lady Curran) in the development of project Window[17] in March 1942; this was called chaff by the Americans. The aluminium strips (used in project Window), which were dropped from bombers to confuse the enemy, resonated with the 50 cm Wurzburg German radar transmissions; on the strips propaganda messages were written. The objects were quite elongated with a length of about 25 cm and a width of only 1–2 cm. Robert Cockburn was the leader of this group at the Telecommunications Research Establishment (TRE). Project Window was first deployed over Hamburg on 23/24 July 1943, causing remarkable confusion among the German defensive anti-aircraft radars. It is striking that Rowe (1948) and Lovell (interview May 2008 with Goss) are more reserved in their assessment of Curran's role in the Window concept , as compared to Jones and Brown.[18] Lovell (1991) also mentions Curran in passing.

Endnotes for Chap. 3: Woodlands School

Glenys Edwards, a volunteer librarian in the Patchell Library Historical Centre at Annesley College in Adelaide, was the facilitator for much of the information received about Payne-Scott at Woodlands School. In March 2007, Miller and Libby Goss visited the school (now St. Peter's Woodlands Grammar School) with Edwards. Wendy Davis of St. Peter's Woodlands Grammar School (Development Officer) provided tours of the school.

In 1998, the original Woodlands Church of England Girls' Grammar School closed; the school had been founded in 1923. In that 75 year period over 7,000 students attended the school. The new co-educational school, St. Peter's Woodlands Grammar School, opened in early 1999 with a staff of 26 and 290 students.

[17] The name was invented by A.P. Rowe, the Superintendent of the Telecommunications Research Establishment (TRE) in the UK.

[18] Another UK woman who worked at TRE was Elizabeth Palmer (letter from Sir F. Graham-Smith, 29 June 2007 to Sullivan). Graham-Smith was in the Test Gear section of the TRE, under the direction of J.A. Ratcliffe. His future wife, Elizabeth Palmer, was a Junior Scientific Officer from 1943 to 1945 in the Trainer Group working on simulators for training radar operators, similar to Payne-Scott's experience in Sydney (Chap. 5). Graham-Smith and Elizabeth were married in 1945, just before returning to Cambridge University where he was to finish his B.Sc. degree. Elizabeth was a research assistant at the Cavendish Laboratory working on the radio solar patrol for Martin Ryle until 1947.

The Woodlands Old Scholars' Association Inc. has an office in the Law Smith Building, maintaining the archives of the previous school. Miller, Libby Goss and Edwards were hosted by the Woodlands Old Scholars' Association Archives Museum Group, including Dawn Geyer,[19] Jill Colley and Mary Carver. The late Jocelyn Pedler (*née* Britten-Jones, formerly Cherry), class of 1939, was also present at Woodlands on this day, providing first-hand memories of Payne-Scott in 1938–1939. Pedler had been a boarder and thus provided additional information about Payne-Scott's role as a "boarding mistress".

[19]Geyer was the Coordinator of the Group in 2007. Information was also provided in letters, emails and telephone conversations with Geyer in 2007 and 2008. Other members of the Archives Museum Group were Christine Ryan, Jocelyn Fuller, Anthea Smith and Margaret Steenvoorde.

Chapter 4
Personnel File from CSIR/CSIRO

The utilisation of Payne-Scott's personnel file is an essential resource in following her 10 year career at the CSIR Division of Radiophysics from August 1941 to May 1949 and then the CSIRO until her resignation in July 1951. By following this detailed record,[1] we cannot only trace her career path as a scientist but also obtain a vivid picture of her difficulties in being a professional woman in the CSIR and CSIRO. This 10-year period began with War time research on military radars (1941–1945) and ended with a productive 6-year period as one of the first solar radio astronomers (Chaps. 7–9); an initial radio astronomy experiment was carried out in March 1944 (Chap. 6). The details of her scientific work during the War are covered in Chap. 5.

Payne-Scott's personnel file reveals many relevant aspects of her career as a scientist and as a woman employee in the CSIR and CSIRO. The conflict that ensued in 1950 about her marriage received some attention in various media reports in Australia; the details have often been reported in a distorted fashion.[2] Based on the personnel file, it is possible to reconstruct the precise events of 1950, as well as details of her resignation in 1951.

Payne-Scott Joins the Radiophysics Laboratory

In the course of 1941, additional scientific staff members were required at the RPL; an advertisement was placed in early 1941 for eight new scientific officers.

[1]National Archives of Australia (Canberra) file, NAA: A8520, PH/PAY/002.

[2]An example was in *The Australian* newspaper (10 March 1999) in an article by Sian Powell concerning Ruby Payne-Watson (sic): "The CSIRO refused to employ married women until sometime after the War and single women who married were politely shown the door ..." Also the ASIO report of 1959 discussed in Chap. 13 contains a distorted version of her conflict with the CSIRO in 1950.

W.M. Goss and R. McGee, *Under the Radar*, Astrophysics and Space Science Library 363, DOI 10.1007/978-3-642-03141-0_4, © Springer-Verlag Berlin Heidelberg 2010

The deadline for the receipt of the application was 28 April 1941. The advertisement stated:

> Applicants should posses a University degree in science or engineering. ... A knowledge of, or experience in, high frequency oscillations would be an additional qualification.

The salary range was to be in the range £A344 to £A588 per annum.

Well after the deadline, on 4 June 1941, Ruby Payne-Scott sent in her letter of application. She got off on the wrong foot by misspelling the name of the Chief of Radiophysics. The letter began:

> Dear Dr. Martin [sic][Martyn was correct]: I should like an opportunity to discuss with you the possibility of my joining the staff of the Radiophysics Laboratory. I am a graduate of Sydney University with first class honours in mathematics and physics, and obtained my MSc. degree in physics in 1936 ... Mr. Macready [sic][in fact McCready], who is a member of your staff, has known me at the University and at AWA and could be referred to for further information. I would be grateful if you would consider this letter as confidential ... [return address listed as 5 Farleigh Street, Ashfield].

On 17 June 1941 David F. Martyn wrote to the CSIR Secretary with a description of eight individuals previously proposed to the Minister of the CSIR for approval; however, two of them could not be released from their current positions. One was from the Patents Office in Australia and the other (Walker) from the New Zealand Department of Scientific and Industrial Research (DSIR).[3] Two late applications were therefore to be considered as replacements for these individuals; these were from A. Richardson and R. Payne-Scott, both from AWA. Martyn had heard favourable reports about both of them and recommended (on 17 June 1941) that an offer be made to Richardson; he planned to interview Payne-Scott personally before making a recommendation. Two days later he met Payne-Scott and was impressed. Then on 20 June 1941, the Secretary of the CSIR wrote to the Minister for Scientific and Industrial Research, Mr. H. Holt,[4] pointing out that two late applicants for the RPL positions should be appointed, Richardson, an electrical engineer, and Payne-Scott, a physicist. Both appointments, as Assistant Research Officer, at starting nominal salaries of £A434 and £A384 respectively, were made "without commitment" – i.e., there was no commitment for future superannuation (pension) rights. In other words, these were not permanent positions. In subsequent years, the issue of superannuation was to be the source of increased conflict between the CSIR/CSIRO and Payne-Scott. In addition, Payne-Scott suffered a slight salary reduction in the new position as her salary at AWA would have been £A400 per annum on 1 July 1941.

[3]Based on personnel file of Payne-Scott. Likely this deliberation was the first occasion that F.W.G. White and Payne-Scott had an interaction. White had just arrived in Sydney a few months earlier (March 1941) on secondment from the Department of Scientific and Industrial Research in New Zealand (see also footnote 25 Chap. 6). In the 17 June letter from Martyn he reported:" Professor White informs me that Dr.[E.] Marsden [the Director of Scientific Developments at the DSIR in Wellington, New Zealand] has told him that he does not propose to release Walker." Thus the position for Payne-Scott opened up.

[4]Holt was the Prime Minister of Australia for a short 22 month period between 1966 and 1967.

On 17 July 1941 Payne-Scott wrote to G. Lightfoot, Secretary of the CSIR in Melbourne, accepting the position which had been offered to her on 26 June; she said that she could only begin the new assignment on 18 August 1941 due to the pressure of "urgent jobs" at AWA.[5] Apparently, she was somewhat late in responding to the offer. She apologized for the delay as being due to her absence from Sydney on a holiday.

Payne-Scott began her new career on Monday, 18 August 1941. This was close to 4 months before Japan's Pacific War was launched against the Allies, including Australia and the USA, in early December 1941. The importance of secrecy was highlighted on her first day at work as Payne-Scott signed the "Secrecy Declaration", promising not to reveal any of the details of the work of RP, except as authorised by the Chief of the RPL. The penalty for any offence would be administered under the Crimes Act of 1914–1937. The signature was witnessed by H.J. Brown, who had been an early appointment (1939) from EMI (Electric and Musical Industries) in England, and by the Assistant Chief, J.L. Pawsey, who was to play a major role in her subsequent career in radio astronomy at RPL.[6]

Payne-Scott joined a cohort who were all to have illustrious careers in disparate disciplines in the coming decades.[7] The work hours were listed as a nominal 36.75 h per week with extra time ("War effort") extended from 17:06 to 17:18 daily and from 9:00 to 12:00 on Saturdays.

1941–1946, Permanent Employment During the War

As stated above, Payne-Scott began at a salary of £A384. By the end of December 1942, the new Chief of RPL, F.W.G. White, recommended that her salary be increased to £A402 per annum; at this time a male colleague on the same

---

[5]The AWA laboratory was located at Parramatta Road, Ashfield.

[6]Pawsey (1908–1962) was originally from Victoria and had studied at the University of Melbourne. He did a Ph.D. degree with J.A. Ratcliffe at the Cavendish Laboratory of Cambridge University in the years 1931–1934. He joined the RPL (also from EMI in the UK) as one of the first scientific appointments from late 1939, starting work on 2 February 1940. Pawsey was the major driving force (along with Bowen) for the development of radio astronomy in Australia at the CSIR/CSIRO starting in 1944. See also footnote 3 (Preface).

[7]Other prominent appointments at RPL in mid-1941 were T.B. (Bruce) Alexander (previously from the Royal Australian Navy and after the War Chief Engineer of Johnson and Johnson in Australia), M. Beard (from Standard Telephones and Cables whose subsequent career at CSIR/CSIRO was working on electronic computers and radio astronomy computing), B.F.C. Cooper (graduate in 1941 from Electrical Engineering, Sydney University, who played a major role in the Darwin radar response to the Japanese attacks in early 1942, and had a long career at RPL working on radio astronomy instrumentation and aircraft landing systems, see also footnote 22, Chap. 5), F.J. Kerr (from Melbourne University and the Radio Research Board with a later career at CSIR/CSIRO and the University of Maryland as one of the pioneers in 21 cm HI research), and Joan Freeman (from Physics, Sydney University – see Chap. 11 for a description of her distinguished career, which culminated as a scientist at the Atomic Energy Research Establishment at Harwell in the UK).

classification but with more experience was earning £A428 per annum. The three highest salaries at the time (not including that of the Chief) were those of H.J. Brown, J.H. Piddington and J.L. Pawsey at £A675. A year later the salary of Payne-Scott was raised to £A420 per annum, to take effect on 1 January 1944.

In July 1944 there was an exchange of letters from the Commonwealth Public Service Professional Officers' Association and the CEO of the CSIR, Sir David Rivett, about the issue of equal pay for women. Payne-Scott was the only female staff member of CSIR who was a member of this professional organisation, a type of trade union. The Women's Employment Board (WEB, see Appendix H) made a ruling in September 1944 that females were to be paid 100% of the male rate for the same class of work.[8] This ruling was clearly made to compensate women for carrying out essential services during wartime; otherwise the women's rate of pay was £A97 per annum *less* than the men's pay scale for Research and Technical Officers.

In mid-1946, Payne-Scott was given a promotion from Assistant Research Officer to Research Officer. The memo from the Chief of RP, John N. Briton,[9] dated 25 February 1946, stated:

> During the past 12 months Miss Payne-Scott has demonstrated her real ability as a research officer. You will recall that earlier we were in some doubt as to her suitability for continuing with the Division. Since then, however, her work, particularly in connection with investigations concerning the visibility of signals and noise measurements, has been quite outstanding and I wish to recommend that she be reclassified as Research Officer, with effect as from 1st July, 1946.

Nowhere in the personnel file do we find any descriptions of the doubts referred to by Briton. A few months later (May 1946), Taffy Bowen[10] became Chief of RPL.

---

[8] As early as the first part of 1941, the Australian Council of Trade Unions (ACTU) had called a meeting of all Federal Unions with women members for a discussion of policy on the role of women in Australian industry. After adopting a resolution that women be paid an equal wage, the full body of the ACTU adopted this position in June 1941 at a time when the average rate of pay for women was only 54% of that of males. The resolution had six conditions including details about creating wage parity for women. Connected to Payne-Scott's problems in 1950 about her marriage (see later in this chapter), there was the important final provision, "... the removal of all restrictions on the employment of married women in gainful occupations and the recognition of their right to economic independence" (Bayne 1943, p. 56).

[9] Briton had succeeded to this position when White moved to the CSIRO Executive in January 1945. Briton had been seconded to RPL during the War from the Grammophone Company in Homebush, NSW (head of manpower from1938 to 1942). He was Deputy Chief for Engineering at the RPL from 1 June 1943. He returned to this employer in May 1946 when Bowen became the RP Chief. In 1952, Briton became the Technical Director of EMI Industries, Australia.

[10] E.G.("Taffy") Bowen (1911–1991) did a Ph.D. degree with E.V. Appleton at King's College, London. He worked with Robert Watson-Watt starting in 1935, and played an important part in the early development of radar in the UK. He went to the US as a member of the Tizard Mission in 1940 and worked at the Radiation Laboratory in Cambridge, Massachusetts (USA) until late 1943, when he joined the CSIR RPL in early 1944. In 1946, Taffy Bowen became the Chief of RPL, where he provided the leadership in the post War research programs of RPL, mainly in cloud physics and radio astronomy. See the obituary by Brown et al. (1992).

Payne-Scott's salary was raised to £A458 per annum at this time; however, a series of events began in July 1946 that were to have far reaching consequences. Bowen realized that many of the more senior RPL staff still did not have "with-commitment" appointments and were thus not eligible for superannuation or a CSIR sponsored pension. Perhaps the exigencies of the War had led to this neglect. In any case, Bowen wrote to G.A. Cook (the new Secretary of the CSIR) insisting that due to these long delays, 19 individuals be given "with-commitment" status, i.e., a permanent appointment with pension rights. Within weeks the ball was rolling; these individuals were called to a medical examination by the Commonwealth Medical Officer at Circular Quay in Sydney to investigate whether the staff member was healthy enough to be expected to "discharge his or her duties until 60 years of age without more than average sick leave …". After some delays, Payne-Scott appeared for her medical examination on 17 September 1946 and was then referred to an eye specialist, Dr. St. Vincent-Welch. Apparently, Payne-Scott's poor eyesight produced some doubts about her long-term health prognosis. On 19 November 1946 Cook wrote to the Minister for Scientific and Industrial Research, J.J. Dedman,[11] with the recommendation that 15 CSIR staff be "gazetted" as "employees within the meaning of Section 4 of the Superannuation Act." Payne-Scott was in this list but recommended for the "Provident Fund" instead of the "Superannuation Fund."[12] The official list was published on 21 November 1946, the date of "gazettal".

A few days later Payne-Scott's classification within the CSIR was changed to Research Officer, Grade III, with a substantial salary increase from £A458 to £A588 per annum.[13] This rapid advance in salary level was continued with an additional increase to £A675 per annum on 1 July 1947; at this point Payne-Scott's salary was comparable with or higher than those of many of her colleagues, e.g., Minnett at

[11]b.1896, d.1973. Minister of Defence and Minister in Charge of CSIRO in the Chifley Labor government during the time of the Kaiser affair (which began in London on 27 July 1949; see footnote 17 this chapter and Chap. 13). John Dedman was the Minister in Charge of CSIR/CSIRO from 1941 to late 1949.

[12]Since Payne-Scott was judged to have poorer health prospects, she was forced to join a "Provident Fund" instead of the superannuation fund. With the superannuation fund, the employer made contributions to supplement the employee contributions. At retirement (for periods in excess of 10 years), the employee received an annual benefit that was a fixed fraction of the last yearly salary. Peter Davidson of the Australian Council of Social Service wrote to Goss on 13 December 2006: "… There was [most likely a] CSIRO Provident Fund and the level of benefit reflected the level of contributions made and the length of membership … CSIRO Staff Provident Fund … would have been owned by the contributing members and the employer (CSIRO) may well have made small contributions but [these] would have been considerably less than the contributions to staff super[annuation]. I assume that the limitations on membership were much less restrictive than those pertaining to the super[annuation] fund. Hence temporary staff and casual staff and staff with some medical conditions would have been eligible for membership. Payments would have been less than the super[annuation] entitlements, even … comparing equal periods of employment". Thus the Provident Fund was a reduced benefit for those individuals not thought likely to attain a retirement age of 65. Payne-Scott lost all her accumulated benefits in 1950 after a refund of her own contributions with no interest (see below).

[13]Memo from J.L. Pawsey who was acting Chief of RPL while Bowen was overseas.

£A725 and both Bolton and Wild at £A500. Both Payne-Scott and Freeman – also now a Research Officer, Grade III, with an annual salary of £A650 – were still being paid at the WEB rates (see Appendix H). In addition, in mid-1947 Payne-Scott also received a substantial windfall; she was paid £A129 in arrears for the WEB award from the period 14 September 1944 to 1 January 1947. This back pay represented an amount of over 2 months' salary. The salary advancements continued on 1 July 1948 with an increase to £A700 per annum, on 1 July 1949 to £A725 per annum, and on 1 July 1950 to £A750, the last full year that Payne-Scott was employed at CSIRO.

Transition from CSIR to CSIRO; Discovery by CSIRO of Marriage of 1944

The proclamation of the law establishing the new Commonwealth Scientific Industrial Research Organisation was made on 19 May 1949, and so the CSIR transitioned into the CSIRO. This transition necessitated a new oath or affirmation. Payne-Scott signed the oath shown in Fig. 4.1, witnessed by Arthur Higgs, the Technical Secretary of RPL. There were two choices for the oath, differing in a key phrase. *So help me, God* was one option, while the other option omitted this sentence. Payne-Scott chose this latter option on 29 June 1949, bearing allegiance to the British King: " . . . do solemnly and sincerely promise and declare that I will be faithful and bear true allegiance to His Majesty King George the Sixth, His heirs and successors according to law." Thus, there is no mention of "God" in this alternative version of the oath signed by her.

The confused and unpleasant events of 1950 began with the official discovery that Payne-Scott had been married since 1944; the consequences were severe for her career at CSIRO. On 8 September 1944 William (Bill) Holman Hall and Ruby Violet Payne-Scott were married. His address was given on the marriage certificate as 6 Stanley Street, Arncliffe, with profession "telephone mechanic", while her address was listed as 5 Farleigh Street, Ashfield, with profession "physicist". The witnesses were Amy Neale, her aunt (Chap. 2), and Mary Pilgrim, Hall's sister.[14] Both of Payne-Scott's parents, as well as Hall's father, were deceased. B.Y. Mills has told us about this day:[15]

> An interesting event occurred early in this work when, quite uncharacteristically, Ruby left early one afternoon without explanation. She later[in fact about a year later] told me that this departure was to marry Bill Hall, something which she found necessary to keep secret . . .

[14]Mary N. Pilgrim (*née* Hall) born in 1908 or 1909, probably in Inverell, New South Wales (Australia). Married to Clifford H. Pilgrim at Kogarah, NSW, in 1933. She was about 3 years older than Bill Hall. Date of death unknown but probably after 1977.

[15]Letter to Goss dated 14 September 1997.

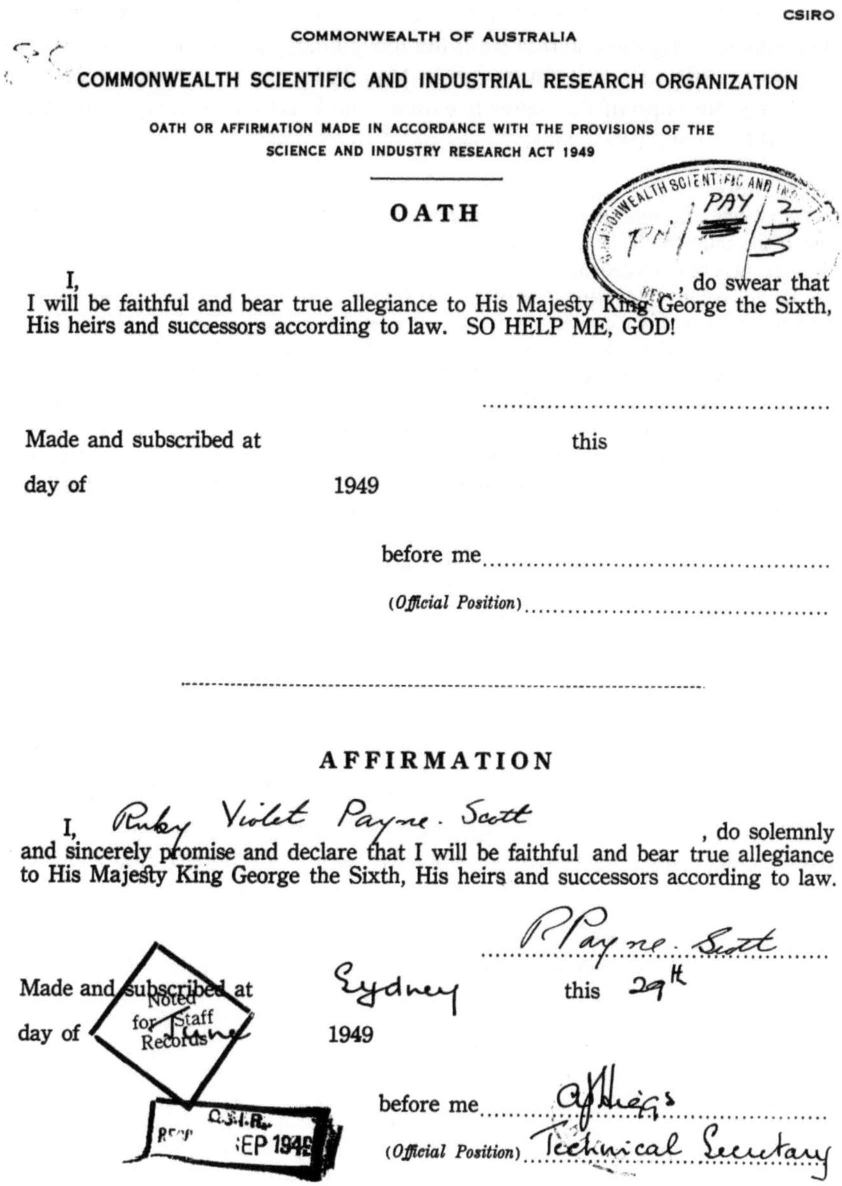

CSIRO 90

COMMONWEALTH OF AUSTRALIA

COMMONWEALTH SCIENTIFIC AND INDUSTRIAL RESEARCH ORGANIZATION

OATH OR AFFIRMATION MADE IN ACCORDANCE WITH THE PROVISIONS OF THE SCIENCE AND INDUSTRY RESEARCH ACT 1949

OATH

I, , do swear that I will be faithful and bear true allegiance to His Majesty King George the Sixth, His heirs and successors according to law. SO HELP ME, GOD!

Made and subscribed at this

day of 1949

before me...

(Official Position)...

AFFIRMATION

I, Ruby Violet Payne-Scott, do solemnly and sincerely promise and declare that I will be faithful and bear true allegiance to His Majesty King George the Sixth, His heirs and successors according to law.

R Payne-Scott

Made and subscribed at Sydney this 29th

day of June 1949

before me A J Higgs

(Official Position) Technical Secretary

Noted for Staff Records

Fig. 4.1 *Affirmation* required by the newly formed CSIRO, signed by Ruby Payne-Scott on 29 June 1949. She did not sign the other option, an oath including the phrase "So help me God". Witness was Arthur Higgs, the Technical Secretary of RPL. From NAA: A8520, PH/PAY/002, page 33 (Personnel file). From the collection of the National Archives of Australia

From a conversation with a friend of Payne-Scott at the RPL in this period of mid to late 1940, we have heard that she wore her wedding ring on a necklace instead of on her ring finger.

Six years later the CSIRO bureaucracy discovered the fact that Payne-Scott was married; this investigation started from the top. In early 1950, Ian Clunies Ross,[16] Chairman of CSIRO since its inception in May 1949, wrote to Payne-Scott about her marriage. No copy of this letter is extant. The handwritten reply from Payne-Scott of 20 February 1950 reads:

> Dear Dr. Ross
>
> Your letter [someone else has written in the word "personal" at the top of the letter to indicate the nature of the letter to Clunies Ross] to me came while I was away on holidays, and in the rush of unprecedented solar activity of the last few weeks I seem to have mislaid it, so I will have to reply from memory. Thanks for your inquiries on my behalf, but when I spoke to you about my marriage I was in effect asking you whether the Executive realises that the customary demoting of women officers on their marriage to the status of 'temporaries' does not appear to be required in the Act and whether the Executive agrees with this procedure or not. Whether or not there are material disadvantages to the women concerned in this procedure, all the married women research officers I have met feel that their classification as "temporary" puts them at a considerable psychological disadvantage in their work. Personally I feel no legal or moral obligation to have taken any other action than I have in making my marriage known. I have never denied to anyone who has asked me the fact that I am married, and it has gradually become common knowledge in the laboratory, particularly as many of the staff are my close neighbours at home. More recently I have stated the fact that I am married, among other information, in a form that I was asked to complete and return to Head Office. I have, of course, also stated it on my income-tax forms . . . I should still be very interested to know whether the Executive is in agreement with the present procedure with regard to married women, which seems to go far beyond the simple statement in the Act that employment may be continued when it seems desirable. I told you my story not in order to implicate you in any way but to demonstrate that the present procedure is ridiculous and can lead to ridiculous results.

The implication in this letter that a personal meeting (". . . when I spoke to you") had occurred is quite likely, given the report in the CSIRO Officers' Association (CSIROOA, also called the Officers Association or OA) *Bulletin* of December 1949 in which Payne-Scott wrote:

> The sub-committee on women's pay took the opportunity of the recent visit of Dr. Clunies Ross to Sydney to discuss the question of women's pay with him. He stated that the Executive favoured the principle of equal pay but might be in a difficult situation if it attempted to set a precedent in the Public Service ... Dr. Ross promised to investigate the position ...[17]

[16]Clunies Ross succeeded Sir David Rivett. Previously he had been Director of Scientific Personnel in the Commonwealth Directorate of Manpower after a career in the field of veterinary medicine at both Sydney University and the CSIR starting in 1926.

[17]Deery (2000) has reported that Rachel Makinson (Chap. 11) wrote to him on 8 November 1998 about this meeting. The meeting was probably in the period September–November 1949 in Sydney. Makinson attended a meeting of the OA in which Tom Kaiser (a CSIR employee on a student assistantship at Oxford) was attacked for engaging in public controversy in London earlier in 1949 (see also footnote 7 Chap. 13 and footnote 78 Appendix H). Clunies Ross even claimed that the stress of this event had led to his heart attack, an assertion that Makinson thought quite "unfair". Makinson was one of the few attendees who spoke ("cautiously") in support of Kaiser along the grounds of freedom in science and even invoked Sir David Rivett (CSIR CEO from 1927 to 1945 and Chairman from 1946 to the creation of CSIRO in May 1949) in support of this view.

Thus it is quite possible that in late 1949 she and Clunies Ross did discuss these issues in person at a meeting in Sydney. Appendix H contains a discussion of Payne-Scott's role in the controversy about the continuation of the WEB (equal wages for women) provision in the post War period.

On 3 March 1950, Clunies Ross wrote a long, scathing response to Payne-Scott's 20 February 1950 letter, in which she was chastised for her behaviour. This letter is reproduced in full in Appendix I. Clunies Ross did admit that in the act establishing the CSIRO, no mention of married women appeared in the legislation itself. But procedures were established under the Act which did imply that "a female officer shall be deemed to have retired from the service of the Council [still operating under the old terms of the CSIR], unless the Minister, upon recommendation by the Council, certifies that there are special circumstances which make her employment desirable". The negative result, as far as Payne-Scott was concerned, was the clause in the Superannuation Act which read:

> A female officer who marries after the commencement of this section shall for the purposes of this Act be deemed to have resigned from the date of her marriage.

Clunies Ross then challenged her statement concerning the issue of a moral obligation to inform the CSIRO about the marriage. He even seems to have doubted that Payne-Scott had submitted the form about her marriage; however, she had in fact done so by submitting a document on 23 January 1950. This form was probably required for all new appointees for the newly formed CSIRO. This form did not reach the CSIRO administration until much later. He wrote:

> In conclusion, I think the simplest way of regularizing the whole affair would be for you to tell us the date of your marriage. We will then look into the matter and tell you what should be done in your own and our best interest.

Then on 3 April 1950, Clunies Ross wrote to Payne-Scott again with a complaint about not having heard from her for a month about "the position created by your marriage. I should be glad if you would let me know without delay whether you intend officially to notify us of your marriage, since I do not feel that we can delay much longer in taking action ...".

Payne-Scott responded on 12 April 1950, pointing out that she had informed the Head Office earlier and that the forms had been delayed in the RPL office awaiting the collection of all other applications (from all RP staff). On 24 April 1950, Bowen did write to Cook, with the dates of the marriage.

At this point, the financial implications of this imbroglio began to be implemented. Cook wrote to both the Superannuation Board and RPL that Payne-Scott was not eligible for the Provident Fund due to her marriage and that a refund of her contributions was required (the CSIR and CSIRO contributions were not to be refunded to her). In the letter to Payne-Scott, the statement was made that she was to

---

Clunies Ross "dismissed us all rather roughly" according to Makinson. At least two other events occurred around this meeting: (1) a discussion about equal pay for women at the CSIRO (Appendix H), and (2) a private conversation between Payne-Scott and Clunies Ross.

be continued as an employee with temporary status, with the requirement for a new appointment each year. The indignity was complete when Cook told the Superannuation Board that the date of the "gazettal" in 1946 was after her marriage in 1944 and thus even the interest on her contributions from 1946 to 1950 was to be disallowed. A dismissive document was prepared by Cook for the CSIRO file that provided a short summary of the whole affair.[18]

In following the chronology of the paper trail, it is plausible that after the CSIR became the CSIRO in mid-1949, the new administration felt the need to survey all of its employees. This survey consisted of filling in the form that every employee had to complete and which Payne-Scott truthfully filled out as a married employee. It was about a month after she filled out the form, and while it was languishing in an administrative inbox waiting for the forms of other employees to arrive, that she questioned Clunies Ross regarding his views on the disparity of treatment between men and women in the workplace, and whether the old penalties for being a married woman still applied under the new codes of the CSIRO. It is likely that during her CSIR years (1941–1949), Payne-Scott had become a part of a scientific community that was happy to overlook unjust, bureaucratic fine points in favour of outstanding performance during War time and the immediate post War periods. Apparently only after the administrative overhaul in the post CSIRO era (mid-1949) did Payne-Scott face renewed scrutiny from CSIRO functionaries, after which her marriage from 6 years earlier became a major issue.

In the end, the result of this brouhaha was the loss of Payne-Scott's pension contributions that the CSIR and CSIRO had made in the period 1946–1950. She had no superannuation funds at all after her 9 years of service to RPL. Although her continued employment was not in jeopardy, the designation as a "temporary" was quite demoralizing.[19] It is quite possible that this negative experience influenced Payne-Scott's perception of her future at RPL in the course of 1951; she never expressed any interest in returning to CSIRO after the birth of her children.[20] Only

[18] In an undated handwritten note for the Payne-Scott personnel file he wrote: "This 'marriage' was kept 'secret': for over 6 years. We gazetted RPS after she was married! We registered a protest but apparently glossed over it from then on, for without doubt we wanted employment to continue. There doesn't seem though, as if there was any disciplinary action of any kind."

[19] In interviews with Goss in August 2003 and February 2007, Rachel Makinson (Chap. 11) stated that her similar treatment as a "temporary" at the CSIRO Division of Textile Physics in the years 1953–1982 was quite demoralizing.

[20] Joan Freeman (Chap. 11) was sympathetic when she wrote in her autobiography (1991): "There was only one case involving sex discrimination, beyond the powers of RP to control, which Ruby [Payne-Scott] went to extraordinary lengths to circumvent. It was in 1944 that she let it be known that she was living with a man (Bill Hall) to whom she was not married. Nowadays little would be thought of such a situation, but in the 1940s 'living in sin', as it was called, was looked on askance". Payne-Scott carried on "unperturbed". "Ruby had hoped, by her deception, to evade what she considered to be an outrageous and discriminatory law. All her RP friends, having developed a strong affection for Ruby as well as respect for her scientific abilities, greeted the story with hilarity, and sympathized with her attitude." Payne-Scott did not see the humour in the consequences of the CSIRO reaction.

in 1966 was the regulation that married women in the Australian Public Service had to resign their permanent employment on marriage rescinded.

Payne-Scott's Resignation in July 1951

Payne-Scott's resignation was quite precipitous as events unfolded in mid 1951. On 18 July 1951, she wrote to Pawsey that she was resigning on the afternoon of 20 July (only 2 days hence) for "private reasons."[21] We know that the reason was her pregnancy with the birth of Peter expected in November 1951. Undoubtedly, Pawsey knew about the upcoming resignation as F.W.G. White, then the CEO of CSIRO, wrote her a letter on the previous day (17 July 1951):

> Dear Miss Payne-Scott,
> I have just seen a letter from Dr. Pawsey which tells me that due to the imminent arrival of your baby, you will not be able to carry on with your research work. This event must be giving you a great deal of pleasure but I can well imagine that you regret having to leave off research, at least for the time being. Unfortunately we cannot give a married woman leave without pay,[22] but I can assure you that I at least would be very pleased to see you return to Radiophysics in due course. I hope the event goes off successfully.

White certainly knew Payne-Scott well as they had both joined RPL at roughly the same time in 1941 and were associated in the research activities of the lab during the War.

Payne-Scott wrote a poignant letter to White from her temporary residence in the home of Richard and Rachel Makinson,[23] while these scientists were on sabbatical in the UK. On 15 August she wrote to White:

> I am writing to thank you for the kind letter that you sent me on the eve of my departure from the Radiophyscis Laboratory. I am, of course, sorry to give up the research work I have been doing and also to leave the laboratory where I have been so happy and have so many friends. If all goes well I do not expect to be returning to Radiophysics at least for some years, but I hope that I may be able to hear the latest on solar noise at the

[21] Payne-Scott was busy in the last weeks of her employment at CSIRO. In early July, Alan Maxwell (a graduate student at Jodrell Bank who came from Auckland) wrote John Bolton a letter which included a series of questions about RPL work on the major sunspot of 14 June 1950 (see Chap. 9 for a description of this Type I solar event). Bolton passed the letter on to Payne-Scott, who wrote back on 11 July. The enhanced radiation was the most intense ever observed at Potts Hill with the swept-lobe interferometer. Payne-Scott gave Maxwell a detailed account of the polarisation state and the position over a period from 13 to 19 June 1950. Maxwell wrote on 31 July with additional questions; from her home in Oatley Payne-Scott wrote to Pawsey asking that he or Alec Little reply to Maxwell. She was doubtful that a reliable angular size of the Type I storm could be determined. Little replied on 1 November, with text largely contributed by Payne-Scott.

[22] This rule was changed in 1973 when Australian Public Servants were given 52 weeks' maternity leave, 12 weeks of which were paid.

[23] 16 Nymbidia Street in Coogee, NSW, an eastern coastal suburb of Sydney. See also Chap. 13.

> A.N.Z.A.A.S. [Australia New Zealand Association for the Advancement of Science] and more particularly the U.R.S.I. [International Union for Radio Science] conference in Sydney next year.

An amusing episode, illustrating the ironic nature of the bureaucratic structure of the CSIRO, occurred as Payne-Scott left RPL in 1951. Pawsey wrote to Cook on 24 July 1951, concerning four items that had been checked out to her some years earlier and could not be found after she left on the previous Friday, 20 July 1951. These items were a soldering iron, a small hammer, some pliers and a side cutter (wire cutter). The total replacement cost was less than two Australian pounds. Pawsey asked Cook: "In view of the length of time Miss Payne-Scott was attached to the Division [10 years] it is recommended that these items now be struck off charge at public expense." Cook agreed to this request on 30 July and Payne-Scott did not suffer the indignity of paying this small sum to CSIRO for the lost items.

The high esteem in which Payne-Scott was held is exemplified by the salary adjustments that were being implemented for the start of the new fiscal year on 1 July 1951, just before her resignation. Already on 27 February 1951, a recommendation had been made that she be promoted to Senior Research Officer, Grade I, at a salary of £A920 per annum.[24] Hers was the second highest salary of any of the scientific staff not in an administrative position; only H.C. Minnett had a higher salary of the scientists who were being reclassified in mid-1951.

At the farewell party held on 20 July 1951, Dick McGee was present as Pawsey described Payne-Scott as the "best physicist in the lab..." to the attendees. John D. Murray has told us about an amusing event at the farewell. Pawsey made reference to the upcoming birth of her baby (Peter Hall was born exactly 4 months later on 20 November 1951). Pawsey began his summary of Payne-Scott's achievements at RPL: "Miss Ruby Payne-Scott ..." and realized that this name was hardly appropriate for someone 5 months pregnant and hastily corrected himself to: "I mean... [long hesitation] ... Mrs. Ruby Hall ..."

Thus the reason that Payne-Scott resigned in 1951 was not a direct result of the controversy from the previous year concerning her marriage. As a 39 year old woman, she was much older than average first time mothers in 1951. During or just after the end of the War, Payne-Scott had had a miscarriage[25] and thus was quite

---

[24] A memo which accompanied the recommendation for promotion lists eight publications and describes her work and that of A.G. Little on the Potts Hill interferometer in detail: "[It] allows the movement of the sources of radio energy emitted at times of solar flares ... to be followed through the solar atmosphere. ... Results of great interest and importance are being obtained."

[25] B.Y. Mills and Rachel Makinson told Goss (in 2003 and 2007) of hearing about the miscarriage based on discussions at the RPL in the mid-1940s. Also see Chap. 7: a possible date for the miscarriage was late 1946, sometime after October 1946. Makinson remembered that the event occurred at a meeting of the Officers' Association; she was told about this tragedy some days later by other colleagues. Fiona Hall was not told of the miscarriage until after her mother's death in May 1981.

probably concerned about beginning a family. The reason for her resignation was simply that maternity leave was not available in 1951.

Certainly the treatment of Payne-Scott by the CSIRO seems draconian in the light of twenty-first century practices. However, her treatment was by no means unusual given the norms of Australian society in 1950. Though Payne-Scott was the beneficiary of increased opportunities for women scientists at the outbreak of World War II, it was her skills and experience as a physicist and mathematician that were the basis of her successes in the remarkably productive decade of 1941–1951.

Chapter 5
War Time Research by Payne-Scott at RPL

The last 4 years of World War II, from 1941 to 1945, were a turning point in the life of Payne-Scott. Her abilities as a radio engineer were perfected; she learned a number of techniques that would serve her well in the post War years as she became one of the early solar radio astronomers. There is no doubt that her association with J.L. Pawsey during these years was decisive in developing her research skills. The ground breaking 10 cm radio astronomy project of March 1944 (see Chap. 6) arose from Payne-Scott's War time activities testing the new S band (10 cm) radar receivers. In particular, this project set the stage for the exciting astronomical endeavours of late 1945 and early 1946. Payne-Scott was prepared for these years because of her experience in radio engineering gained from War time research on radar.

Payne-Scott's research activities during this period at RPL can be illuminated by summarising a number of internal reports written by her between 10 February 1942 and 6 August 1945, as well as the two page report of a colloquium held on 30 January 1945. All these reports were originally classified as "most secret", "secret" or "restricted", but were later marked "unclassified" with the latter word written in pen on the original documents that were located in the RPL archive in late 2006; reports PD 30, RP 233, RP 252/1 and RP 252/2 have not been found.[1] The RP series of documents were in the RP, PD and TI[2] classes (see Appendix B for a complete list of Payne-Scott's publications).

[1] Barnaby Norris of the Australian Telescope National Facility located these internal RPL documents in late 2006 in the RPL archive in Sydney. The unlocated documents RP 233 and RP 252/1 and /2 are dated 10 January 1945, 20 May 1945 and 6 July 1945 respectively, and are earlier versions of their publication in the *Proceedings of the IRE* (Institute of Radio Engineers) in 1948. The unlocated document PD 30 is titled "S Band Signal Generator", 8 November 1943, and is probably related to the later document (described below), RP 211.

[2] The designations refer to different internal memo series during the War at RPL. RP (for "Radiophysics") were more polished documents that evolved into the famous RPP ("Radiophysics Publications" after 1945) series. TI is probably "Technical Information" and represented a preliminary document, while the PD series is of an unknown provenance.

W.M. Goss and R. McGee, *Under the Radar*, Astrophysics and Space Science Library 363,
DOI 10.1007/978-3-642-03141-0_5, © Springer-Verlag Berlin Heidelberg 2010

Payne-Scott's War time research can be divided into two categories: (1) establishment and description of accurate standards for radio engineering test equipment used at RPL during the War, and (2) a description of the theory and practical applications of the detection of weak radar signals. Based on Payne-Scott's article in 1948a, "The Visibility of Small Echoes on Radar PPI Displays",[3] we can reconstruct project (2) in some detail. In addition, B.Y. Mills has written two summaries of his collaboration with Payne-Scott during the years 1944–1945.[4] The initial research projects at RPL continued the radio engineering investigations that Payne-Scott had carried out at AWA in the period 1939–1941 (Chap. 3).

In addition to the decisive work on the LW/AW 200 MHz radars (Chap. 1), the RPL was heavily involved in higher frequency radar research through out the War. Interactions with both British and US groups (the MIT Radiation Laboratory) were frequent and led to testing and prototyping of higher frequency radar systems (Brown 1999; Minnett 1999). Payne-Scott's research was mainly involved with testing of various S band (10 cm or 3,000 MHz) components. When Mills joined the RPL in December 1942, he found that Payne-Scott was already establishing accurate standards for much of the test equipment. While he was working on receivers for radar systems with McCready, Mills was given the task of determining the sensitivity of these systems using hot loads. "This naturally led to comparisons with signal generators calibrated by Payne-Scott and much discussion about differences which showed up".[5]

The first report (10 February 1942, RP 138/1 – classified as "most secret") was written by T.N. Basnett and Payne-Scott: "Oscillator Frequency Required for Sh.D. [Shore Defense] and G.L. (Gun Laying) Calibrator". This report contains a calculation of the frequency of the oscillator used to drive an "A type" display (deflection modulated with a determination of range only; the radar blips were displayed as deflections perpendicular to the time axis on the cathode ray tube) at intervals of 1,000 yards. The rationale for the calibration was to provide accurate distance marks on the display. In this way the operator could report the range of the enemy ship to the nearby artillery battery. A fascinating aspect of the calculation is the realisation of the uncertainty in the value of the speed of light (c) existing in 1942. Brown (1999) has a thorough discussion of the determination of the speed of light during the War years. The accepted value in 1941 was $299{,}776 \pm 4$ km/s, compared to the modern values of $299{,}792.458 \pm 0.0012$ km/s. Basnett and Payne-Scott adopted a value of $299{,}780 \pm 40$ km/s.[6] A discussion of the appropriate value

[3] *Proceedings of the IRE* (Institute of Radio Engineers, USA).

[4] In the summary of the 1999 conference at the University of Sydney (*The Boffins of Botany Bay: Radar at the University of Sydney, 1939–1945*), Minnett has described the activities of Mills and Payne-Scott in detail in the chapter, "Light-Weight Air Warning Radar". He also wrote to Goss on 14 September 1997.

[5] Letter from Mills to Goss, 1997.

[6] As an example, for a target at a range of 20 km, this fractional error would cause an error of about 5 m. No explanation was given for the use of a slightly different value compared to the accepted value in 1941.

of the index of refraction ($n = 1.00036$) for standard temperatures and pressures was also required. The setting of the calibrator unit frequency to provide the correct distance scale was then found to be 163.86 kHz with an uncertainty of only one part in 6,000.[7]

A series of internal reports followed in the period 8 November 1943 to 6 August 1945, all descriptions of microwave components, mostly at S band (10 cm). In the TI /80/2 report of 23 February 1944, she discussed the effects of extreme daytime temperatures on the performance of the CMH (centrimetric heightfinder for aircraft warning) S band experimental station at Collaroy and the 272 Mark I Bondi naval radar station, where summer temperature variations inside the radar building in excess of 20°C were observed. The heated crystals caused a severe deterioration in the noise factor (sensitivity) of the radar receivers and thus a major decrease in the sensitivity of the overall radar system. The report contained an analysis of the level of increase in the noise factor and the manner in which the receiver recovered after the crystals were removed from the receiver. The concern was the rapid deterioration of the radar system in the tropics (the South West Pacific Area, SWPA, north of Australia) during the final years of World War II. The report contained almost no discussion of the causes of the effects other than a passing reference to "some effect" on the surface layer of the silicon crystal. Payne-Scott probably suggested a remedy for removing (and then remounting) the crystal during periods of intense heat. This procedure would then have become a part of the operating instructions to avoid serious degradation in sensitivity during hot weather. Apparently, this recipe for repair was effective.

Two classified reports were written in 1943 and 1944 describing an S band signal generator and an S band noise "thermal noise generator", the latter (RP 211, Payne-Scott, 1944a) dated 29 May 1944.[8] In modern terminology the latter device would be called a "noise tube". This report described an elegant device to produce a known amount of S band radio frequency power that was then used to determine in a precise manner the "noise factor"[9] (current terminology is "noise figure") of S band receivers. Knowledge of the sensitivity of the S band radar receiver for an active radar station was always essential. A rapid deterioration in the noise factor indicated that something was malfunctioning in the receiver; in this case the ability

[7] The report was typed by Sally Atkinson, RPL administrator from the early 1940s to the 1970s (see Preface).

[8] Robert Hayward (Electronics Engineer at the NRAO, Socorro, New Mexico) has summarised these reports in a document dated 24 January 2007: "[These documents] provide a good description of the type of radio frequency test equipment that was available shortly after World War II. [The techniques] owe their genesis to radar systems developed in ... radar labs . . . during a time of . . . global conflict."

[9] This term was invented by Harold Friis of Bell Laboratories in 1944. Friis was the supervisor of Karl Jansky (Sullivan 2009). In the current era the common usage is "noise temperature" to describe the sensitivity of the radio astronomy or radar receiver. As an example, the typical 200 MHz receiver of the 1945 era would have a noise figure of about 6 dB or 860 K. A modern 3 GHz receiver in 2009 would have a noise figure of only about 0.22 dB, a noise temperature of 15 K.

to detect incoming aircraft would be seriously impaired. The resultant errors in the location of incoming aircraft could be substantial.

During this period, receiver noise factors were usually determined with what is now referred to as the "signal generator twice-power" method.[10] The technique required two pieces of laboratory test equipment. A signal generator was used to feed a continuous wave (CW – a fixed frequency) signal of known frequency and strength into the receiver. A power meter was then used to measure the level of the amplified signal at the output of the receiver.

As the name implies, the twice-power method was a two-step process. First, the output power of the receiver was measured with a matched "room temperature" load connected to the input. This presented an effective noise power signal of 290 K at the input. Next, the load was replaced by a signal generator providing a CW signal within the measurement bandwidth of the receiver. The power being injected was adjusted so that the output level of the receiver doubled (i.e., a 3 dB increase). The power coming from the generator was thus equal to the input noise power. If the actual generator power level being injected as well as the bandwidth of the receiver were known, the noise factor could be calculated using a simple and straightforward formula.

While the method was advantageous in that it did not require the gain of the receiver to be known, the absolute power level of the signal generator had to be determined. Because the receiver was designed to deal with signals of exceptionally low power levels, this usually meant that precision attenuators were needed to reduce the CW signal to a very weak level before being injected into the receiver. This level was far too low to be measured by a power meter and thus was inferred by the amount of attenuation added to the signal path. This procedure could result in substantial errors in determining the signal strength. The method also required that the frequency response of the receiver be accurately determined so that the effective noise bandwidth could be calculated. Additionally, the output level was to be measured with a true power detector since the receiver has a mixture of both noise and CW signals. Some detectors would respond differently to these types of signals. Thus the final result could be quite uncertain.

A significant improvement over the "signal generator twice-power" method was possible if a high level of noise power – much stronger than the 290 K room temperature load – could be used. Payne-Scott developed a noise source ("noise tube") based on the Johnson noise produced by a hot resistor of known temperature. The elegant setup (Fig. 5.1 shows the waveguide apparatus, probably using RPL made waveguide components at 2 GHz) used a tungsten lamp at 2,100 K (Fig. 5.2) embedded within a section of S band rectangular waveguide. A great deal of the design effort was spent in ensuring that the lamp was well matched to the characteristic impedance of the waveguide, with the conclusion that the noise power delivered into the waveguide agreed to within a few percent with the predicted power.

[10] The following eight paragraphs have been provided by R. Hayward (2007).

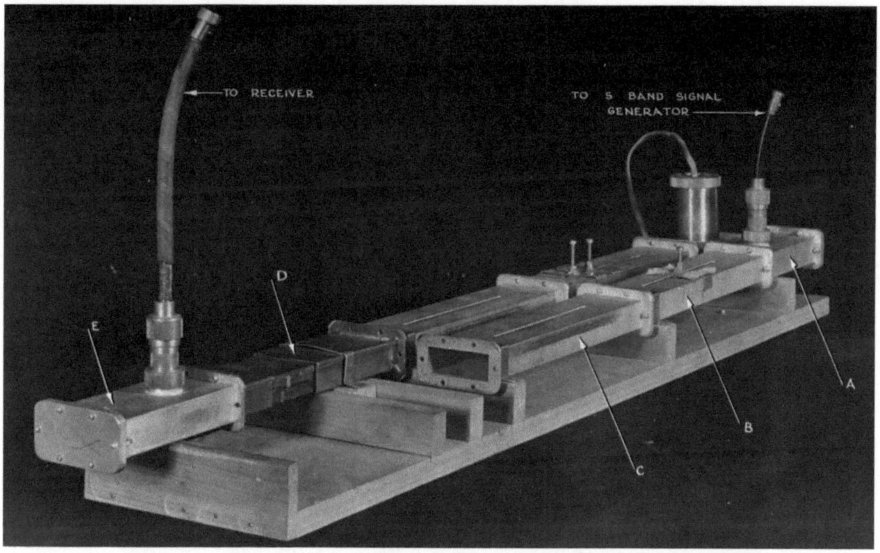

Fig. 5.1 A S band (3 GHz or 10 cm) set-up to determine the sensitivity of radar receivers. Designed by Payne-Scott and published in a secret memo from 29 May 1944, RP 211. Note the officer in charge was J.L. Pawsey. "A" is a length of waveguide into which the hot or cold load resistor was coupled. "B" is a length of slotted waveguide with adjustable tuning screws serving to match the hot or cold resistor to the characteristic impedance of the waveguide. "C" is a length of slotted waveguide along which a standing wave detector could be moved to check for matching. "D" is a length of telescopic waveguide set to a length that would keep the crystal current of the receiver the same when either the hot or cold system was connected. "E" is a standard waveguide to coaxial cable coupling. From RP 211 29 May 1944. (A "secret 'document:' A Thermal Noise Generator for Absolute Measurement of Receiver Noise Factors at 10 cm" by Payne-Scott, 1944a. (CSIRO, ATNF Archive)

The availability of this high power noise source allowed the RPL to carry out what is now known as the "Y-factor" method of measuring the "noise figure" of a receiver. By presenting two loads at different temperatures to the receiver, both the gain and the noise factor of the receiver could easily be determined. In this case, the 2,100 K noise tube acted as the "hot load", while a standard room temperature 290 K load was the "cold load". This combination resulted in the hot load having nearly ten times more noise power than the cold load. "The increase in noise on switching from the hot to the cold load circuit depends on the ability of the receiver to distinguish small signals above the normal noise, i.e., on its own noise factor".[11] As seen in Fig. 5.1, the two paths (noise generator and the signal generator) were completely interchangeable; the waveguides were simply unscrewed and interchanged.

---

[11] From RP 211, "A Thermal Generator for Absolute Measurement of Receiver Noise Factor at 10 cm", 29 May 1944a. See Appendix B for a list of Payne-Scott's unpublished reports.

Fig. 5.2 The tungsten lamp used as a noise source for the hot resistor. See Fig. 5.1 for source of this image. (CSIRO, ATNF Archive)

The cold load was actually a 10 dB attenuator, as seen in Fig. 5.1.[12] When connected to the input of the receiver, the noise power from the attenuator acted as a 290 K termination. The cold load portion of the setup could also be fed with a CW signal from a signal generator, with its power being reduced by a factor of 10 by

[12] Surprisingly the cold load was just the high loss cable that went from the signal generator into the waveguide "A" in Fig. 5.1. This provided a load at room temperature or 290 K for comparison with both the thermal noise generator and the signal generator.

the attenuator. The signal generator could be used to line up the receiver in frequency and to determine the bandwidth (although this requirement was not needed for the actual Y-factor measurement). The signal generator could also be used to carry out a twice-power measurement for comparison.

The noise factors based on the noise source and the signal generator methods agreed to better than 1 dB (about 25%). Thus a method of providing a fundamental check on the twice power method, which was commonly used in the field, was available. The cumbersome noise tube method was not suitable for field tests due to its complexity and size; various receivers could be checked out in the laboratory using this new and unique noise source. As a procedural step, the more portable signal generator setup could be calibrated against the noise tube back in the laboratory.

The S band noise tube designed and constructed by Payne-Scott in 1944 had an effective noise power of 2,100 K. Today, noise diodes bought off-the-shelf are capable of achieving noise temperatures of 290,000 K or more. The twice power method was commonly used before high power noise sources became available; however, the method is still used for systems with large noise figures where the Y-factor method results can be inaccurate. For modern cryogenic (cooled to liquid He temperatures close to 20 K) receivers, which have noise temperatures of much less than 100 K,[13] the Y-factor method is the preferred measurement technique. Currently in the early twenty-first century, the hot load remains at room temperature while the cold load is typically a cryogenic load immersed in liquid nitrogen (i.e., 77 K).[14]

The report RP 211 was followed within a week by the document TI 121/1 ("The Present Position of Low-Power S-Band Measurements in the Radiophysics Laboratory", 6 June 1944, Payne-Scott, 1944b), providing a summary of the full range of power measurements at S band from 10^{-4} to 10^{-13} W; the methods, calibration and precision of a number of methods were summarised.[15] The noise factors for S band crystals (available in 1944) at the power level of 10^{-13} W were in the range of about 10 dB or a T_{sys} of 2,600 K.

---

[13] A typical receiver (10 cm or 3 GHz) would have a receiver temperature of less than 15 K.

[14] An engineering colleague of Payne-Scott, D.F. Urquart, wrote a memo on 28 May 1947, "Operating Instructions for the 200 Mc/s Receiver" with copies to Mount Stromlo and McCready. Detailed step-by-step instructions were provided to determine the noise factor of this system using the method developed by Payne-Scott in 1944. The signal generator method was described with the warning "results obtained in this way are liable to be 1 or 2 dB out owing to the difficulty in getting an exact measurement of the bandwidth ... The most accurate and direct way of measuring the N.F. [noise factor] is by the use of a Thermal or saturated diode noise generator ..." All the equations were provided to determine the N.F. Four general tips for operation were then provided. An example was, "When the receiver is in use or when measurements are being made on it, both the preamplifier and the receiver proper should be firmly screwed down to the base-board". RPL file: A.1/3/1(a) from the Sullivan archive.

[15] Bowen (1947, p. 7, 1954 edition) has remarked on the remarkable dynamic range required by radar systems. The transmitted pulse was typically 10^5 W while the received energy from the target was roughly 10^{14} W. "The overall operating efficiency is therefore 10^{-19} and it is a great tribute to the pioneers of radar that they persisted in their efforts to attain apparently impossible ends".

The final technical report (TI 191/1 – "restricted"[16]) was written by Payne-Scott and dated 6 August 1945, nine days before the end of World War II. This report is the most comprehensive of these documents and is entitled "Present Position of Fundamental R.F. [radio frequency)] Measurements in the Radiophysics Laboratory" (Payne-Scott 1945a). This 14 page document provided a description of RF engineering knowledge and practice at the end of World War II. The document provided a primer for the determination of frequency (or wavelength), power at various levels, attenuation, impedance, cable loss, etc. at P band (10–500 MHz), L band (1–2 GHz), S band (2–4 GHz), X band (8–12 GHz), and K band (18–27 GHz). Many of the references were to RPL publications (9 of the 22 or 41%), while MIT Radiation Laboratory publications provided a comparable fraction of the cited publications.

The report concluded with a number of sensible recommendations concerning practices in the running of the "test room", the section of RPL responsible for the use of standard equipment. As an example, "at present some standard measuring equipment is the property of individual sections ... much of it is liable to wander out to field stations and return full of sand and with impaired accuracy ... Standard measuring equipment should be kept in a specified place and not removed under any circumstances". The final page of the report consisted of five concrete recommendations for establishing and maintaining primary and secondary standards.

A cryptic and significant paragraph appeared as Section 3.2.2(c): "Other Methods (for noise factor determinations)". The report stated:

> Two other methods for measuring the noise factor of a receiver have been considered in the laboratory; one involves using the radiation from the *sun* [our emphasis] as a standard source, and the other comparing the receiver noise when the aerial points alternatively at "free space" and an enclosure at ambient temperature. Neither effect has yet been sufficiently well investigated to be used to provide a standard noise voltage, but in particular the second, which does not require a directive aerial, has attractive possibilities.

As discussed in Chap. 6, Pawsey and Payne-Scott carried out S band observations in March 1944 of the sky and the Milky Way, but did not observe the sun. Sullivan (2009) has pointed out that Pawsey and Payne-Scott were not aware at this time of Southworth's secret reports of 1942 at Bell Laboratories, describing the microwave detection of thermal radiation from the sun. However, in August 1945, Bowen and Pawsey became aware of the detection of microwave radiation from the sun based either on rumours of Southworth's observations that reached RPL or on receiving the "secret" Bell Laboratory reports (see Appendix G).[17] But they could not have

[16] As Harry Wendt has pointed out to us (March 2009), this lowest level of security classification seems surprising given the far reaching scope of this report. Perhaps the realisation of the impending end of World War II influenced the censors at RPL. This low level of security classification probably had an effect on the use of "secret" documents from the US in RPL internal publications. See footnote 17 this chapter.

[17] As described in Appendix G, Bowen wrote Richard Woolley (Director of the Commonwealth Solar Observatory at Mount Stromlo) on 20 March 1945 with copies of the two Bell Labs reports. The complex chronology of these events is described in Appendix G. Woody Sullivan (6 May 2009) has proposed a plausible answer to the question: "Why did Payne-Scott not refer to the two

seen the early 1945 Southworth article (published in the open literature), since this was not cited even as late as 23 October 1945 by Pawsey et al. (1946, see Chap. 7 and Appendix G) at the time of the first publication of RPL solar noise observations (Chap. 6). In the above quotation from Payne-Scott's report, there is the suggestion that she knew about the Southworth results at 10, 3.2 and 1.25 cm of 1942–1943 and she made the prescient suggestion that the detection of the sun at microwave frequencies might be a useful standard source. As we now know, the quiet sun was unlikely to be used as a precision standard source at microwave frequencies, due to the time variations of the intensity and the large angular size (about 30 arc min) of the solar disk at cm wavelengths.[18] For large antennas the huge flux density of the sun would present problems in the calibration of sensitive receiver systems. The first high frequency observations of the sun at 1.25 cm at RPL were only made in early1948 (see Piddington and Minnet 1949; Pawsey and Yabsley 1949); lower frequency detections of the quiet sun at RPL were carried out by Lehany and Yabsley at 50 and 25 cm in August to November 1947 and published in 1949. Based on the statement by Payne-Scott in 1945 in this technical report (TI 191/1), the delay of a few years in the detection of the quiet sun at RPL remains surprising.

The major project on which Payne-Scott worked from 1944 to the end of the War was the new 25 cm LW/AWH ("light weight aircraft warning – height") MkII radar. This advanced radar was developed as a prototype in 1945; the first contract for the construction of 47 was cancelled only in August 1945 as the atomic bomb explosions in Japan brought at end to World War II. One of the radar prototypes at Middle Harbour, Sydney, is shown in Fig. 5.3. In early 1944, the Japanese had developed the ability to jam (transmit a confusing signal at the same frequency as the radar transmitter and receiver) the 200 MHz LW/AW workhorse radars in the SWPA (Appendix C). In addition the height determination precision of the 200 MHz radars (LW/AW and the LW/AW GCI series, "ground control of interception") was limited, in addition to the absence of low angle coverage (Briton 1947; Minnett 1999). The 25 cm magnetrons had been designed by the group from the University of Melbourne in the previous years; in 1945 they were being manufactured by AWA. The new prototype produced narrow beams and experienced a complete lack of ground-reflected radiation, in contrast to the 200 MHz radars. Thus there was

Southworth documents which described the 1942–1943 microwave detections of the sun?" The suggested resolution is that the RPL document was only "restricted", the lowest level of security classification, while the Bell Lab documents were "secret", a higher level of security. Possibly this restriction implied that the RPL group was not able to quote the Southworth results until the Bell Labs paper of 1945 was published in a scientific journal. (See also Chap. 6.)

[18] In RPR 111 from July 1950 ("Possible Use of Solar Radiation to Check Sensitivity of Microwave Radio Receiving Equipment"), Pawsey (1950b) proposed to use the sun as a "standard source" at wavelengths of 1–60 cm. As an example, at wavelengths of 3–1 cm the flux density of the quiet sun is in the range 10^6 to almost 10^7 Jy. Pawsey pointed out the major problems in determinations of the absolute calibration of receiving equipment with accuracies of better than 25–50%. With small radar aerials the sun was, in fact, too faint and time variability of the solar radio emission was a severe limitation. The use of the sun to compare the relative sensitivity of two aerial systems was more feasible, with accuracies of a few per cent for a relative comparison.

Fig. 5.3 LW/AWH (Light Weight Aircraft Warning, Height) prototype 25 cm radar with size of antenna 25 × 12 ft (7.6 × 3.6 m), located at Georges Heights, Middle Harbour, Sydney (note South Head in the distance). The magnetrons were designed at the University of Melbourne and manufactured by AWA. Pulse lengths were either 4 or 1 μs, with a power of 500 kW. Weight was only 7 t compared to 35 t for comparable UK and US radars. This radar was the major technological achievement of RPL during World War II. The end of the war brought a cessation to development activities; only a few prototypes were built. The other antenna to the left is the communications aerial for control of the radar. This radar spearheaded RPL's postwar drive to solve civilian problems for control of aircraft traffic. ATNF Historical Photographic Archive: B1362

no interference pattern due to a Lloyd's mirror effect (see Appendix C, Fig. C.1) as was observed at 200 MHz. The beam widths were determined at the 1.5 dB points or at the 3 dB points for the return echo with values of 1.3° in elevation and 3° in azimuth. The required volume of space (Fig. 5.4) was searched once per minute with a flat helical scan by the radar beam. At 25 cm, the peak power of about 500 kW was transmitted with a detection range of 70–100 miles for small aircraft up to heights of 35,000 ft. Measured heights were accurate to ± 2,500 ft at ranges of 30–100 miles, representing a vast improvement in precision compared to the 200 MHz radar sets. This 1 GHz (L band) project represented the most challenging and successful technological achievements of RPL in the era 1939–1945.[19] Both Mills and

[19] This advanced radar system remained the only military project at RPL after the War. In August 1946, Bowen (NAA: C3830, D1/2, "Programme of the Division of Radiophysics", 8 August 1946) presented a report on the future activities of RPL. A detailed description of the AWH Mark II was presented: "The Radiophysics Laboratory was formed to develop radar equipment for the Armed

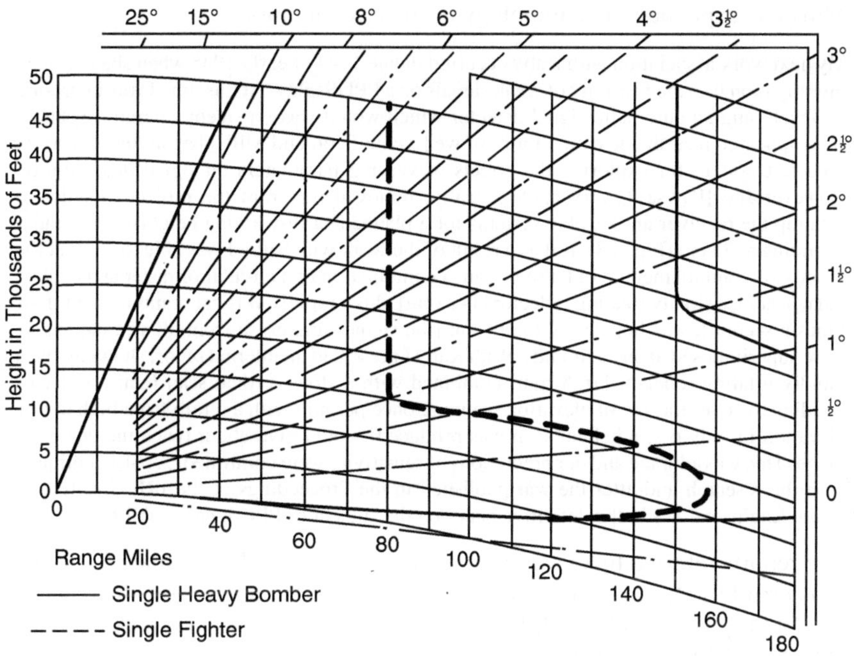

Fig. 5.4 The angular coverage of the of the LW/AWH 25 cm radar system. With this radar, objects at greater height could be detected than previously. Errors in determining the height of incoming aircraft were less than 2,500 ft at distances up to 100 miles. Figure 15 from Briton, *The Journal of the Institution of Engineers Australia*, 1947, vol 18, page 121. Figure obtained from CSIRO, ATNF, RPR 41. 'LW/AWH MkII Light-Weight Air-Warning and Height -Finding Ground Radar.' O.L. Wirsu, 1 October 1946. Used by permission of CSIRO ATNF

Payne-Scott developed skills that were to serve them well in the explosive growth in radio astronomy in the post World War II period.

B.Y. Mills and Payne-Scott played the major roles in the planning of the scanning method and the design of the advanced displays for this sophisticated radar. A complex of three displays was available: (1) an A type display (*x* axis was the radar echo range, while the strength of the echo was displayed on the *y* axis) for monitoring and IFF (Identification Friend or Foe), (2) a PPI (Plan Position Indicator with the position angle of the display representing azimuth of the echo and the radial displacement of the display indicating range) for detection and location, and (3) a RHI (Range Height Indicator, a modified A type display).

Forces and during the War years its programme was determined by their requirements. ... This development was started during the War and was so nearly completed at the end of hostilities that under pressure from the Air Force it was decided to complete the construction and perform field tests. [The new radar] ... is approaching the test stage at our field station at Georges Heights [Middle Harbour] ... It is expected that the design will form the basis of peacetime radar sets for the Air Force and for various civil purposes."

Mills has explained the chronology of their collaboration:[20]

> My next work association with Ruby occurred in late 1944 or early 1945 when she began an investigation into the visibility of weak signals on a PPI display. At this time I was interested in developing an automatic signal detector which would wake a sleepy operator when an echo appeared near the detection limit, so we [Payne-Scott and Mills] began together to set up a test system for exploring the effects of varying parameters on signal detectability. However Joe [Pawsey], very apologetically, removed me from the project in order to develop the receiver and display systems for a planned early-warning height-finding radar operating at 25 cm (Fig. 5.3). But as a result of the work with Ruby I was able to see that the proposed scanning method [a Palmer scan, a combination of a circular or raster and conical radar scans], a slowly rotating reflector [1–2 rpm] plus a rapidly vertical scanning feed [four times a second], was hopelessly inefficient [the signals and noise from all vertical angles were superimposed at any azimuth]. With Ruby I presented an internal report showing that a rapidly rotating reflector [16.5 rpm] combined with a slow vertical scan [entire reflector nodding to cover the 11° of elevation scanning once per min with a return from high to low elevation in only 6 s; a helical scan was produced] would give about twice the detection range. This was adopted and our predictions of sensitivity were confirmed.... She continued with the research and after the war published in the Proceedings of the IRE (Institute of Radio Engineers, USA) the definitive paper on the subject of PPI signal visibility.[21]

The reports referred to above were: (1) Payne-Scott, "The Ultimate Visibility of Signals on a PPI Display and the Effect of Electrical Parameters on Visibility", RP 252/1, 20 May 1945, and (2) Mills, "Scanning Considerations in LW/AWH Mk II", RP TI 137/4, 22 June 1945. In addition Payne-Scott had given a research colloquium on 30 January 1945, presenting preliminary results of the research.[22] The two reports each had only one author. It is clear, however, that there was a great deal of collaboration between the two; for example, much of the discussion of the scanning considerations in the Mills report was based on calculations carried out by Payne-Scott.

---

[20] Letter to Goss, 14 September 1997.

[21] In the article by Minnett (1999), Mills was quoted as pointing out that the first Palmer scan method shook the antenna so excessively that this structure fell off due to the high acceleration. The nodding feed was driven up and down four times a second with a linear scan. The helical scan was much superior since at any one time the scan information on the display was collected over a much smaller vertical range. Mills wrote: "I see the rapid and successful development to have depended on the foresight of Joe Pawsey in setting up a program to study the basics of signal visibility. Without it months would undoubtedly have been wasted struggling with the poor performance of the first scanning method, before the light dawned."

[22] "Notes on the Research Colloquium held on 30/1/45" (NAA: C3830, D4). For the first time visibility was defined: "Visibility is in terms of the reciprocal of the power of the minimum visible signal", i.e., the smaller the detected signal (for a more distant object) the higher the visibility. The presentation seems to have been controversial based on a number of questions from the audience. The advantages of the newer PPI displays compared with the more traditional A type display were discussed by Payne-Scott. The main conclusion was that the visibility was proportional to the square root of the pulse repetition frequency (PRF) up to a critical PRF and thereafter was independent of the PRF. Brian Cooper (1917–1999, well known scientist at RPL since 1940 who later played a major role in the instrumentation and use of the Parkes 64 m radiotelescope starting in the late 1950s, see footnote 7 Chap. 4) asked Payne-Scott about the nature of the signal to noise ratio of the return echo.

In late 1945 and early 1946, Payne-Scott was encouraged to convert her research on PPI visibility issues into a publication; at this time there were no restrictions since the War time restrictions on security were no longer in place. The PC (Propagation Committee, Appendix A) minutes of 15 January 1946 reported that she was working on this publication. The 17 page paper was submitted to the *Proceedings of the IRE* (Institute of Radio Engineers, USA) on 13 January 1947, with a revised manuscript appearing on 6 August 1947 (probably after a peer review report). The paper was published in February 1948 – "The Visibility of Small Echoes on Radar PPI Displays". A preprint (RPR 40, Payne-Scott 1948a) was issued on 8 October 1946.

The introduction provided a summary of the goal of the research:

> Detection of an object by radar depends ultimately on the ability of an observer to pick out a small change, in brightness or position, of part of the pattern on the screen of a cathode-ray tube. This ability depends on the one hand on physiological and psychological factors, and on the other, on the parameters of the whole radar system, which determines the nature of the change to be detected. In order to design systems of predictable performance or to compute the effect of any proposed change in a given system, we need to know the laws governing visibility.

She then presented a general theory of visibility with a rigorous derivation of an equation for visibility on a PPI display. The details of the various parameters affecting the response of the total system were analysed in detail. Examples of the parameters that were relevant were: radar pulse duration, pulse repetition rate, rate of the base sweep, bandwidth of the receiver, and antenna beam size. A number of nomograms (graphical calculating devices such as a slide rule) were presented that provided useful calculations for detectability as a function of numerous variables. ("A series of nomograms are provided, from which the least visible signal under any set of conditions can be read off. ... The fact that signals up to 18 dB below noise can be seen under favorable conditions is most striking.") As an example, the angular spot diameter (in degrees on the screen) was presented as a function of the screen type for various antenna rotation speeds (revolutions per min). A number of recommendations were presented to optimise the visibility of the display. Examples were: (1) use the maximum available peak power in the pulse, (2) use as long a pulse duration as possible, (3) use an overall-bandwidth of the system equal to the reciprocal of the pulse duration, (4) maximise the antenna gain, (5) use the maximum possible beam-width consistent with this gain, (6) use the slowest feasible antenna rotation rate, and (7) use the highest possible transmitted pulse repetition frequency, combined with the chosen pulse duration, that would not exceed the display tube dissipation. A detailed appendix to the paper contained the derivation of mathematical formulae for visibility of the minimum signal required to produce a detectable echo on the display.[23]

[23] The parameters of the equation were the available power at the antenna output, the peak signal power, the output voltage from the detector, the brightness of the screen for zero spot size and screen persistence, the area of the spot size on the screen, the effect due to the long persistence of the display screen, and the effect of a limited number of returned echoes (e.g., a moving target or a change in elevation of the antenna between successive rotations).

A novel aspect of the testing was the creation of an artificial radar station in a small darkened room. In current terminology, this simulation would be called an analogue computer. "The apparatus allowed a wide range of values of system parameters to be arranged in any desired combination". The observations were obtained by three people, Payne-Scott and two Royal Australian Air Force radar operators. Each observer could be trained in a few hours because various observers were found to have similar responses to the system. The operator was subjected to a simulated attack of enemy aircraft at a random bearing. The next step was a detailed comparison of the measured visibility on the screen as each of the parameters was changed, one at a time; comparisons were then made with the theoretical predictions. In most cases the observations fitted the theory, as was demonstrated in the publication with many illustrations showing numerous combinations of parameters. The illustrations indicated that the empirically determined observational parameters agreed well with the theory developed by Payne-Scott.[24] As Mills has pointed out, this paper remained the definitive publication on PPI displays for many years.[25]

At the end of the paper Payne-Scott acknowledges "many conversations with B.Y. Mills". As has been discussed in Chap. 16, it is likely that this experience with detectability of weak signals led to Payne-Scott's interest and expertise in discussions of radio astronomical source "confusion" in the post War era (see footnote 5 Chap. 16). Payne-Scott's research during the War led to a perfection of her research skills; her reputation among her peers was strengthened and probably increased her acceptance by her colleagues in the 1945–1951 era, as radio astronomy in Australia experienced rapid growth.

[24] McCready (in "Receivers", Chap. 10, *A Textbook of Radar*, edited by E.G. Bowen [first edition 1947 and second edition 1954]) has described the results of the Payne-Scott research. He pointed out that the signal to noise of a radar receiver was not limited by the noise factor of the receiver but by the sensitivity of the cathode ray tube. He wrote, "Payne-Scott has shown that under these conditions we can detect a signal whose power is 15–18 dB below the noise power, depending upon the type of detector ..." He summarised the 1948 publication of the *Proceedings of the IRE* paper and concluded: "Although many existing radar systems can detect signals whose powers are of the same order as noise in the input circuits, it is preferable to calculate the sensitivity at the cathode ray tube making use of the basic theory and charts in Payne-Scott's paper". See footnote 40 Chap. 6.

[25] Interview with Goss, 1 April 2007.

Chapter 6
Payne-Scott – The First Woman Radio Astronomer and the Transition to Peacetime – 1944–1945

Payne-Scott's career as a radio astronomer began in 1944 and extended to her retirement in July 1951. Her remarkable career, which led to many of the early ground breaking discoveries in solar radio astronomy, can be roughly divided into four phases. The earliest phase began in March 1944 and extended to late 1945, a period of transition from War time radar research to early solar noise research. In this period Payne-Scott, under the leadership of J.L. Pawsey, together with others, was laying the ground work for the beginnings of solar radio astronomy. The next period from October 1945 to late 1947 marked the ground breaking solar work at Dover Heights and the publication of their first important research papers on radio studies of the sun. Her contributions to Fourier radio astronomy imaging were crucial at this stage. The third period was in 1948, during which time Payne-Scott had an interlude working on her own at the Hornsby field station; here the detailed properties of Type III bursts were elucidated. Then, starting in 1949 up to the end of her career at RPL in July 1951, she was mainly involved in the building and use of the high resolution swept-lobe interferometer at Potts Hill Reservoir in Sydney, in collaboration with Alec Little. There was a brief coda in August 1952 during the URSI International Assembly in Sydney.[1]

Propagation Committee of the Radiophysics Laboratory: 1944–1954

The minutes of the "Propagation Committee" (PC) of the CSIR RPL from 14 September 1944 to 7 March 1949 and the renamed "Radio Astronomy Committee" from 11 April 1949 to 9 April 1954, have provided an invaluable source of

[1] Orchiston, Slee and Burman (2006) have discussed early RPL contributions to solar astronomy in Australia from 1945 to 1948, in addition to work at the Commonwealth Observatory at Mount Stromlo and in the Physics Department at the University of Western Australia.

W.M. Goss and R. McGee, *Under the Radar*, Astrophysics and Space Science Library 363, DOI 10.1007/978-3-642-03141-0_6, © Springer-Verlag Berlin Heidelberg 2010

information about the research program of the RPL in these years. The committee had probably been instigated during World War II under the leadership of Pawsey (the Chair) for bi-weekly or sometimes monthly discussions. At these meetings, a new topic would often be introduced by Pawsey asking; “Ruby what do you think?”[2] In Appendix A, we provide some details concerning the invaluable 332 page record of contemporary accounts of the progress of research for this nine year period at RPL.

Neophyte Radio Astronomer: Ruby Payne-Scott

Payne-Scott’s first radio astronomy experiment[3] was performed in March 1944. The observations were carried out with J.L. Pawsey at the short wavelength of only 10 cm (3,000 MHz). Payne-Scott played a key role in these observations based on her experience with the calibration of 10 cm receivers (see Chap. 5) and her thorough understanding of the thermodynamics of radio frequency noise. The results were described in a RP classified memo by J.L. Pawsey and Ruby Payne-Scott (RP 209) dated 11 April 1944.[4] The equipment undoubtedly was experimental since most of the radar work being done at this time was in the 1.5 m (200 MHz) range, although another group at RPL was working on the LW/AWH Mk II 25 cm (1,200 MHz) system (Minnett 1999).[5] The report RP 209 had the unspectacular title of “Measurements of the Noise Level Picked Up by an S-Band Aerial”.[6] The short report of five pages, including an appendix with the calculation of the noise factor or noise temperature of the receiver, contained a number of prescient conclusions. The first experiment consisted of determining the response of a 20×30 cm microwave horn connected to a receiver with a noise temperature of about 2,900 K (a factor of 200 less sensitive than modern receivers, i.e., those with a system temperature of 15 K or less) as the horn was pointed at various positions within the

[2] We have received a comprehensive letter from John D. Murray, dated 24 January 2004, about the format of the PC meetings. The late Chris Christiansen wrote in October 1977 with a description of the role of Pawsey and Payne-Scott at the PC meetings. At one memorable meeting, Pawsey made a decision when Payne-Scott was not present. Pawsey initially said: “Well that is settled.” Then suddenly he hesitated and added: “We had better ask Ruby before we proceed.”

[3] See Sullivan (2009) for a possible earlier Sydney solar observation by Martyn and Piddington (1939) at 60 MHz. No written record of this non-detection observation is extant.

[4] W.T. Sullivan III has provided the authors with his annotated version of this report. His notes, written on 14 December 1983, have provided valuable background information. RP 209 was mentioned in passing by Lovell (1983), who was impressed: “I had all the documents from RP and this most interesting paper was amongst them. I hope it is still there.” Sullivan (1988, 2009) has also described RP 209 in detail.

[5] See Chap. 5 for a description of Payne-Scott’s role in this 25 cm project.

[6] S band is the obscure name given to the 10 cm microwave band (now well known as the frequency of microwave ovens) during World War II. “S” stands for “short” wavelength. S band radars were decisive in the later stages of World War II, as described by Brown (1999).

room and out of a window towards the eastern sky. Unfortunately, no detailed description of the exact test set up was provided. The detailed thermodynamics of the noise from the receiver and various sources of this noise were discussed.[7] Various contributing factors were mentioned, including the intrinsic noise of the receiver, the temperature of the room ("ground"), clouds and "matter in space". The authors were apparently surprised by the low temperature of the sky (between 0 and 140 K) and seemed to detect the increase due to clouds or even rain on 20 March 1944 at 10 a.m. in Sydney. The most precise sky measurement of 20 K was a surprisingly accurate value for the brightness of clear sky conditions. Pawsey and Payne-Scott were surprised that inserting attenuation between the horn and the receiver *increased* the output as the radio horn was positioned looking out of the window, in contrast to pointing the horn within the room.[8] This result[9] would be of no surprise to radio engineers today; the result showed that the sky had a low temperature compared to the room temperature of about 300 K. Pawsey and Payne-Scott wrote:

> Despite the theoretical and practical interest attaching to measurements of ultimate noise level the authors are not aware of any reported measurements of received noise powers in the centimetre range of wavelengths. Those described here are of a preliminary nature, and the authors hope to extend them further.[10]

This prediction turned out not to be the case, as no follow-up work was reported in subsequent years. The concluding paragraph of the report also suggested a new way to calibrate the noise factors of receivers (at least in the cm bands); the use of cumbersome signal generators (see Chap. 5) could be avoided if it could be established that the temperature of the clear sky was constant. If this were to be the case, the receiver could be calibrated by pointing alternately at the sky and a room temperature enclosure. This technique is often used by radio engineers in the modern era as a component of the calibration procedure of cm and even sub mm radio astronomy systems.

Pawsey and Payne-Scott attempted one astronomical observation. They used the same receiver and attached it to a 4 ft diameter paraboloid (the report is filled with mixed units, some in feet and inches, some in centimetres and metres). They observed a region near the galactic plane in Centaurus (new galactic longitude

---

[7]In Appendix G a discussion is presented of a detailed unpublished report that Payne-Scott sent in late 1945 to Southworth at Bell Telephone Laboratories in New Jersey. In this report a similar treatment of the radiometer noise was presented.

[8]The attenuator reduced the strength of the signal from the sky but added considerably more noise power from the ohmic resistance at room temperature.

[9]In the jargon of current engineering, this effect was due to noise arising from the excess loss in the inserted cable, a loss of 20 dB or a factor of 100.

[10]In fact Southworth and Mueller (Appendix G) had carried out a similar experiment in 1943, but this was apparently unknown in Sydney in early 1944. Later in 1944, Southworth was aware of the work by Pawsey and Payne-Scott (clearly having read RP 209), while the Australians may not have been aware of his earlier investigations until 1945. See Appendix G and Chap. 5 for a description of the complex chronology.

near 300°) and reported no signals greater than 10 K,[11] "very much less than that observed by Jansky and Reber and so small as to have no observable effect... No attempt was made to observe radiation from the sun".[12] Sullivan (2009) has made the startling calculation that had they done so, the sun would have been detected easily with a large signal to noise. Pawsey and Payne-Scott were apparently not aware of either the Southworth 3.6 cm solar detection of 1942 (see Appendix G and Chap. 5; Southworth 1945) or Hey's 5 m solar detection in the same year (Hey 1946). Both results were still classified results in 1944, although RPL was receiving a number of classified documents from the UK and the US during this period as described in Chap. 5. Harry Minnett told Sullivan in 1986 that the March sun may not have been visible from the chosen window at the National Standards Laboratory.[13] The Pawsey-Payne-Scott report contained a number of astronomical references. There were four references to Jansky's papers in the *Proceedings of the IRE* from 1932 to 1937, a reference to Reber's paper in the same journal in 1940a (there was no reference to the ground breaking 1940b "Cosmic Static", *Astrophysical Journal* publication), and a reference to a paper by A.S. Eddington in the 1926 *Proceedings of the Royal Society*. Pawsey and Payne-Scott were struck by the divergent predictions of Jansky and Reber concerning the likelihood that the galactic background would have been detectable at 10 cm. Based on Jansky's suggestion that the radio background arose from a large scale distribution of matter at high temperatures, the 10 cm radiation might well have been observable. On the other hand, if Reber's claim that the radiation intensity scaled as wavelength, then there would have been negligible emission at 10 cm. The interpretation given to Eddington's predictions by Pawsey and Payne-Scott was, however, a misinterpretation that others made throughout the twentieth century:

> It appears likely that all noise received is of thermal origin and comes from regions of very low temperature. Eddington, working from the measured intensity of starlight, calculates that the radiation from the matter in space is equivalent to that of a black body of 3.2 K. If most of the noise power received by the aerial originates from space, this would account for the very low noise temperature measured.

As Sullivan has said, Eddington had only pointed out that the average energy density of starlight in the interstellar medium near the sun is comparable to that of a black body with temperature of 3.2 K. The Pawsey-Payne Scott interpretation,

---

[11] The expected value based on current knowledge would only be a few K antenna temperature, which would have implied an impressive total power stability of almost 0.1%, likely an unattainable value in 1944. The detection of 20 K from the sky only implied a more plausible stability of somewhat better than 1%.

[12] RP 209.

[13] In 2007 and 2008, Goss visited the two inner courtyards at the Madsen Building of the University of Sydney in an attempt to locate the possible room used for the 1944 observation. Several candidate windows were detected with eastern facing exposures with visibility to the north blocked by the building (the sun is in the north as viewed from Sydney); but the exact location remains a mystery.

which suggested that the 10 cm radiation field in the interstellar medium would be characterised by a black body temperature of 3.2 K, was not correct.[14]

In summary, there is a strong likelihood that the March 1944 observations at 10 cm by Pawsey and Payne-Scott were the first radio astronomy observations in Australia and even in the southern hemisphere. We agree with Sullivan (1982) that Ruby Payne-Scott is probably the first woman radio astronomer. The 1944 observations described in RP 209 predate the solar observations of Elizabeth Alexander (Orchiston 2005) by about a year. The major importance of RP 209 was the determination of the upper limit on the sky brightness at the short wavelength of 10 cm. This limit was an unexpected result in 1944. The fact that these data were never published is a surprise. The paper was a remarkable contribution, showing an understanding of the thermodynamics of radio receivers that is taken for granted six decades later. The techniques that led to an understanding of the absolute calibration of the instrument would serve the RPL group well in the next few years as the Australian radio engineers participated in the rapid post World War II growth in radio astronomy.

Post War Activities at RPL, 1945: How This Affected the Role of Payne-Scott

Before the end of World War II on 15 August 1945, the leadership of RPL began planning for post War activities. The nature of their discussions has been treated by numerous authors.[15] The existence of an intact group of scientists was decisive for the emergence of Australia as one of the two major players (along with the United Kingdom) in the post War development of radio astronomy. As Sullivan (1988, 2009) has pointed out, only in certain fields of medical science did the international reputation of Australian scientists rival that of their fellow Australian radio astronomers. As both Sullivan (1988 and 2009) and Wild (1968) have emphasised, the fact that the Australian radio astronomers had the Southern sky to themselves played only a minor role; the experience and expertise of the scientists were the decisive factors.

As will be shown, Payne-Scott was destined to play a major role. Her radio engineering skills acquired during the War, her physics background and especially

---

[14]The remarkable coincidence with the cosmic microwave background of 2.7 K, discovered by Penzias and Wilson (1965), is just a coincidence. When Goss was a graduate student at the University of California, Berkeley in the 1960s, a well known astronomer (in an interstellar medium course) suggested that the Eddington value was consistent with the previously known 2.3 K excitation of the optical interstellar lines of the CN molecule; in fact the excitation of this line is due to the cosmic microwave background.

[15]Sullivan (1988 and 2009), Wild (1965, NAA: C3830, D12/1/5, "Origin and Growth of Radio-Astronomy in C.S.I.R.O", delivered at the Division of Plant Industry, 15 October 1965), Wild (1968, 1987), and Bowen (1984, 1988).

her mathematical skills, were all contributing factors to her success in the years 1945–1951. Clearly Pawsey relied on her experience and judgment as he became the force behind the rapid growth of radio astronomy in Australia in this post War period. We present a review of the evolution of RPL from a War time radar laboratory to a major radio astronomy institute within the short period of 1945–1947. (We do not discuss the other major activity of RPL in this era, cloud physics.) This survey gives the boundary conditions for Payne-Scott's role in the evolution of solar radio astronomy from 1945 to 1951.

J. Paul Wild[16] (1987) has summarised the main ingredients of success in the transition from War time to new programmes in 1945. Major components of the success were the presence of two key individuals: (1) Taffy Bowen (see footnote 10, Chap. 4), who was the newly appointed Deputy Chief (Research) of RPL since January 1944, and who was to become Chief (May 1946) of RPL, and Joseph L. Pawsey, the leader of the radio astronomy group and later Assistant Chief of RPL (see footnote 6, Chap. 4). Bowen wrote in 1984:[17]

> What were the ingredients which led in 1946 to the development of radio astronomy? The first and by far the most important of these was the decision by the Chairman of CSIR, Sir David Rivett [1885–1961], that at the conclusion of World War II, CSIR would be devoted only to peace-time research, and that defence research would be carried out by other agencies. This meant that a highly developed laboratory with a superlative staff became available for a wide range of researches and practical developments in a peace-time environment.[18] ... The next ingredient was that the staff, about two hundred strong, was already highly skilled in electronic research and development.[19] They ranged from professors of physics to practical engineers from industry. Many of them had spent months, if not years, at the best overseas laboratories and were saturated with the most recent electronics techniques. In view of later events, it is also rather remarkable that there was not a single astronomer on the staff, nor, for that matter, anyone who had done a university course in astronomy ... Next in importance to the people was the store of components of all kinds which had been accumulated during the war years ... It is clear that another important factor was morale ... Our policy was to try anything that gave promise of useful scientific or practical applications; if successful, we poured in manpower and resources. Radio astronomy was to become one of the most productive of these.

As the War came to a close in 1945, discussions within the CSIR began concerning the details of the post War future of RPL. In an unpublished manuscript,[20] Marjorie Barnard (RPL Librarian and famous Australian author) has foreshadowed a number of post War developments that built on the foundation of radar research during the War. She discussed the future use of radar in meteorology

---

[16] J.P. Wild (1923–2008), future Chief of the Division of Radiophysics, 1971–1978.

[17] In Sullivan (1984).

[18] In Appendix H we summarise the role that Payne-Scott played in discussions of possible secret research in the transition period from the CSIR to CSIRO, from 1947 to 1949.

[19] There was some downsizing at the end of the War, as many of the professional staff took up positions in industry, went back to universities or started graduate degrees (for a Ph.D. the only option was to go overseas, in most cases the UK).

[20] See Chap. 11, footnote 4, *One Single Weapon.*

and air navigation as well as in studies of the nature of clouds in the terrestrial atmosphere. There was an extensive discussion about the use of radar in surveying the vast land area of Australia. In passing, Barnard wrote: "Radar may also be used for detecting and measuring the distance from earth of some astral bodies such as planetoids and meteorites (sic)." There was no mention of additional uses of radio astronomy in order to study the universe. In a final comment, she summarised the future of radar: "It is an extension of man's senses and knowledge . . ."

The most far-reaching discussion of the future programme of the RP division occurred in an extensive memo written by E.G. Bowen on 2 July 1945, a month before the end of the Pacific War. The memo was an agenda item for the 35th Session of the Council of the CSIR to discuss the "Future Programme of the Division of Radiophysics".[21] The future programme consisted of nine possible areas of research: (1) Propagation of radio waves including the ionosphere, radio noise and super-refraction, (2) Vacuum research including the generation of power at millimetre wavelengths and the use of radar techniques to accelerate elementary particles (programmes directed by Pulley and Gooden),[22] (3) General radio and radar research including antennas and receiver systems, (4) Radar aids to navigation including long range navigation (programme directed by V.D. Burgmann, a future Chairman of CSIRO,[23] (5) Aids to ground survey, (6) Atmospheric physics research (this became the Cloud Physics research group within RPL), (7) Research and development for the armed forces, (8) Industrial co-operation and (9) Co-operation with other divisions of CSIR. The future radio astronomy was hidden in a Section (1.2) with the obscure title of "Study of Extra-thunderstorm sources of noise (thermal and cosmic)". Bowen wrote: " . . . a type of noise appears which is thought to originate in the stars or in interstellar space . . . Little is known of this noise and a comparatively simple series of observations on radar and short wavelengths might lead to the discovery of new phenomena or the introduction of new techniques." Bowen then suggested that it might be possible to calibrate the radar receivers by pointing the antenna at the sky and then at an object near the antenna which was at room temperature.[24] No mention was made of Jansky's and Reber's results, which in fact were well known at the time. This meagre text was the basis of the formation

[21] NAA: C3830, D1/1 (1945 Programme of the Division of Radiophysics) and D1/2 (1946 Programme of the Division of Radiophsyics).

[22] On 22 June 1946, a short paper by Bowen, Pulley and Gooden, "Application of Pulse Technique to the Acceleration of Elementary Particles" appeared in *Nature*. Joan Freeman Jelly (see Chap. 11) worked in this "Vacuum Physics Laboratory" until early 1946, ably assisted by a young Alec Little (Chap. 9), who worked with Payne-Scott from 1948 to 1950 at Potts Hill.

[23] Starting in 1942, Burgmann (1916–1991) had been at the Australian Scientific Liaison Office in Washington and later at the Radiation Laboratory at MIT. He was at RPL from 1943 to 1949 working with Bowen on air navigation (later DME, distance measuring equipment) and then at the Division of Textile Physics, 1949–1969 (Chief, 1958–1969). See also Chap. 13 for a discussion of Burgmann and the ASIO file of Payne-Scott.

[24] Note the connection between this application of radar techniques to the descriptions by Payne-Scott in the technical report TI 191/4 (6 August 1945, Payne-Scott, 1945a in Chap. 5) and also RP 209 of 11 April 1944 (see above in this chapter).

of one of the major components of the RPL research program for many years to come. In addition there was no mention of radio radiation from the sun in the Bowen report, possibly due to the fact that uncensored reports of solar detections had not yet reached Sydney.

However, within a matter of days on 1 August 1945, Bowen wrote to F.G.W. White,[25] at this time a member of the CSIR Executive in Melbourne:

> ... these results [the "Norfolk Island effect" of March 1945, as reported by Orchiston (2005); see below for a discussion of this March 1945 detection of the sun at 200 MHz by the Royal New Zealand Air Force] are remarkable in that while one would expect to receive solar noise radiation on S. or X. band [10 or 3 cm wavelength], a COL [Chain Overseas Low-Flying] antenna and receiver at 200 Mc/s is quite unlikely to do so. I have heard rumours of the same thing in England, but as far as I am aware, the subject has never been followed up. We are therefore going to attempt to repeat the observations here in Sydney to see if we can track down the anomaly.[26]

Following Sullivan's (2009) conclusion, this quotation indicates that Southworth's detection of microwaves from the sun in 1942, which he published in 1945, was known at RPL in 1945 (see Appendix G and Chap. 5 for Payne-Scott's role in the evaluation of these results), whereas Hey's detection of solar bursts in 1942 (published in 1946) was the source of the "rumours". Ironically, J.S. Hey has written in his book, *The Evolution of Radio Astronomy* (1973):

> I well remember Sir Edward Appleton's [then Head of the Department of Scientific and Industrial Research in the UK] astonishment at a meeting in 1945 when I remarked that I was contemplating publishing in a scientific journal my 1942 paper on solar radio emission for, by some mischance, no-one had informed him of the 1942 episode.

Sullivan (2009) has given a detailed description of how this misunderstanding may have contributed to the conflict between Appleton and Hey over priority for the discovery of solar radio bursts.[27] Thus poor communication of the 1942 results had occurred not only in the UK, but in the far reaches of the Dominions.

---

[25] Professor (later Sir) F.W.G. White (1905–1994) arrived in Sydney in March 1941 from the University of Canterbury in Christchurch, New Zealand, initially seconded from the New Zealand DSIR for 3 months to provide assistance with the CSIR radar work. This period was extended to 9 months and then indefinitely as the RPL was reorganised. White first became the Acting Head of the Radio Research Board and later in October 1942 became the second Chief of the RPL as the deficient leadership qualities of D.F. Martyn were recognised by the CSIR (Minnett and Robertson 1996). In January 1945, he became the Assistant Executive Officer of CSIR and then CEO from 1949 to 1956, finishing his career first as Deputy Chairman and then Chairman from 1957 to 1970. White played a major role in the success of the radar effort in World War II with numerous contacts in the UK and the US (Evans 1970). Later on he was a major proponent for the growth of radio astronomy in Australia in the post War period. In the mid to late 1950s, he was a major backer of the RPL plan to build the Parkes 210 ft (64 m) radio telescope under the leadership of E.G. Bowen (whom White had recruited during an extensive visit to the US in mid-1943). (see also footnote 3, Chap. 4)

[26] NAA: C3830, A1/1/1, Part 1.

[27] Sullivan (2009) has described in detail the controversy generated by the Appleton paper ("Departure of Long-Wave Solar Radiation from Black-Body Intensity") in the 3 November

The New Zealand results had been sent to Pawsey on 20 August 1945 in a letter from Dr. Elizabeth Alexander of the Radio Development Laboratory of DSIR 9 (Department of Scientific and Industrial Research) in Wellington, New Zealand; the letter was dated 1 August but there was a postscript dated 8 August. The letter summarised the RNZAF (Royal New Zealand Air Force) results and included the report (RD. 1/518 "Report on the Investigation of the Norfolk Island Effect"[28]). (See Orchiston 2005, for the details of this report and a biographical sketch of this pioneering scientist.) Bowen, Kerr, McCready, Payne-Scott and Briton (Chief from January 1945 to May 1946) all received copies of this document from New Zealand. An enigmatic statement in the Alexander letter to Pawsey shows that the Hey results were not yet known: "I think the main differences between Southworth's latest results and ours are first, his work in the centimetre band fits more or less with black body theory and ours shows definitely too much energy on 200 Mc/s for theory. Sir Edward Appleton has also taken measurements on 200 Mc/s and confirms our finding. He suggests that at times of increasing sun spot activity there is an increase in energy at both ends of the sun's spectrum, and has encouraged us in our efforts". Orchiston (2005) has suggested that Appleton may well have been disingenuous in this claim because it is not substantiated in the Appleton (1945) publication, which in fact included no new solar data of his own. Another possible interpretation is that Alexander may have confused this claim with the "rumour" reported by Bowen above, based on the Hey report of 1942. Given the slowness of communication in 1945, it is quite possible that these types of misunderstandings might have arisen.

1945 issue of *Nature*. The publication made use of radio data provided by D.W. Heightman, a prominent radio amateur. The 10-40 MHz noise ("hissing") from near sunspot maximum in 1936 was detected only during the day and was the precursor of a low frequency fade-out. Appleton calculated that at these frequencies the received noise could not be thermal radiation from the sun; an enhancement of at least a factor of 10,000 was indicated. Heightman was upset and wrote a public letter claiming that his "work had been wronged". But the most serious aspect that Sullivan has reported was the fact that many of the ideas in the *Nature* paper were in fact due to R.E. Burgess (Chap. 8) who had written a letter to Appleton 5 days before the *Nature* paper was submitted on 24 September 1945.

[28]Norfolk Island is an Australian island between Australia and New Zealand; the Royal New Zealand Air Force was responsible for its air defence in World War II. Using a COL radar, the RNZAF personnel detected the sun on a number of days, starting on 28 March 1945, at 200 MHz. Elizabeth Alexander (1946) has given a popular account of the early New Zealand solar radio astronomy, the only publication of these results. A preliminary report from New Zealand had been sent to Bowen (9 July 1945, NAA: C3830, A1/1/1 Part 1) by E. Marsden (footnote 3, Chap. 4) who reported on obervations his son, E.D.L. Marsden, had made of the sun, confirming the "Norfolk Island effect." The younger Marsden's data was obtained from the radar station at Whangaroa on the North Island from 14 to 18 April 1945. The initial New Zealand detection from Norfolk Island was made on 27 March 1945 by Flying Officer Hepburn. Bowen wrote back to Marsden on 27 July 1945 indicating that 200 MHz observations were to start in Sydney in the near future. Bowen told Marsden that: "while.... [thermal noise] is actually received on 10 and 3 cm equipment, one would not expect to be able to detect it on COL equipment at 200 mc/s." Clearly Bowen was quoting the Southworth detections from 1942 to 1943.

RPL's First Observations of Solar Noise: Role of Payne-Scott

As described in the Preface and Chap. 1, J.L. Pawsey was well positioned to become a leader in the post War activities. His leadership in the War time activities of RPL and his experience in antenna design were to serve him well. Pawsey made two informal key appointments to his group, an action that would have far reaching affect in the next years; Lindsay L. McCready[29] was to head the receiver developments, while Ruby Payne-Scott was to become an overall advisor for scientific issues.[30] On 14 September 1945, a little over a month after the Alexander letter arrived from New Zealand, the "Propagation Committee" (see above and Appendix A) of RPL was reconstituted; present were Drs. Pawsey (Chair), Pulley, Piddington, Messrs. Kerr (secretary), Wood, Iliffe, Yabsley, Parker, Wing Commander Taylor, SLDR Hall and Flying Officer McDonald.[31] The minutes state that "the Committee's functions are to review progress in propagation work, and plan new work. The peacetime method of working will be closer to the typical University research system, one of individual responsibility and we hope to gradually work into this method". Within the purview of this committee, five topics of research were proposed. The list elaborated on the proposed research topics described by Bowen 2 months earlier. The subjects were: (1) troposphere propagation (super-refraction), (2) ionospheric propagation (Loran - navigation), (3) scattering from clouds, (4) scattering from the middle atmosphere and (5) radio noise levels. In the minutes Don Yabsley described the 2 and 6 MHz receivers which apparently were recording the noise level of the ionosphere. Of remarkable significance was the note that Miss Payne-Scott "is going to look for 200 Mc/s signals from the Sun at sunrise and sunset. Such signals, at a level greater than suggested by black body theory, have been reported on COL [Chain Overseas Low – at 200 MHz] sets in New Zealand".

At the 12 November 1945[32] meeting of the PC, the following discussion was noted under the rubric of solar noise:

> The programme of future work prepared by Dr. Pawsey was discussed. Present observations will be continued. Mr. Yablsley will shortly take on the development of special

---

[29]McCready (1910–1976) joined RPL in 1940 from AWA (as did Payne-Scott). He moved to CSIRO Applied Physics in 1962 and retired in 1971. In Fig. 6.1 McCready is the first person on the right in the first row.

[30]B.Y. Mills interview, 1 April 2007.

[31]The minutes also stated that an English mathematician, Mr. T. Pearcey, would be joining the group in the near future. Also footnote 96 in Appendix L. Pearcey was the leader of the group that subsequently built the CSIR automatic computer, CSIRAC [valve or vacuum tube electronic computer with 2,000 valves, a 2 ms cycle time and mercury acoustic-delay line memory], in the years 1948 to 1956. See *Annals of the History of Computing*, 1984, a publication by Maston Beard and T. Pearcey ("The Genesis of an Early Stored-Proram Computer: CSIRAC.").

[32]The minutes of the meeting of 15 October 1945 contained the statement that "reports on this topic [solar noise] will continue to be made to the Research Committee [probably the PC] for the present". Note that the first observations at Collaroy had already begun on 3 October 1945.

> equipment for the investigation. He will study the time variation of solar noise, and collect information where possible on the wavelength variation. Miss Payne-Scott will write a report on work to date in T.I. [technical report – see below. It is likely that this report evolved into the comprehensive summary report published in December 1945] form. Steps will be taken to start observations on the Sun and Milky Way from the Dover Sh.D. [shore defence] station.

This text was followed by a detailed two page report by Pawsey (B.51/4, in the PC file given to us by Yabsley), with outlines of possible cosmic and solar programmes. The subsections were prescient: study of solar intensity variations at 200 MHz, and the study of spectra of both solar and cosmic emission at a number of wavelengths above and below 1.5 m, polarisation of the radiation at the high frequency of S band (3 GHz or 10 cm), and the precision determination of the direction of arrival of the solar radiation using the Sh.D. in a beam "swinging" mode.[33] A number of possible new solar observations at higher frequencies were proposed (wavelengths 50, 25, 10, 3, 1.2 cm). A major new development was proposed to build antennas which were equatorially[34] mounted, so the sun could be observed continuously, not only at sunrise and sunset, as was the case with the radars at Collaroy and Dover Heights, which could only be moved in azimuth at a fixed elevation near the horizon. Also the proposal was made to move some of the radio equipment to Mount Stromlo to "carry out observations in close liaison with Stromlo personnel with visual equipment". The remainder of the report described the radio equipment in some detail. A fascinating detail was mentioned at the conclusion. "For subsequent work it is desirable to obtain a large, say 30 ft diameter, paraboloid suitably mounted". Had this proposal been implemented, the course of radio astronomy development in Australia in the post War era would have been vastly different.[35] It would be seven years before the 36 ft transit telescope was available at Potts Hill.

At the next meeting on 10 December 1945, the solar noise report stated: "Observations by the R.A.A.F. station at Collaroy are continuing. Miss Payne-Scott has completed a survey of the subject, to be issued as an internal report. Mr. Yabsley is starting on the development of equipment."

---

[33] Piddington and Minnett (1951) described this method of beam "swinging" in their observations of a few radio sources at 1,210 and 3,000 MHz. This method is a crude version of the common on-off technique used with current single dish radio telescope. The major problem that limited the sensitivity of the observations in 1951 was the large variable radio background arising from the galactic plane of the Milky Way.

[34] An example was the 4.9×5.5 m radio telescope used for centimetre observations of the sun, first at Georges Heights and then after the move in the late 1948 to Potts Hill (Wendt 2009). The initial use of this antenna was with a cumbersome altitude-azimuth drive, making it difficult to follow the sun during the day. At Potts Hill, a polar mount was installed with a tracking capacity to follow the sun continuously for many hours during daytime.

[35] A cryptic comment appears at the end of the Pawsey report (B.51/14). In a rubric number 4 (Personnel) the statement is made: "Yabsley to take over in a few weeks with assistance on equipment design from Receiver Section [McCready's group]. In the meantime continue exploratory work on catch-as-catch can basis." In the course of 1946, Yabsley was a major participant in the radio astronomy enterprise but was never "in charge". In late July 1950, he left the radio astronomy group.

The First Summary Paper in Radio Astronomy, December 1945" Author, Payne-Scott

In a document dated simply December 1945, but probably written before 10 December (SRP 501/27), "Solar and Cosmic Radio Frequency Radiation; Survey of Knowledge Available and Measurements Taken at the Radiophysics Laboratory to Dec. 1, 1945" (hereafter "the December 1945 summary paper Payne-Scott, 1945b"), Payne-Scott provided a detailed chronology of events during the course of 1945 that led to the first Australian solar observations from October, 1945 (see Chap. 7). She wrote that the initial solar observations "... were inspired by the almost simultaneous arrival of three reports in the laboratory ..." in mid-1945. Two of these were the classified reports referred to above: (1) the Hey report of 1945 describing the 1942 detections of bursts with the British radars on the south coast of England,[36] (2) the report from New Zealand of a large increase in solar noise from Norfolk Island from late March 1945 using a COL radar (see above). The final decisive report was G. Reber's article, "Cosmic Static", showing the galactic plane at 160 MHz with a resolution of about 12.5° (Reber 1944). This paper appeared in the November 1944 issue of the *Astrophysical Journal*, having been submitted on 8 May 1944. The detection of the sun was presented in the publication almost as an afterthought. Reber wrote:

> It has been suggested that this long-wave radiation could be set up in the corona of the sun. Until recently no positive evidence was available ... In any case the sun had the rather surprising intensity ...

The galactic centre scans carried out on 12 December 1943 showed the sun and the galactic plane in Sagittarius superimposed, with comparable intensity. Reber did not indicate that the implied brightness temperature was, in fact, in excess of 10^6K, if the emission were to arise from a disk the size of the optical sun.[37] Due to the slow speed of sea-mail, it is not surprising that the *Astrophysical Journal*, published in late 1944 in the US, did not arrive in Sydney, Australia until June or July 1945.

---

[36] This was the Army Operational Research Group (UK) Report No. 275, 13 June 1945, describing the 27 and 28 February 1942 detections of the sun using GL stations at 55–85 MHz (see Sullivan 2009 and above).

[37] An example of an erroneous interpretation of the Reber observation of the sun appeared in the publication by Appleton and Hey (1946a). These authors suggested that the 1.9 m data of Reber implied a brightness temperature of only about 6,000 K, roughly the black body temperature of the optical sun. In fact, Reber made no claims about the intensity of the solar radiation based on his detection of late 1943. As Sullivan (2009) has pointed out, Charles Townes (1947) published a paper: "On the Interpretation of Radio Radiation from the Milky Way" in which an earlier version did include a statement about the large solar intensity (hence non-thermal) implied by the Reber data at 1.9 m. However, the referee of the paper insisted that this conclusion be dropped; the final Townes publication contains no reference to the anomalous solar intensity at 160 MHz.

In Chap. 7, the ground breaking solar radio astronomy that was carried out by Payne-Scott and her colleagues at RPL in late 1945 will be described in detail. The December 1945 summary paper provided a valuable source of the details of the research and motivation in these first months after the War.

A remarkable additional non-solar observation was described in the December 1945 summary paper; the North Head antenna had been used. The Milky Way was observable in the daytime in these months and a 200 MHz map of the galactic centre at right ascension 17 h 30 min and declination –33° was presented. (Unfortunately this figure is missing from the copy of the Payne-Scott report located in the CSIRO Division of Radiophysics library by Wayne Orchiston in the early years of 2000.) With this system the angular resolution was some tens of degrees:

> It will be apparent that, in addition to the radiation from the sun, there appears to be radiation from a more diffuse area centred approximately on the centre of the galaxy.

At the end of November 1945 the Collaroy antenna was also used to scan the centre of the Milky Way at sunrise. The observations consisted of scans covering about 20° in azimuth and roughly centred on the galactic centre. The summary of these results followed:

> There are not yet sufficient results to produce a clear picture, and a number of puzzling variations have been observed; it is possible that some of these are due to absorption in the clouds of matter that cause the dark patches observable in the Milky Way.

Probably Payne-Scott was not aware that the interstellar dust was completely transparent at these radio frequencies. The final figure in the December 1945 report (also missing in the existing copy of the December 1945 summary paper) was a reproduction of the first contour maps of the sky which had been published in 1944 by Reber. This map of the northern Milky Way at 160 MHz with a resolution of about 12.5° would most likely have been far superior to the lower sensitivity and incomplete map presented in the Payne-Scott report. Improved maps of the radio continuum of the southern Milky Way were not available until 1950 when Bolton and Westfold (1950) and Allen and Gum (1950) presented maps of the galaxy at 100 and 200 MHz, respectively.

The final paragraph of the Payne-Scott report was a plan for the future. Here a number of prescient predictions were made concerning several aspects of the future of radio astronomy in Australia:

> It is hoped to soon begin here a programme of more exact work, in conjunction with the Stromlo Observatory. Among questions to be investigated are the frequency dependence of the [solar] radiation, its polarisation, further study of the long-term variations and an investigation of the short-period fluctuations. There is also hope to make a survey of the Southern sky; Sydney is almost at the antipodes of Reber's stations (sic), so that we can survey areas inaccessible to him; in particular it will be interesting to see whether radiation can be detected from the Magellan Clouds.

In fact the Magellanic Clouds (the nearest galaxy neighbours of the Milky Way and only readily observable in the southern hemisphere) were detected a few years later in Sydney by RPL scientists in both the HI line at 21 cm and in the radio continuum (HI by Kerr et al. 1954; continuum by Mills and Little 1953).

Bowen was impressed with the quality of the December 1945 summary paper and commented on it in a letter to E.G. Appleton in early 1946:[38] "Miss Payne-Scott, who with Pawsey and McCready has been largely responsible for the work here in Radiophysics, has written an internal report [the December 1945 summary paper] summarising latest ideas on the subject of solar and cosmic noise." Bowen asked Appleton's advice about publication of the Payne-Scott paper: "After adding some further experimental results we propose publishing it in one of the journals. So do you think this is a good thing and would you suggest the Proceedings of the Physical Society or the Astrophysical Journal?" Unfortunately the paper was never published;[39] this omission remains a serious loss for early radio astronomy in Australia since the Payne-Scott report so thoroughly documented the early thinking of the Sydney group with regard to both solar noise and cosmic noise. No record has been found that explains the reasons for not publishing. We can only guess that the frenzied pace of research and publications in the following year (and Payne-Scott's probable absence in late 1946 and early 1947, see Chap. 7) may have played a role. More important tasks were at hand.

Symposium on Radar, 5–7 December 1945: Payne-Scott was Present

A final event in this transition period from War time radar research to post War radio noise research occurred in late 1945 at the School of Tropical Medicine in the grounds of the University of Sydney.[40] At this "Symposium on Radar", which took place from 5 to 7 December, the techniques of radar and its military and non-military uses were described to a number of armed service personnel and some representatives of industry, as well as numerous RPL personnel. The symposium began on Wednesday with a short opening presentation by J.N. Briton (Chief of RPL), followed by "Introduction to Radar" by E.G. Bowen (Assistant Chief)

---

[38] See Chap. 7 for additional details of Bowen's description of the early RPL solar noise research in 1945 in this letter to Appleton, NAA: C3830, A1/1/1, Part 1.

[39] If the paper had been published, it could have been considered the first *review* paper in radio astronomy since the results of Jansky and Reber were discussed as well as new solar data at radio wavelenghts.

[40] The late Ron Bracewell has kindly given us a copy of the program for this conference. Joan Freeman has described the event in her biography (see Chap. 11). She writes that the conference was such a success that "Bowen decided that our papers should be published in a book, called A *Textbook of Radar* ... [which] appeared in 1947". The second edition was published by Cambridge University Press in 1954. Freeman's contribution was titled "Local Oscillators" while Bracewell's was titled "Microwave Transmission and Cavity Resonator Theory." See footnote 24 Chap. 5.

Fig. 6.1 Symposium on radar: 5–7 December 1945. Sponsored by RPL at the Lecture Theatre in the School of Tropical Medicine, the University of Sydney. The concluding talk was given by Bowen ("radar research"), and final remarks by the RPL Chief Brition. The front row (*left to right*), is E.G. ("Taffy") Bowen, J.N. Briton, J. Eagles, J.L. Pawsey, H.C. Minnett, L.U. Hibbard, T. Kaiser and L.L. McCready. Note personnel from armed services in the audience. Ron Giovanelli is sitting immediately to the extreme right in the third row. Photo sent to Goss by Minnett in March 2000. Photo used by permission of the late Harry Minnett

and "Fundamentals of Radar" by J.L. Pawsey. Presentations were given by Joan Freeman (Klystrons and Triodes) and Ron Bracewell (Resonator and Transmission Theory). On the second day the talks were by Harry Minnett (Transmit and Receive Switches and Mixers), Tom Kaiser (Chap. 13) (Aerials), Lindsay McCready (Receivers and Displays) and B.Y. Mills (Chap. 11) (Time Base and Marker Circuits). In total there were 19 presentations. The morning of the final day (Friday) was dedicated to demonstrations in the nearby RPL laboratory and concluded with a presentation on "Radar Research" by E.G. Bowen.

Harry Minnett (1999) has described the development of the Light-Weight Air Warning Radar during World War II at the RPL in the compilation of articles edited by MacLeod (1999). Minnett had found a photo of the assembled audience which had been stored for years in a filing cabinet, and later in a cardboard box, in the Minnett home. This photo is shown in its original form in Fig. 6.1, with a number of well known RPL staff sitting in the front row. When the original publication arrived in Socorro, New Mexico in early 2000, one of the author's colleagues, Loretta Appel, recognised Ruby Payne-Scott in the background. Minnett had an enlargement made at the RPL photo laboratory and the resulting image is shown in Fig. 6.2. A number of additional RPL and NSL personnel can now be seen, including W.H. Steel, Rachel Makinson, and Gordon Stanley. The names provided by the late Harry Minnett are given in the figure caption. Minnett (1999) writes: "It was fitting that RPL's most advanced Wartime radar should spearhead the Laboratory's post-War drive to apply radar techniques to civil problems."

Fig. 6.2 Enlargement of previous figure, showing Payne-Scott and colleagues toward to the back of the lecture theatre (pointed out to the authors by Lori Appel). Payne-Scott is in the next to last row (blond with glasses). Rachel Makinson is three rows down. Several well known CSIRO colleagues are near Ruby Payne-Scott. To our left of Ruby is Noel Thomdike. On her other side in the back row is Stuart Dryden of the National Standards Laboratory (NSL) and to our right of Dryden are Gordon Wells and Gordon Stanley, both of RPL (extreme right of photo, back row. See Chap. 11). David Holloway is in front of Dryden (row in front of Payne-Scott). Holloway was in charge of microwave standards at NSL. Behind Makinson and to our right is Mel Thompson of NSL. In front of Rachel (with glasses) is the well known expert in optics W.H. ("Beattie") Steel (see Appendix F) from NSL. Identifications provided by Harry Minnett 14 April 2000. Permission, see Fig. 6.1

Thus in late 1945, Payne-Scott was poised to begin her participation as a major member of the early team of physicists, engineers and technicians who were to lead to the rapid growth in Australian radio astronomy in the next decade. In Chap. 7 we will describe in detail the exciting years from late 1945 to the end of 1947 as the research endeavours were concentrated at Dover Heights, Sydney.

Chapter 7
Payne-Scott at Dover Heights 1945–1947: Discovery of Type I, II and III Solar Bursts and the Introduction of Fourier Synthesis in Astronomy

In early October 1945 the stage was set for the burst of energy within the radio noise group at RPL that led to a rapid growth in the fields of solar physics and radio astronomy in the following 5 years; the initial participants were Pawsey, Payne-Scott, McCready and Yabsley. Why did Ruby Payne-Scott play such a major role and why was she accepted as a key member of the *team* that produced major advances in solar physics? A number of these factors have been outlined in Chap. 6. A major factor was that she, unlike many of her colleagues, was a *physicist*, with a strong background in mathematics. Many of her colleagues were engineers, who had made significant contributions to the perfection of radar equipment during World War II. In addition, Payne-Scott was fascinated by the electrical engineering aspects of radar; this combination of physics and electrical engineering must have ensured that J.L. Pawsey, the leader of the nascent radio astronomy group, had complete faith in her. As she became a member of the team, she had the trust of the other members because of her experience and insights. For a few years she had no rivals in her scientific leadership as Pawsey's partner in setting the direction of solar radio astronomy research.

In September 1946 this leadership was strongly challenged by the arrival of John Bolton (Chaps. 8 and 11). Their sharing of facilities at Dover Heights produced endless conflict in 1947, eventually, in late 1947-early 1948, leading to her departure to the site at Hornsby (Chap. 8). The intellectual dominance of Payne-Scott in these first years trumped the difficulties that she experienced as a woman in the CSIR/CSIRO (Chap. 4); her colleagues accepted her and probably covered up for her for a few months in late 1946 when she may have had a miscarriage (see below). Certainly admiration for her remained until her resignation in 1951 and continued throughout 1952, during the time of the URSI international conference in Sydney (Chap. 10). In this chapter, we describe the genesis of solar radio astronomy in Australia, with particular emphasis on Payne-Scott's role. Only with the arrival of J. Paul Wild at RPL in 1947[1] was Payne-Scott's dominant role at RPL diminished;

[1] Wild joined RPL in 1947 and the "solar noise group" in February 1948. See footnote 16, Chap. 6. Wild's version (1987), that in February 1948 he be transferred to the solar noise group at the RPL

W.M. Goss and R. McGee, *Under the Radar*, Astrophysics and Space Science Library 363, DOI 10.1007/978-3-642-03141-0_7,

the day-to-day leadership of the solar group was taken over by Wild in the 1950s after Payne-Scott's departure.

The nature of RPL in 1945 was also critical for Payne-Scott's success; the research environment developed with the stimulation provided by the leadership of Bowen and Pawsey and the strong support of White within the CSIR/CSIRO administration. Thus fundamental astronomical research was fostered at a period when there was little astronomical research at Australian universities, apart from that at the Mount Stromlo Observatory in Canberra. In the years 1945–1951, Payne-Scott played a major role in the foundation of solar radio astronomy. Important contributions were the introduction of "Fourier Synthesis" into radio astronomy and the discovery of Type I, II and III solar bursts.

The Collaroy Campaign, October 1945: First Solar Radio Astronomy in Australia

These initial observations were obtained on a COL (Chain Overseas Low-flying) 200 MHz radar antenna. This was similar to the British built COL sets that had been used by the Royal New Zealand Air Force with which the sun had been detected earlier in 1945 (March and April, Chap. 6). The initial Australian detections were obtained at a Royal Australian Air Force location at Collaroy, a seaside suburb of Sydney about 24 km north of the city centre. The site was located about 120 m above sea level with the coastline running NNE to SSW. The sun was observed at sunrise and sunset since the COL antenna could only be moved in azimuth. The December 1945 summary paper (Chap. 6) gave the characteristics of Radar Station Number 54:

> The aerial is a broadside array of four horizontal rows each of ten half-wave dipoles with a reflector, having a gain of 80 relative to a half-wave dipole (i.e., $G = 130$) and a horizontal beamwidth (to half power points) of 10°.[2]

The first detection was on 3 October 1945 at 0531 EAST (Eastern Australian Standard Time), with a signal increase of about 27% over the general receiver level

site at Georges Heights cannot be correct. Wild has suggested that both Lehany and Pawsey took him to Middle Head for an initial visit. But Pawsey was in the US at that time on his world tour. Probably Fred Lehany was involved with Pawsey's approval at a distance.

[2]Observations with the RAAF (Royal Australian Air Force) Radar Station 54 were to continue to 15 February 1946. A series of letters from J.N. Briton (RPL Chief) in October and November 1945 to Air Commadore L.V. Lachal of the Eastern Area Headquarters of the RAAF at Point Piper (Sydney) elicited promises of cooperation for sunrise and sunset observations of the sun. Flight Lt. Katz, the Commanding Officer of Station 54, was quite helpful. On 15 November 1945, the Air Officer Commanding Eastern Area, promised to continue providing the use of this antenna. Briton had written on 7 November 1945 that "if you find that the observations seriously interfere with the normal operation of the stations you will immediately discuss with us the desirability of cancelling … them". Earlier Briton had promised to return the station to operation status (as a radar station) within about half a minute in case of an emergency. All these letters had as subject matter, "*Radar* Observations on the Sun at No. 54 Radar Station."! Probably RPL did not want to confuse the RAAF personnel with the fact that only reception, not transmission, was involved using their aerial (NAA: C3830, A1/1/1, Part 1).

at a bearing (azimuth) of 94°, essentially in the direction of the eastern horizon. Nine minutes later the noise power on a bearing of 93° was 4.5 times the normal noise power. The level corresponded to an equivalent temperature of the sun (size 30 arc min) of 15×10^6K or a flux density of order 10^6Jy (Jansky). The publication of these data in 1946 (*Nature*, see below) also summarised short terms fluctuations:

> We observed, from the direction of the sun, a considerable amount of radiation having the apparent characteristics of fluctuation "noise" when observed on a cathode-ray oscillograph or head-phones. However, the output meter reading fluctuated considerably, a characteristic which is not typical of normal thermal agitation "noise". The variation of apparent azimuth of arrival and of intensity with horizontal rotation of the aerial and the sun's elevation was qualitatively consistent with the assumption of radiation from the body of the sun modified by the known directional characteristics of the aerial ...

The December 1945 summary paper (Payne-Scott , 1945b) had given addition details about these "kicks":

> One feature ... is the short period fluctuations; the noise from the sun causes a fairly steady meter deflection on which are superimposed at intervals of perhaps a few seconds kicks which may be of the same order as the steady deflection; the relative magnitude and frequency of occurrence of these kicks seems to be independent of the elevation of the sun over the hour or so during which it is observed.

The 1946 publication (in *Nature*) described the short term fluctuations, named for the first time as "bursts", with no mention of the burst time scale of some seconds (which were discovered in subsequent observations in 1946–1947). The paper concluded:

> Furthermore, because of the very high levels relative to expected thermal radiation ... and the observed short-period meter fluctuations, it seems improbable that the radiation should originate in atomic or molecular processes, but suggests an origin in gross electrical disturbances analogous to our thunderstorms.[3]

The December 1945 summary paper contained three plots that illustrated the detection of the sun in October 1945. In Fig. 7.1, we show Fig. 2 from that publication, illustrating the observed vertical pattern of the sun on 4 days in October and November 1945; all the measured sizes were consistent with a point source (less than about 10° in diameter) of radiation arising from the sun. In the 1946 publication by the RPL group, there was no indication that the sea-cliff interferometer technique (see below), to determine either the location or the size of the emitting regions, had been employed.[4] There was a remark (with no detailed justification) that the magnitude of the sunrise intensities were reduced by a factor of three and the sunset records by a factor of two, to take into account reinforcement due to reflection from the sea and ground, respectively.[5]

---

[3] As seen above, it appeared that the consensus at RPL changed in the next months; at the later period the likely origin was thought to be from the solar interior ("crater") or even the corona.

[4] The McCready et al. publication of 1947 stated that no interference pattern was observed in these October 1945 data; this absence was probably due to a wide distribution of sunspots over the sun at this date, in contrast to the period in early 1946, when a compact sunspot group formed.

[5] Based on later publications, these correction factors were indeed plausible.

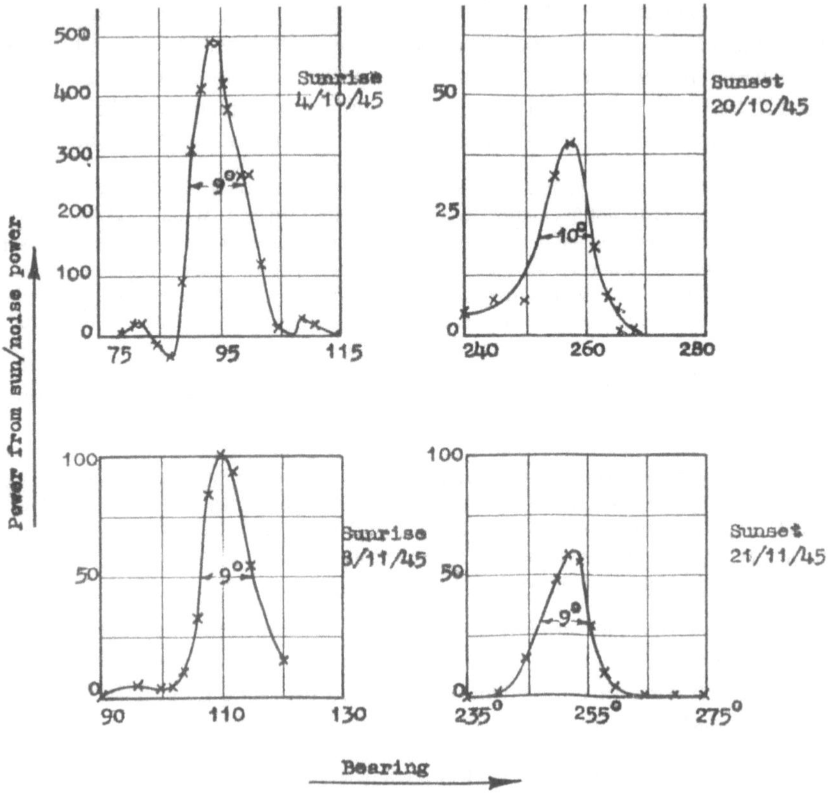

Fig. 7.1 From Payne-Scott "solar and cosmic radio frequency radiation; survey of knowledge available and measurements taken at radiophysics laboratory to 1 December 1945." (SPP 501/27 Fig.2 in her publication). Scans of the sun with Collaroy 200 MHz radar in late October and early November 1945, showing that the solar radio radiation was not resolved by the 10° beam of the antenna. Note that the signal intensity of the sunrise (over the sea) data were more intense than the sunset (over land) data. Thus the radio radiation arose from the sun. Observations were only possible at sunrise and sunset since the antenna was only steerable in azimuth or bearing at an elevation close to the horizon (CSIRO, ATNF archive)

The short term events ("kicks") were completely unexpected. Possibly the reason that these features were not mentioned in the publication of February 1946 was the uncertainty that was emphasised in the December 1945 summary paper:

> The meter fluctuations observed over a period of a few seconds ... may be due either to absorption or scattering of the radiation in the earth's atmosphere or to genuine fluctuations in the solar radiation. There is so far little evidence one way or the other, but this will be one of the first points to be investigated in future work, as it is critical in deciding the origin of the radiation.

As we will see below, work carried out in 1946 did elucidate the nature of the "kicks"; in later years these would initially be called "storm bursts" and later, in the mid-1950s, Type I bursts.

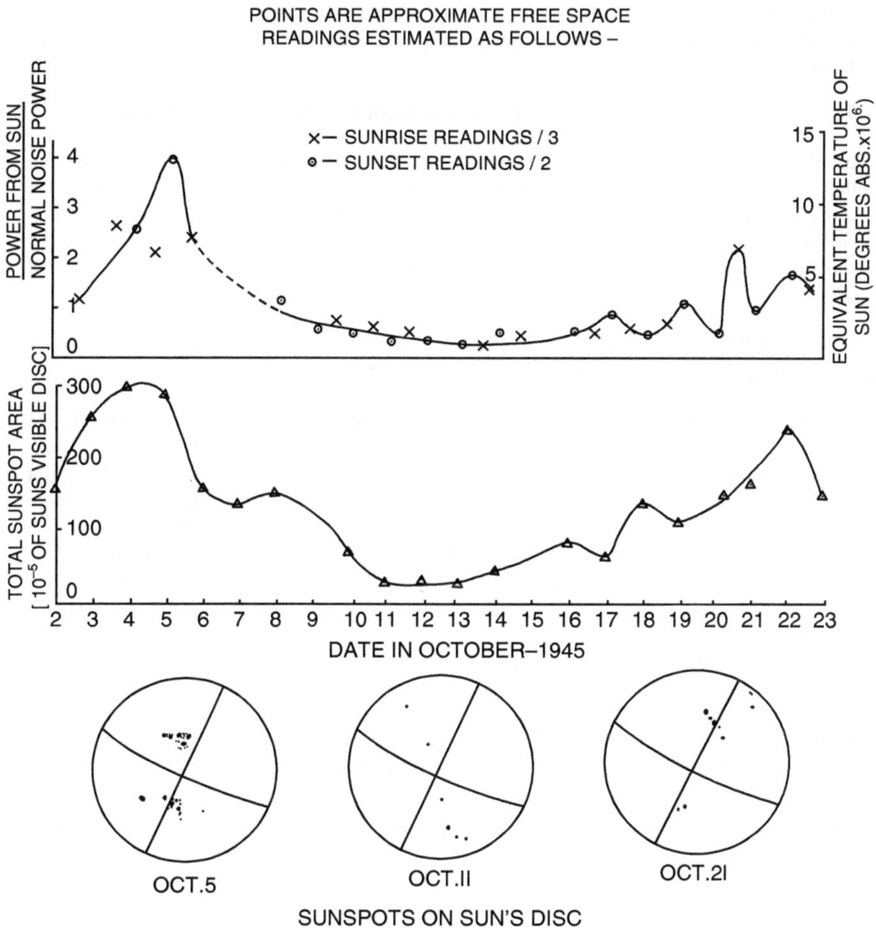

Fig. 7.2 The main result from the first publication in radio astronomy from Australia, submitted to *Nature* on 23 October 1945 and published 9 February 1946. For dates in October 1945 the radio power from the sun at 200 MHz was plotted vs. date in October 1945 and compared (below) with the total sunspot area (from the Commonwealth Solar Observatory at Mt. Sromlo). Major sunspots on the solar surface are shown in the bottom of the figures from three dates in October 1945. Reprinted by permission from Macmillan Publishers Ltd: *Nature*, vol. 157, p. 158, 1946. Figure 1 in "Radio-Frequency Energy from the Sun", by Pawsey, Payne-Scott and McCready

The main results of the October 1945 campaign at Collaroy were shown in Fig. 7.2 (from the 1946 publication); the sunspot data were provided by C.W. Allen[6] of Mount Stromlo:

[6]Clabon ("Cla") W. Allen (1904–1987) was on the scientific staff of the Commonwealth Solar Observatory (CSO) near Canberra, which became the Mount Stromlo Observatory of the Australian National University. Later on Allen carried out an independent radio solar patrol consisting of

> It is apparent that the peaks of 1.5-m radiation coincide with peaks of the sunspot area curve and with the passage of large sunspot groups across the meridian (*Nature* publication, 1946).

The December 1945 summary paper contained additional conclusions:

> It will be seen that there is good correlation between the two curves, particularly between their peaks, the peaks of the radiation curve being sharper than those of the curve for sunspot area. It is possible that the radiation may emerge from say, a *crater*[7] on the sun's surface, and so be highly directional. This suggestion is borne out by the sketches of the sun's disc, which show that the peaks coincide with the passage of large optically visible spots across the meridian.

The meaning of the word *crater* is puzzling; probably this concept was related to the *holes* discussed below. At about this time Bowen reported:[8]

> Within a few weeks of the end of the war, the phenomenon [solar noise detected by Hey in 1942] was investigated afresh in Australia, and a few weeks' continuous observation was sufficient to show a very close correlation between excess noise from the sun and sunspot activity. It was shown that the existence of active sunspots corresponded to an average noise temperature at the sun of 2 or 3 million degrees, and on occasions as much as 25 million degrees absolute. One theory of the production of this noise is that sunspots correspond to holes in the surface layers of the sun through which energy corresponding to the internal temperature can escape in the direction of the earth. The fact that little or no energy comes out in the visible spectrum is readily explained by the heavy absorption of these frequencies which is bound to occur. Alternatively, someone is letting off atom bombs in the sun and has been doing so for some considerable time.

Perhaps the *holes* referred to above are related to the *crater* described in the December 1945 summary paper;[9] a probable connection was being made between the hot internal temperatures in the solar interior and the cooler solar surface at 6,000 K. No explanation for the bizarre statement concerning atom bombs was

---

200 MHz data from Mount Stromlo (antenna and receiver provided by RPL) during a 6 month period starting in April 1946. The PC (Propagation Committee) meeting of 16 April 1946 mentioned that the RPL personnel were to improve the 60 and 200 MHz instrumentation. It is not clear if any 60 MHz data was obtained by the CSO group. The RPL group's appreciation of the importance of optical collaboration had already been mentioned at the PC meeting of 12 November 1946 where it was proposed for the first time to place a 200 MHz receiver at the CSO in order to ensure "... close liaison with Stromlo personnel with visual equipment". The impressive solar events of late July–August 1946 were observed at CSO; the publication was prepared by Allen (1947). The 200 MHz data provided to Payne-Scott et al. from the March 1947 event was also obtained at Mount Stromlo. Allen later moved to University College, London, where he was Director from 1951 to 1972. Allen was the author of the first editions of *Astrophysical Quantities*, a reference used by most astronomers (edition one in 1955 up to edition three in 1972). He published one additional solar radio paper in 1957 ("The Variation of Decimetre-Wave Radiation with Solar Activity".).

[7] See the letter from Bowen to Appleton, described below.

[8] "Radar in War", *Australian Journal of Science*, 22 October–21 December 1945.

[9] In this early attempt to make sense of the properties of solar bursts, the holes could have had walls which could have led to some degree of directionality.

provided! As far as we can ascertain, this novel explanation was never discussed again. Within a short period, there was no discussion of craters on the sun being involved in the emission of the excess solar radio emission; the RPL group realised that solar bursts, in fact, arose higher in the solar corona.

The December 1945 summary paper and the publication of early 1946 also mentioned additional observations that were noteworthy. Probably using the higher frequency antenna at Georges Heights at Middle Harbour (see Orchiston et al. 2006), a marginal detection was reported of the sun at 25 cm on 4 and 5 October 1945, with an equivalent temperature (the average values of brightness over the 30 arc min diameter solar image) in the range of 6,000–20,000 K. Based on current knowledge about the properties of the sun at 20 cm, the low value of 6,000 K is implausible and must indicate the large uncertainty in these first observations in the cm range. At 50 cm, the sun was not detected with an upper limit of about 60,000 K. (We have used the more conservative values from the report.) Since the Collaroy data were only available at sunrise and sunset, additional 200 MHz observations were carried out with a lower gain Yagi system at North Head (Sydney); this antenna could track the sun during the day, since the elevation of the aerial could be varied. Using a system with reduced sensitivity compared to the Collaroy antenna, a marginal detection of the sun was reported in the December 1945 summary paper; from 1,400 h to sunset on 31 October 1945, the sun was not detected.

Both the December 1945 summary paper and the 1946 publication contained an attempt to compare the solar radiation with the previously known properties of the "cosmic static" reported by Jansky, and Reber from the 1930s and 1940s as well as Fränz (1942). The conclusions from the two sources are somewhat different. The 1946 publication states:

> In view of observations of such intense bursts of radiation from the sun at the wave-length at which "cosmic static" is known, it appears desirable to question the suggestion that the latter originates in the interstellar space. It seems more reasonable to attribute it to similar bursts of radiation from stars which, because of their large number, could yield an approximate constant value for any one area in the sky.

The December 1945 summary paper contained three relevant conclusions concerning the comparison of the radio radiation from the Milky Way and the sun:

> (8) Similar radiation [i.e., the solar radiation] is obtained from the collection of stars in the Galaxy; it is most intense in the direction of the centre of the Galaxy, and has a distribution corresponding approximately to that of the stars in the Galaxy.
> (9) The effective temperature of the Milky Way increases with decreasing frequency [in modern terminology, has a non-thermal spectrum].
> (10) Because of their similar nature and behaviour, it seems likely that the radiation from the sun and the Milky Way have a similar origin.

As is now well known, these parallels between the solar emission and the galactic emission did not stand the test of time. Within a few years the realisation that the filling factor or dilution factor (i.e., the fraction of the sky covered by the stellar disks, a value of about 10^{-14} for stars similar to the sun) was tiny and the low luminosity of the sun implied that a collection of "suns" (stars similar to our sun) in the Galaxy could not explain the non-thermal emission of the Milky Way.[10] By the late 1950s astronomers accepted that the Milky Way extended emission arose in the interstellar matter via synchrotron emission from relativistic electrons distributed throughout the Galaxy (see Alfvén and Herlofson 1950; Kiepenheuer 1950; Ginzburg 1951; also footnote 16, Chap. 10).

The final conclusions in the December 1945 summary paper concerned the physical nature of the solar radiation. There is a prescient discussion of the importance of the solar corona, influenced by conversations with D.F. Martyn.

> The radiation [the metre-wave non-variable emission] may come from the corona, which has been recently shown to have a much higher temperature than the photosphere, and which, although transparent to visible light, may well be opaque to long radio waves. Dr. D.F. Martyn, of the Mount Stromlo Observatory, Canberra, has suggested this origin.[11]

Payne-Scott pointed out that Martyn's theory did not seem to account for the greatly increased solar radiation associated with sunspots. "It would [also] not give a continual increase in temperature with decreasing frequency, as appears to occur for the Milky Way radiation. There is not yet enough evidence to determine how the solar radiation varies at frequencies below 160 Mc/s."[12]

Other possibilities for the origin of the solar radiation were discussed, such as non-thermal emission similar to atmospherics in the earth's atmosphere, as well as plasma oscillations in the ionised atmosphere of the sun. Payne-Scott then pointed out that that the previous theory of the Jansky-Reber observations of the Milky Way had postulated that the origin for this emission was emission from the ionised matter in the Galaxy. "This would not account for the observed solar radiation", presumably due to the vastly different physical scales and densities in the interstellar medium of the Milky Way.

On 17 October 1945, Bowen wrote to White with an enthusiastic assessment of the first solar noise campaign: "... Pawsey and Miss Payne-Scott's noise measurements are bearing considerable fruits". He summarised the results, in particular the correlation of noise level with sunspot number. A letter to *Nature* was planned in a

---

[10] Sullivan (2009) has emphasised that Payne-Scott did not point out a problem mentioned by Reber and Jansky: if the sun and the radiation of the Galaxy had the same origin, then the fact that the ratio of sun to Galaxy for the optical regime was much higher than for the radio was unexplained. The optical sun is much brighter than the optical Galaxy as compared to the radio; at intermediate frequencies of 200 MHz the sun and the Galaxy are, in fact, comparable in intensity. Greenstein et al. (1946) also pointed out the serious discrepancies due to the small dilution factors of stars in the Milky Way which would "vitiate this interesting new suggestion" (from Pawsey et al. in the 1946 *Nature* paper) in order to explain the galactic radiation as arising from the collective effect of emission from stars in the Galaxy.

[11] This statement certainly strengthens the assertion that Martyn deserved much of the credit for the association of the hot corona with the radio frequency quiet sun. The controversy concerning the adjoining papers by Martyn and Pawsey in *Nature*, 2 November 1946, describing the theory and detection of the radio radiation from the million degree solar corona, has been summarised by Orchiston et al. (2006) and Sullivan (2009).

[12] Many of these questions would be sorted out in the next few years as the nature and frequency dependence of both the Milky Way radiation and solar radiation became apparent.

few days. The entire operation happened quickly. The first observations from the Collaroy site were obtained on 3 October 1945 and the experimental work continued to 23 October. "Radio-Frequency Energy from the Sun" by Pawsey, Payne-Scott and McCready, was submitted to *Nature* on 23 October (including data from this date!) and published on 9 February 1946.[13]

A number of authors (e.g., Sullivan 2009) have commented on the long publication delays that the early Australian radio astronomy papers, in particular the first two submissions from Sydney, were subject to after submitted to British publications. Two publications from the UK predated the Australian paper of 1946: (1) the Appleton paper, "Departure of Long-Wave Solar Radiation from Black-Body Intensity" (submitted on 24 September 1945 and published on 3 November 1945 in *Nature*) and the Hey publication as described above ("Solar Radiation in the 4–6 m Radio Wave-Length Band", submitted on 17 October 1945 – a week before the Sydney submission – and published in *Nature* on 12 January 1946).

The references in the RPL paper of 1946 were striking. There was no reference to the Southworth paper of April 1945 (see Appendix G); the references that

[13]This paper was RPP 1 (Radiophysics Publication No. 1); this series ran from 1945 to 1997 and produced 3,934 publications. We have copies of the first notebook from the Publications Office which ran from RPP 1 to RPP 667, ending in May 1961. The notebook consisted of a series of handwritten records providing the publication history of the publication, e.g., dates of submission, publication, etc. There were two additional RP series of internal reports. The RPR series was a series of preprints. The RPL series was a collection of laboratory reports, some of which were later converted into polished reports or publications. In NAA: C3830, A1/1/1, Part 1, the entire publication of RPP 1 has been found as a single letter to the editor (Dear Sir ...) with the signatures of the three authors. The published version is essentially identical to the submitted paper. Five days before the publication, Pawsey sent a telegram (costing 19 shillings and 4 pence) to the editor of *Nature*, complaining about the delay:. "[D]ue to unfortunate and incorrect reports of our work originating in the US would appreciate early publication". Bowen even characterised this as: "... irresponsible sources in the US". The Australian High Commission in London had sent a telegram to RPL on 31 January 1946, with a confused report of press reports claiming that Piddington had made *radar* contact with the sun and moon. "We are being pestered for details ..." Pawsey tried to kill this story in a return telegram to the High Commissioner on 6 February 1946, pointing out that they had only detected "noise from the sun ... apparent temperature over one million degrees ... and shown close correlation with sunspot activity". Appleton was also involved in trying to quell the adverse publicity in the UK with respect to radar echoes from the sun: "Fortunately 'The Times' has a good scientific correspondent (Andrade) and they refused to touch it. But you can imagine what the less responsible papers did with it." (4 March 1946, NAA: C3830, A1/1/1, Part 1). This clear cut story with the blame laid at the footsteps of the US press is, however, inconsistent with a letter from Pawsey to Southworth 10 days later; the latter had sent the US press newspaper clippings to Australia. According to Pawsey, "The Australian press got it thoroughly mixed up, they were even claiming echoes from the sun ... The New York Post seems to have copied a false Australian press report." The two press reports (NAA: A1/1/15 Part 1 and D9/4H Part 1) indicate that at least the New York Post of 29 January 1946 only reported the false claim that the CSIR group were the first to detect radio emission from the sun. The 30 January 1946 article in the Sydney Sun (!) reported incorrectly that radar echoes had been detected from the sun. The hyperbole was extreme: "It is likely that many .... new discoveries will follow which will undoubtedly affect the lives of everyone." Bowen wrote in the margin: "What is this about?"

appeared were Reber (1944), Jansky (1933), and the restricted reports from Hey and the New Zealand group.[14]

Bowen (then Deputy Chief of RPL and soon to become Chief) was clearly proud of this first publication. He wrote to Appleton on 23 January 1946, after receiving the Appleton paper of November 1945:

> This laboratory obtained what I think is the first direct experimental verification of this effect [solar bursts] during October 1945 when over a period of 3 weeks we measured solar noise on 200 Mc/s and obtained a close correlation with sunspot activity. A letter to *Nature* was immediately concocted and you may have seen the published version by now . . . We are now speculating on the source of the noise, and have departed somewhat from the suggestion in the letter [the February 1946 publication] that it was of electromagnetic origin. Personally I feel it must be of thermal origin either from the depths of the sun or in some way from the corona.[15]

As 1946 began, plans were being made to extend the observations of the sun and the Milky Way. At the 15 January 1946 meeting of the Propagation Committee a "Cosmic and Solar Noise Section" had been established under the leadership of Pawsey and McCready. The members of the group included Payne-Scott, Yabsley (see footnote 36 this Chap.), a technical officer (Don Urquart) and a technical assistant. The observations were still being carried out at Collaroy; in December and January, observations were difficult, however, because the sun was then in the centre of the Milky Way and there was confusion with the intense radio noise emanating from there. The Solar and Cosmic Noise Section planned to concentrate on equipment development. Plans were also announced to carry out simultaneous "fading" observations at two sites to determine if the fading (pronounced decrease of intensity in the solar radio radiation) was due to solar or

---

[14]The use of these still classified reports caused some trouble back in the UK; although the results of both reports were then widely known in the ionospheric and radio noise field, the reports had not been cleared with various committees in the UK responsible for the declassification of World War II documents. Appleton and E.V. Hill (1886–1977, Nobel Prize, 1922, for Physiology or Medicine; member of the Tizard Committee for air defence in the mid-1930s and Chair of the Committee of Post-War Publication, dealing with declassification in 1945) were involved. Appleton wrote to White with a complaint about the citation of two classified reports. Both Bowen and White wrote letters of apology to the UK. Neither could really see that major damage had been done. White wrote to Bowen on 30 April 1946, that "... I do not think it matters a great deal – it is only a formal point". Bowen had a more biting response in a letter to White on 26 April 1946: "I am sorry that Appleton is making a song and dance about our letter [*Nature* letter of 1946] . . . but I suppose he is just expressing his well-known ownership of all radio and ionospheric work." (NAA: C3830, A1/1/1, Part 1).

[15]Bowen's claim that the RPL detection was the first observation of the excess radiation associated with sunspots was not correct. As Pawsey, Payne-Scott and McCready pointed out in their 9 February 1946 *Nature* paper, earlier reports were cited from the British AORG (Army Operations Research Group, Hey et al.) and the New Zealand group. In addition, the discussion by Bowen about the timing of the *Nature* paper is ironic, since the prevailing view held in Australia was that Appleton (as the probable referee for the *Nature* paper of the Australians) may well have held up this first RPL solar publication (Pawsey, Payne-Scott and McCready) in favour of the Hey paper. As far as we are aware, there is no definite proof of this claim.

atmospheric (also ionospheric) effects. If the fading was observed to be dissimilar at the two displaced sites, then the probable explanation was that the intervening medium (e.g., the atmosphere) imposed the fading and the fluctuations would not arise from an intrinsic process on the sun.

Ground-Breaking Developments in Solar Noise Research and Techniques of Interferometry: February 1946, Dover Heights

The major impact of the two 1946 research campaigns at Dover Heights was based on a number of factors. The team of Pawsey, Payne-Scott and McCready were building on the experience gained from the ground breaking results from the previous October. Relevant scientific questions were posed and revolutionary instrumental techniques were applied: "directional observations, based on interference phenomenon"[16] yielded an association of the radio bursts with sunspots. This technique was built on experience gained with the anti-aircraft radars during World War II. This experience led directly to the concept of radio Fourier synthesis, a technique applied in the construction of numerous radio telescopes during the remainder of the twentieth century. Each member of the team made major contributions to the ultimate success of the endeavours. Finally, there was an amazing piece of luck; this period was close to the time of a major sunspot maximum, with major sunspot groups occurring in February and July 1946. The RPL team had the relevant expertise and equipment and they were present at the "right" time.

After the October 1945 campaign was complete and the *Nature* paper submitted at the end of the month, the daily monitoring at the Collaroy site (the air force radar installation) with the COL antenna continued to 15 February 1946. In late January or early February 1946 a new site was inaugurated at the ShD site (a Shore Defence radar) at Dover Heights RPL field station at the Army Reserve (see Fig. 7.3), in the eastern suburbs of Sydney about 17 km south of the Collaroy site (see Fig 1.4).[17] The cliff site was slightly lower in elevation with an eastern horizon 85 m above the sea. The antenna has been described by Bird (1993); the original design of the broadside array was by Minnett (1943) and the individual elements by Pawsey et al. (1940). The 200 MHz antenna consisted of 36 half wave elements,

[16]The 'directional observations' referred to the interferometer described below; with this technique it was possible to determine both the position of the radio burst emissions and limits on the angular size of the radio sources.

[17]This site is now Rodney Reserve, Dover Heights. A memorial plaque was unveiled on 20 July 2003 by the Governor of New South Wales, Her Excellency Professor Marie Bashir. On the same day, a model of the 100 MHz sea-cliff interferometer used in the early 1950s by Bolton, Slee and Stanley was inaugurated.

Fig. 7.3 RPL field station, Army Reserve, Dover Heights; the Fortress Fire Command Post was in this reserve, 2.2 km south of the Macquarie lighthouse and 2 km north of Bondi beach. (Also 7 km east of the centre of Sydney.) Photo, 18 February 1943. Possibly the two men near the Shore Defence Radar (ShD – 200 MHz) are John Worledge (head of the NSW railways radar structures group) and Fred White (then Chief of RPL; see Minnett 1999 "Radar and the Bombing of Darwin"). This 200 MHz antenna was used for the solar radio astronomy experiments in 1946. The view is to the north, North Head is clearly visible. The antenna to the north is the experimental array on top of the small building. ATNF historical photographic archive, B81-1

arranged in a plane, 33 cm in front of a wire-mesh reflector. A number of improvements in the receiving apparatus were made, such as the use of continuous paper chart recorders (instead of visual reading of meters).

In addition several tracking antennas (200 MHz Yagi) were installed at North Head, Mount Stromlo and even Dover Heights which could follow the sun throughout the day; these simple Yagi antennas had a lower gain than the radar aerial (ShD). The ShD and COL antennas could only observe at sunrise and sunset due to the lack of elevation mobility. Calibration was a major issue with a claimed precision of the intensity scale of only 40%; the calibration of the systems was to improve in subsequent years. The fringe spacings (see below, the effective angular resolution of the sea-cliff interferometer) were about 30 arc min at Dover and 21 arc min at Collaroy. The primary beam was about 9–10° (December 1945b report).

The sun was detected on practically every day on which observations were carried out in early 1946. As a number of authors have pointed out, an event of great good fortune occurred in early February 1946. The largest sunspot group ever detected up to this time was observed to cover a fraction of 0.0052 of the solar area[18] (Figure 7.4 shows an optical photograph from the Royal Greenwich Observatory of 5 February 1946). C.W. Allen from Mount Stromlo contacted the RPL group by telephone and a series of intense observations began. Examples of the data from 7 and 8 February, as well as from 2 months later, are shown in Fig. 7.5.[19]

The bursts were mainly Type I bursts (using the modern nomenclature); the 5 April 1946 bursts were isolated and might be Type III bursts. The rapid bursts observed were the "kicks" referred to in the Payne-Scott report of 1945b. The intensities ranged up to 12×10^6Jy (North Head, 7 February 1946, at about 1600 EAST). There were two types of variation: (1) a relatively slowly varying type with time scales of many days; these were probably Type I "enhanced radiation" or "noise storms", with intensities in the range $0.05–10 \times 10^6$Jy (a variation of a factor 200), and (2) intense bursts with time scales of seconds ("the kicks", probably Type I bursts) "of widely varying frequency of occurrence". The intensities of the bursts could be an order of magnitude or greater than the mean level. Later on in the 1950s, the classification of these events was characterised as: "thousands of short-lived spikes [seconds duration] superimposed on a slowly varying continuum", continuing for periods from a few hours to a few days (Kai et al. 1985).

Two important pieces of data indicated that the rapid variations were intrinsic to the sun and did not arise in the atmosphere; a comparison was made to the twinkling of stars in contrast to the more steady appearance of planets in the night sky. There were no systematic changes in the character of the rapid variations as the sun rose from the horizon to the zenith; also the variations always consisted of increases above a mean level with no decreases. The major reason that the RPL group claimed that the rapid variations were due to intrinsic processes on the sun was the near identity of the intensities, as observed from Collaroy and Dover Heights, a distance of 17 km.[20] In addition, the similarity of the Mount Stromlo and Dover Heights data (from later observations of April–May 1946) at a distance of about 260 km was decisive: "It is

[18] This sunspot was the second largest in area in the years 1874–2009; the sunspot of April 1946 was even larger. Newton (1955) has summarised major sunspots from 1874 to 1952. The largest sunspot in this period was on 7 April 1947 (see below, 6,132 millionths) followed by one on 6 February 1946 with a maximum area of 5,202 millionths of the sun's hemisphere. He pointed out that the 1 March 1942 sunspot was number 51 in the ranking of the most prominent sunspots (2,048 millionths) in this 78 year period from 1874 to 1952; this was the sunspot group ("a minor giant") that led to the discovery of solar bursts by Hey on 27 and 28 February 1942. The list of major sunspots can be followed to at least 2001 using the NASA web site, "http://www.science.nasa.gov/spaceweather/sunspots/history.html". Four of the largest ten sunspots since 1874 occurred in the years 1946–1947. Newton gave little credit to RPL for the later development of solar radio astronomy, the only references being to Appleton and Hey.

[19] At the 14 February 1946 meeting of the PC, there was a lengthy discussion about the major solar event of the previous week. The report stated: ". . . a series of records obtained at sunrise on the 200 Mc/s Dover and Collaroy sets showed a strong correlation with the actual visible sunspot area . . . These records indicated that the activity originated from an area small compared with the total area of the sun although the analysis had not yet located the source with respect to the spot. The effective temperature on 200 Mc/s of the active area was of the order of $10^{1\circ}$K."

[20] Astronomers would use the term "spaced receivers".

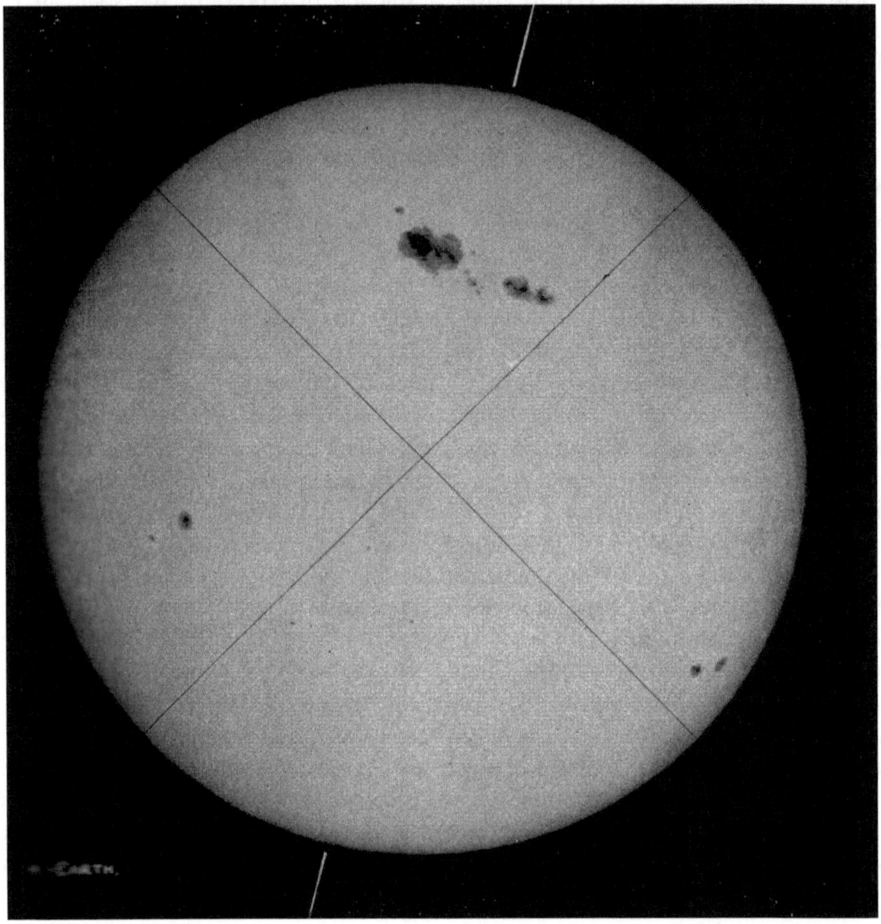

Fig. 7.4 An optical image of great sunspot of February 1946. Note the orientation with north to the top (see Fig. 7.9 for the orientation with respect to the eastern horizon as observed from Dover Heights). Adapted from Appleton and Hey 1946a, *Philosophical Magazine (Ser 7),* vol. 37, p. 73, Fig. 1 in their publication. (Used by permission of Taylor and Francis Ltd., http://www.informaworld.com). This was one of the larger sunspot groups ever recorded. Major radio solar events were associated with this sunspot group. Observed at the Royal Greenwich Observatory 5 February 1946. The intense radio emission from this sunspot group was observed in the UK and Australia

highly improbable that variations having such a high degree of correlation at widely separated sites should be due to any effect in the atmosphere, and it seems that most of them are extraterrestrial, and presumably solar, in origin."[21]

[21] At the meeting of the Propagation Committee on 16 April, 1946, a report was given summarising the coordinated Mount Stromlo and Dover Heights data at 200 MHz. The main result was that the bursts occurred simultaneously at the two places, suggesting that the bursts were not due to the atmosphere or the ionosphere of the earth. The publication of the results was being prepared.

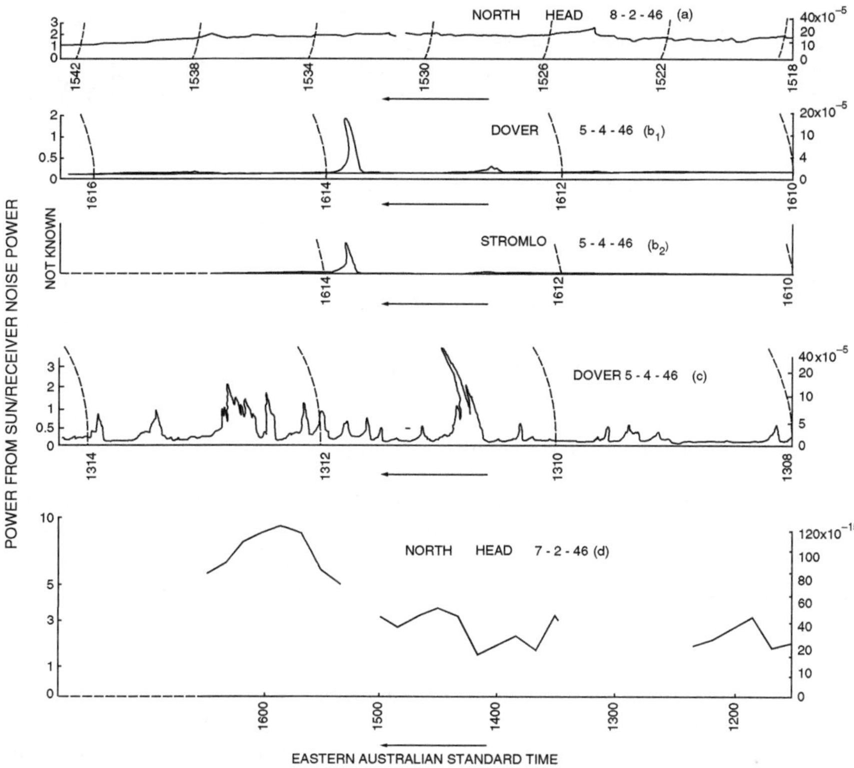

Fig. 7.5 Radio observations by the RPL group of Pawsey, Payne-Scott and McCready from February to April 1946; these were the second systematic series of solar noise observations made in Australia. Various sites were used at 200 MHz: North Head, Dover and Mount Stromlo near Canberra (see Fig. 1.4). Time increases to the left. The units of intensity are 10^{-15} W/m^2/MHz or 10^5 Jy. Figure 1 of the 1947 publication of McCready, Pawsey and Payne-Scott (MPP) in *Proceedings of the Royal Society, Series A*, vol. 190, p. 357, "Solar Radiation at Radio Frequencies and its Relation to Sunspots." Used by permission of the Royal Society

At the subsequent meeting on 13 May 1946, the announcement was made that the paper was almost complete and was to be submitted "in a few weeks". This prediction was repeated in subsequent meetings in June and early July (the paper was in fact submitted on 22 July 1946). At the 13 May 1946 meeting, the topics that were to be discussed in this new paper were: correlation of solar noise with sunspots, the short burst phenomena and their simultaneous nature at the two observing sites, and the location of the bursts on the solar surface. A new development, that the Mount Stromlo scientists were attempting to carry out a correlation of the radio observations with visual data on sunspots, was also described. Pawsey reported that "visual observations are very difficult. No correlation had yet been obtained at Stromlo between radio and visual phenomena."

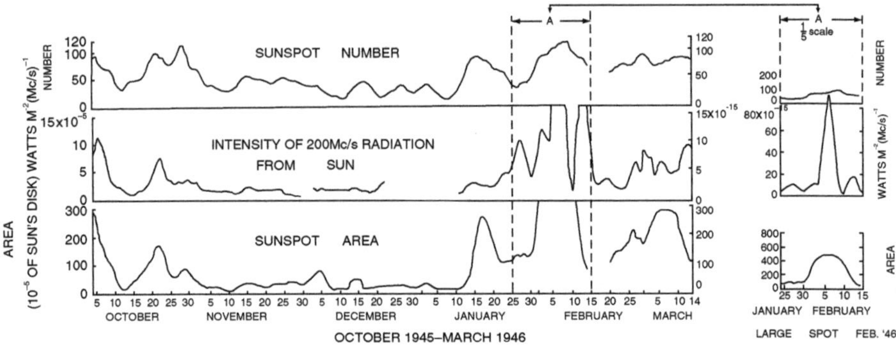

Fig. 7.6 Day-to-day variation of the radio noise at 200 MHz compared to sunspot number and sunspot area; the connection of radio noise from the sun and sunspot activity was established (Figure 3 of MPP). *Proceedings of the Royal Society, Series A*, vol. 190, p. 357, "Solar Radiation at Radio Frequencies and its Relation to Sunspots." Used by permission of the Royal Society

In addition to the 200 MHz data, a few additional observations were obtained at 3,000, 1,200 and 75 MHz.[22] The latter frequency represented an extension of solar metre wave observations to a considerably longer wavelength by the RPL group. During disturbed periods the 75 MHz intensity was much higher than at 200 MHz. The two high frequency observations agreed roughly with the Southworth data (1945, from Bell Laboratories) at 3,000 MHz.

The correlation of the flux density of the solar 200 MHz emission and sunspots that had been found in October 1945 was now extended using the observations from 3 October 1945 to 15 February 1946; the Collaroy data were still being provided by Air Force personnel. The collection of data was then continued up to early May based on observations from Dover Heights, the first of which was made in late January or early February 1946 and continued to about 6 May 1946. A partial summary for a portion of this period is shown in Fig. 7.6; during the great sunspot of early February an expanded scale is used on the right hand side of the figure. The correlation suggested earlier between sunspot area and radio intensity was once again confirmed. During the period the mean flux density of the 200 MHz sun was 13×10^5Jy, 1.2×10^5Jy, 10×10^5Jy, and 106×10^5Jy on 5 October 1945, 11 November 1945, 26 January 1946, and 7 February 1946, respectively. Data on a few days in early February from a simple Yagii at North Head was utilised, while the Mount Stromlo 200 MHz system was available after early April.

The major result of the 1946 February campaign was the interferometry which began on 26 January 1946. This was an historic occasion as it represents the first use of interferometry in radio astronomy. Sullivan (2009) has provided a thorough historical summary of this event. As he points out, the first use of interferometry

[22]The PC of 14 February also mentioned attempts to observe at 60 MHz; the observations were compromised by interference.

in radio astronomy used only a single radio telescope! The record is not clear which of the two stations (Collaroy or Dover Heights) was used. Payne-Scott may have been present in person while she carried out the observations on 26 January 1946. In any case, it is certain that she did analyse the data after this historic event.

The method made use of the sea-cliff interferometer,[23] a radio analog of a "Lloyd's mirror".[24] Payne-Scott's unpublished RPL 9 Report of August 1947 (day unknown, see below) provided a succinct description of the method:

> Briefly, at dawn both direct radiation from the sun and that reflected from the surface of the sea are received, and as the sun rises the receiver output varies sinusoidally as the phase difference between the direct and reflected rays increases; from the times of minima relative to the known time of sunrise, the elevation of the radiating source relative to, say, the centre of the sun can be calculated, while the ratio of the minimum to maximum power gives a measure of the width of the source.

The interference arose from the portion of the incoming wavefront that arrived directly from the source to the antenna and from the reflected radiation from the sea; the latter wavefront would travel an additional distance (twice the cliff height times the sine of the elevation angle about the sea); the source is above the sea. This is equivalent to having another virtual antenna located at a cliff height below the base of the cliff, Fig. 7.7.

The Dover Heights antenna had a lobe separation of about 30 arc min at 200 MHz, as compared to the primary beam size of 9°; this lobe separation implied that sources could be located with a precision of a few arc min. The upper limits for the angular sizes of the widths were typically 8–13 arc min. The use of this type of interferometer was well known from radar experience during the War. In Appendix C we describe a number of aspects of the history of this "Lloyd's mirror" type of interferometry as it developed in both Britain and Australia during World War II.

As Kellermann and Moran (2001) and Wild (1955) have pointed out, the sea cliff interferometer had a number of advantages. It had twice the sensitivity of an interferometer, consisting of two similar antennas connected as an interferometer, no interconnecting cables or preamplifiers were required, the sharp horizon could be used to eliminate confusion due to sources which had not risen, and there were no lobe ambiguities in identifying the maxima on the records with their corresponding lobe numbers. The obvious disadvantages were the limited observing time as the source was rising or setting and the low elevation of the observations with resulting large (also uncertain) ionospheric and tropospheric refraction corrections. The roughness of the sea due to wave action also caused a loss of coherence, especially

---

[23] We follow Sullivan (1991) in calling this system a "sea-cliff" interferometer; in 1946 the interferometer was called either a "sea interferometer" or a "cliff interferometer". The paper referred also to "Lloyd's mirror" via an optics analogy and also "lobes", as with a radar set.

[24] At the PC meeting of 12 November 1945 an impressive list of research projects for the fledgling radio astronomers was presented by Pawsey. The use of interferometry was anticipated as a method to study small scale structure if the "noise originates in single small area". The method was described as the use of precision directive finding techniques, obviously influenced by World War II radar practices, as discussed in Appendix C.

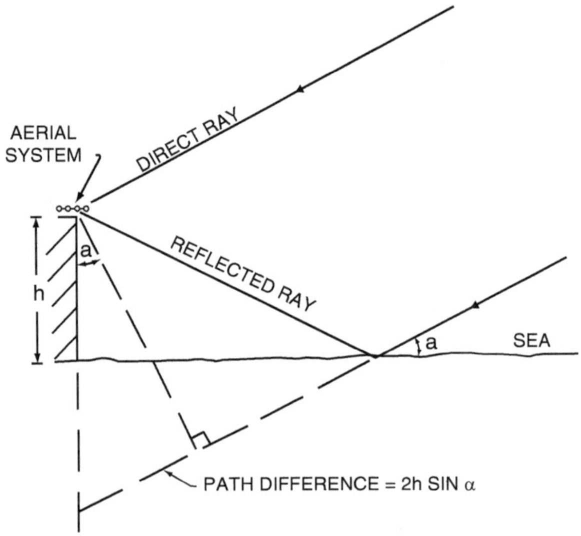

Fig. 7.7 A schematic diagram of a sea-cliff interferometer. The effective baseline of the virtual interferometer is twice the cliff height. The direct ray from the radio source and the reflected ray interfere to form the interferometer. From Eastern Australia, the sun and "radio stars" could only be observed as the source rose in the East. ATNF historical photographic archive: B1639-4

at shorter wavelengths. In addition, no delay compensation was possible as in a conventional interferometer of the Michelson type (which is much more flexible since the beam can be steered at any time of day, not just source rising or setting).[25]

The first fringes on the active sun were observed on 26 January 1946 with deep minima; the modulation of the 200 MHz intensity was almost complete implying small scale structure in the distribution of the solar radio radiation. Fringes were obtained from the "sunspot radio radiation" (the enhanced radiation or noise storms) and even from the quiet sun,[26] with angular size of about 40 arc min. It was possible to determine the absolute elevation of this radiation while no positional information was obtained for the short time scale bursts. The observers immediately realised that a large fraction of the 200 MHz solar radiation arose from a source(s) much smaller than the sun itself. During the great sunspot activity of the first days of February, data were obtained at Collaroy on 6–9 February, and on 7 and 8 February at Dover Heights. The data in Fig. 7.8 shows the deep minima on 2 days in February from Dover and Collaroy; fortunately sea-cliff interference data existed at both antennas on both 7 and 8 February. The comparison was encouraging as the positions agreed from both sites. "In each case the radiating strip has a width

[25] Stanley and Slee (1950) described a number of details of the sea-cliff interferometer, such as the effect of a finite bandwidth on the interference fringes and the decrease of the power received from the reflected beam due to curvature of the earth, thus resulting in incomplete interference.

[26] The radio quiet sun represents the thermal bremstrahlung (free–free) radiation from the million degree corona. The concept of a quiet sun arose simply due to the fact that the sun is a hot body, with the expected thermal emission. The intensity depends on the electron temperature and the total number of electrons along the line of sight. The quiet sun at radio wavelengths is not, however, a simple black body. See Sheridan and McLean (1985).

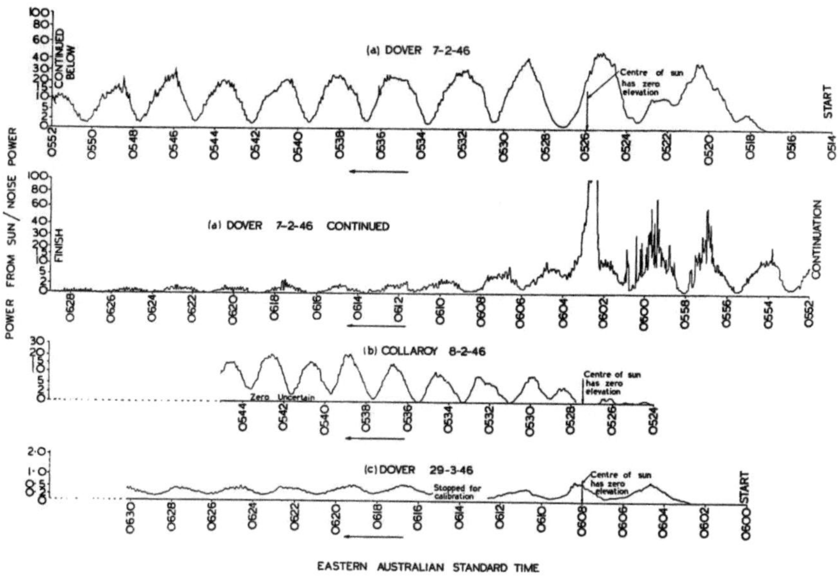

Fig. 7.8 Interference patterns at 200 MHz at sun rise for the great sunspot of February 1946; note the radio emission is observed before optical sunrise (centre of sun has zero elevation) due to increased radio refraction compared to optical refraction (Appendix L). Panel "a" from Dover shows deep minima indicating a small source size. Panel "b" is the next day at Collaroy; where a higher cliff height makes the period of the interference fringes shorter. Panel "c" shows a "normal" day when the shallow minima indicates a larger angular size. See Fig. 7.9 for the appearance of the radio and optical sun on these days. Time increases to the left. Optical sunrise is about 0524 EAST. Figure 5 of MPP. *Proceedings of the Royal Society, Series A*, vol. 190, p. 357, "Solar Radiation at Radio Frequencies and its Relation to Sunspots." Used by permission of the Royal Society

considerably less than that of the sun's disk, being of the order of the size of the sunspot group, and passes through the group". The fringe periods at the Collaroy site were shorter, because of the increased height of the cliff at this site. Due to the short duration of the intense Type I bursts observed from 0556 to 0604 EAST (compared to the fringe period of about 2 min), it was impossible to determine the angular size of these events; the fact that the maxima occurred at the same time as the steady radiation implied that the two originated in the same vicinity, as expected for Type I bursts. The inferred positions and widths of the equivalent radiating strips are show in Fig. 7.9. Thus for the first time there was direct evidence that the bursts and the Type I noise storms originated from a region close to a sunspot group.

In summary, the new data extended the results of the previous campaign of October 1945: the radiation at 200 MHz consisted of possibly two types of processes: (1) a slowly varying type of emission, and (2) bursts of duration with time scales of about a second. The emission originated from different positions on the sun and could usually be associated with a compact source near a sunspot. Finally, there was no evidence to make a detailed association with particular optical

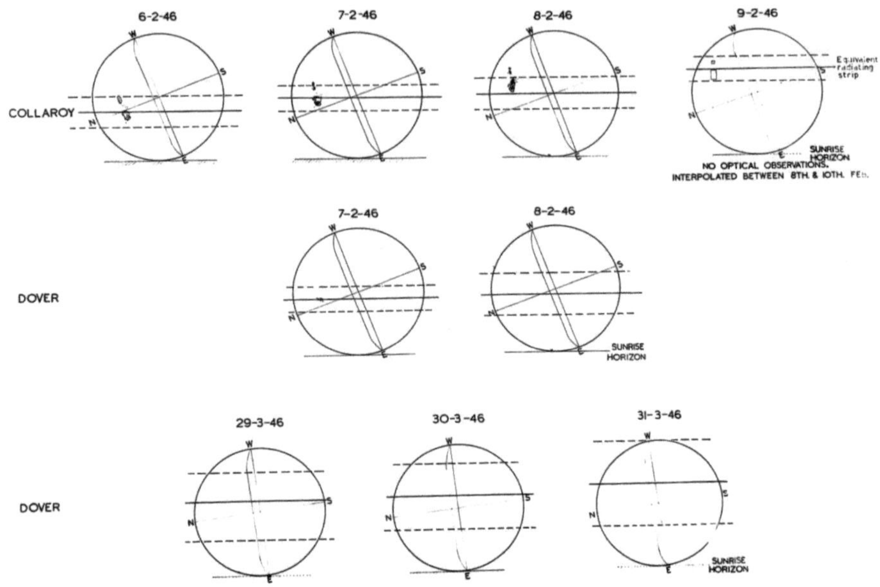

Fig. 7.9 Sketches of the sun with the position and the width of the radio noise source for a number of days in February and March 1946; the sea-cliff interferometer was used to derive the radio noise position and equivalent size. The positions were uncertain to a few arc min (solar diameter is about 30 arc min). The widths were upper limits. The top panel refers to Collaroy data while the bottom two refer to Dover observations. Figure 7 of MPP. The sun is shown with the horizon to the east shown as a hatched area. The equator of the sun is shown (E–W) as well as the solar poles (N–S). The great sunspot of February 1946 is sketched. For the top two panels the radio emission was localised, arising for a region near the sunspot. In the bottom row, the sources of radio emission are scattered, although the emission arrises mainly from one sunspot group. The intense radio emission arose from an area much smaller than the optical sun, close to the optical sunspot. ATNF historical photographic archive: B779-1.

phenomena (e.g., flares) and radio emission. For the first time a derivation of an approximate brightness temperature for the storm radiation became possible. On 7 February 1946, the radio flux density was about 10^7 Jy with an angular size less than 6.5 arc min (the optical disk of the sun is about 30 arc min). The equivalent temperature was thus greater than 3×10^9 K, much larger than any conceivable temperature on the sun. As was the case for the 1945 data, gross electrical disturbances (similar to electrical storms in the atmosphere[27]) were suggested as possible causes. A parallel was made with the generation of noise in radio valves (tubes). There was discussion about the uncertainty concerning the association of the two types of emission:

[27] There were qualifications, however: "It must be borne in mind that the detailed mechanism in the partially conducting ionised atmosphere of the sun will be different from the sudden breakdown occurring in lightning flashes." (McCready et al. 1947).

> The appearance of bursts superimposed on a more slowly varying background suggests that there may be two separate mechanisms, though it is possible that the background may consist of a large number of overlapping bursts (McCready et al. 1947).

We now realise that the radio emission observed by McCready et al. (1946) was mainly "enhanced radiation", later described by solar radio astronomers as Type I storm radiation. This radiation appears to have a background component upon which Type I bursts are superimposed. The detailed nature of Type I bursts was elucidated by Wild (1951), who found that the bursts have a short duration (about 1 s), with a bandwidth of only some 10 s of MHz. These can be associated with noise storms lasting hours to days. This emission originates from the plasma fundamental radiation with high degrees of circular polarisation. The polarisation is left handed when the magnetic polarity of the dominant sunspot is positive and right handed when the field is negative. This pattern indicates that the Type I emission bursts are polarised in the ordinary wave mode (Kai et al. 1985).

The suggestion in the above quote by McCready et al. that the background (the Type I storms) might consist of the superposition of numerous weak storm bursts was considered in detail by Wild (1951). He attempted to show that "the background level may be the resultant of a large number of bursts, only the larger ones being observed as distinct phenomena". His analysis was not conclusive, but he did favour the superposition hypothesis.[28]

A major achievement of the early 1946 research was the introduction of Fourier synthesis in radio astronomy. McCready, Pawsey and Payne-Scott suggested at this time that measurements of the amplitude and phase of observations taken at a large number of antenna spacings would provide the complete Fourier spectrum from which the true distribution could be derived. This suggestion was based on the result of their analysis of the response of the sea-cliff interferometer as the source rose over the horizon. Twice the cliff height was the effective spacing (in metres) of the interferometer. They wrote (in the 1947 publication, see below):

> Since an indefinite number of distributions have identical Fourier components at one frequency, measurements of the phase and amplitude of the variation of intensity at one place at dawn cannot in general be used to determine the distribution over the sun without further information. It is possible in principle to determine the actual form of the distribution in a complex case by Fourier synthesis using the information derived from a large number of components. In the interference method suggested here delta [the phase difference between the direct and the reflected ray from a distant source] is a function of *h* [height of the cliff] and lambda [the wavelength of the radiation], and different Fourier components may be obtained by varying *h* or lambda. Variation of lambda is inadvisable, as over the necessary wide range the distribution of radiation may be function of wave-length. Variation of h would be feasible but clumsy.[29] A different interference method may be more practicable.

---

[28] Kundu (1965, p. 451) discussed counter evidence suggesting that storm bursts may well be distinct from the background continuum. This question remains unresolved even in 2009 (Tim Bastian, email, 19 April 2009).

[29] Examples would be sea-cliff interferometers at locations with different cliff heights or even observations from an aircraft at different heights, as suggested by Ryle in an unpublished memo of

The more practical method that has been adopted in the last 60 years has been interferometry with separable and movable antennas. Payne-Scott and collaborators also pointed out that for an arbitrary distribution the response of the interferometer consisted of a steady term and another term that had the form of a Fourier cosine series. For the first time Fourier synthesis was described in radio astronomical context, a milestone acknowledged by Ryle in 1952.[30]

We have been able to determine the origin of the concept of Fourier synthesis within the RPL group.[31] There remains little doubt that Pawsey was the originator of the concept, with major contributions from Payne-Scott. The association of Fourier synthesis and interferometry can be ascribed to Pawsey and Payne-Scott, a major highlight in the development of twentieth century astronomy. Since Pawsey was averse to mathematics (Sullivan 2009, quote from Steve Smerd), the equations were certainly formulated by Payne-Scott.[32] Kevin Westfold[33] was a newly arrived staff scientist in 1946. He wrote:[34]

> I joined the Lab just at the time the McCready, Pawsey and Payne-Scott paper had been submitted to the Royal Society. In order to familiarise me with the subject Joe [Pawsey] had put me on to reading all about the sun, particularly the chromosphere and corona. I was given the MS [manuscript] of their paper to read and, as a mathematician, I thought I should check the derivation of (4) [to determine *R*, the ratio of the minimum to maximum power of the interference fringes] and the approximation (5) for a spherical earth. Their derivation had assumed a plane earth. When I put the earth's radius into the calculation I came to the conclusion that there was a significant error in the formulae they had used. I reported this to Joe, who immediately referred me to Ruby. I think Joe's action established that Ruby had been responsible for these derivations, but I am pretty sure that it was Joe's idea to use the sea as a Lloyd's mirror interferometer. We can assume Lindsay[35] had been responsible for the design of the equipment. The sequel: Ruby was, of course, correct. What I had done,

August 1946, discussed by Sullivan (2009). Sullivan writes: "... in essence a sea-cliff interferometer with variable fringe spacing – simply observe a low-elevation sources from an aircraft flying at a different heights above the sea!" As far as is known, this novel experiment was never attempted.

[30]In 1952 Ryle ("A New Radio Interferometer and its Application to the Observation of Weak Radio Stars") wrote: "The relation between the magnitude of the varying component of power intercepted by an interferometer and the Fourier transform of the distribution across the source was first pointed out by McCready et al. (1947)." Sullivan (2009) has reminded us that Ryle too was thinking of Fourier synthesis in 1946; he has pointed out that in the same memo of August 1946 (footnote 29, this chapter). Ryle suggested the principle of what we would now call aperture synthesis for solar radio observations using multiple aerial separations.

[31]Correspondence with Christiansen et al. (1999–2005).

[32]In 1946, Fourier transforms were not as well known to radio engineers as is the case today. In 1944, Payne-Scott attended a course of lectures on the related topic of Laplace transforms given by J.C. Jaeger for RPL and NSL (National Standards Laboratory) scientific staff. The lecture series was held at the School of Tropical Medicine on the University of Sydney campus (Bracewell interview, January 2007).

[33]Westfold (1921–2001) was a Professor of Mathematics and later Astronomy at Monash University in Melbourne, Australia, as well as Deputy Vice-Chancellor, 1982–1986.

[34]Email, 5 November 1997.

[35]McCready's role had been to provide the 200 MHz receiver and electronics. There is little evidence suggesting that he played a major part in the data analysis.

being all clued up with my research on the sun, was to estimate orders of magnitude by inserting the radius of the sun into my formulae instead of the radius of the earth. The factor of ~100 was not insignificant! This was a salutary lesson for a tyro theoretical radio astronomer. Joe, of course, never forgot the fright I had given them, though I am not sure that he was aware of exactly how this had come about. Ever after he would speak of me thus: "Westfold is a good mathematician, but he makes mistakes."

For cases for which the minima were quite deep, the radio power was concentrated in a compact source; it was then possible to derive both a position and width of the equivalent emitting strip. The width (in elevation) was derived from the value *R* and the elevation from the timing of the minima compared to the known time when the sun had zero elevation. For a number of days the emission arrived from small regions and some of these are shown in Fig. 7.9. For the days in February 1946, the radio emitting strip was much smaller (about 1/5th) than the size of the sun. The radio emitting region was quite compact and roughly corresponded with the major sunspot. The strip moved across the sun (whose rotation rate is 27 earth days) with the sunspot; the sea-cliff interferometer only provided information about the elevation (up-down coordinate) of the sunspot radiation, but little information on the azimuth of the radio signal. The detailed structure of the radio emission was unknown with this crude interferometer; however, the position determination and size information were decisive in making the identification of the compact radio source with sunspots. Thus the inferred association of these two that had been suggested by Appleton, Hey and the earlier RPL paper was now clarified. In addition, the bursts of one sec duration had the same times of maxima as the steady noise storms; thus both roughly arose from the same location on the sun.

The bottom row of images in Fig. 7.8 shows the inferred radio properties in late March 1946, when the intensity of the radio emission was much weaker, with fewer and less prominent sunspots. The radio minima from the dawn records were then less pronounced and the inferred width of the radio emitting strip was larger. The radio storm emission was at this time a few times 10^5 Jy while the quiet sun emission (angular size of about 40 arc min) had a flux density of about 10^5 Jy. Thus the centroid of emission was pulled to the centre of the sun as the storm radiation weakened, because about half of the intensity then arose from the disk of the quiet sun.[36]

[36] During the exciting observing campaign of early 1946, the weekly progress can be followed in the minutes of the PC. Examples mentioned in these reports were correlation of solar noise with sunspots, the short burst phenomena and their simultaneous nature as observed at two observing sites, and the location of the bursts on the solar surface. On 19 February 1946, Don Yabsley wrote a short document that might have been discussed at the next PC meeting (19 March 1946) (NAA: C3830, A1/1/1, Part 1). Yabsley had become part of the "Cosmic and Solar Noise Section" in early January 1946, joining Pawsey, McCready, Payne-Scott, a new Technical Officer and a Technical Assistant. The Yabsley plan indicated that already a great deal was known about the strong bursts, only a few weeks after starting observations at Dover Heights and Collaroy. The list of projects included six items (e.g., relation of sunspot area and daily solar noise level, and interference pattern "noise area small compared with sun's area" and comparison of fading on different wavelengths). See footnote 35 Chap. 6.

In the midst of these reports, a very intriguing report by McCready and Payne-Scott appeared from the 19 March 1946 meeting: "Some explanatory investigations of stellar noise have been conducted with inconclusive results." No additional details were provided.[37]

A number of major questions had been raised by the RPL group and it was at this time that the publication of these results from early 1946 occurred. The novice Radio Noise Group had learned many of the techniques and was becoming familiar with astronomical concepts. The publication that resulted from this early 1946 work, McCready, Pawsey and Payne-Scott, "Solar Radiation at Radio Frequencies and its Relation to Sunspots", appeared in the *Proceedings of the Royal Society* on 12 August 1947, having being submitted by Dr. David Rivett FRS almost 13 months earlier, on 22 July 1946.[38] This paper will be referred to as MPP. Sullivan (2009)

---

[37]The first detection of Cygnus A in Australia was made by Bolton and Stanley only in June 1947 from Dover Heights. Bolton (1982) claimed that initial futile attempts had in fact been carried out by Pawsey in September 1946. Hey, Parsons and Phillips reported the discovery of Cygnus A in the 17 August 1946 issue of *Nature* (submitted 4 July 1946); these observations were carried out in late May and early June 1946.

[38]Bowen had written to Sir David Rivett (Fellow of the Royal Society, Chairman of CSIR) on 10 May 1946 with the suggestion that Rivett should submit the paper in person to the *Proceedings of the Royal Society* during his upcoming trip to the UK. An outline of the publication was enclosed. Rivett facetiously responded that he was willing to do this but insisted that he "be assured beforehand that no one will be so tactless as to subject me to a viva voce examination on it!" On 31 July 1946, Fred White (then Executive Officer of CSIR) wrote to Bown that Rivett had informed him that the paper had been submitted in London to J.D. Griffith Davies, the Assistant Secretary of the Royal Society of London (NAA: C3830, A1/1/1, Part 1). As was the case for the February 1946 *Nature* paper (footnote 15, this chapter), the suspicion was that Appleton was also the referee of this publication. Ironically Pawsey had written to Appleton (NAA: C3830, A1/1/1, Part 1) a few days later (5 August 1946) with a copy of the new publication. He told Appleton that Rivett was in the process of submitting the paper in person in London and "we are hoping it will be accepted". In subsequent years, the long delay was blamed on Appleton by the RPL group, again with no definite proof; this delay may have been beneficial to the astronomers at the Cavendish Laboratory of the University of Cambridge, as their first paper was published on 7 September 1946 (Ryle and Vonberg, "Solar Radiation on 175 Mc/s.", 1946) after being submitted on 22 August 1946 – a delay of only 2 weeks. The Cambridge paper described interferometer observations with a fringe spacing of 24 arc min of the large sunspot of 27 July–3 August 1946 (see later this chapter for Payne-Scott's data from this period). Based on the upper limit of the angular size of 10 arc min (thus a brightness temperature $> 10^9$K, clearly arising from a non-thermal process) the authors suggested, "Since the value obtained does not greatly exceed the diameter of the visual spot, it is reasonable to relate the source of this radiation with the visual spot itself, or a region closely associated with it". This is a surprising statement since the sun spot had a diameter of at most a few arc min (Newton 1955); also the position of the radio source was not determined, in contrast to the Dover Heights data from the preceding February which showed that the enhanced radio radiation did coincide (in elevation) with the major sunspot. In 1948, Pawsey and Appleton exchanged letters (NAA: C4695,8) about who should receive priority for the first discovery of the association of bursts of radio emission with large sunspots. Appleton was preparing URSI Publication No. 1, "Solar and Galactic Radio Noise", published in 1950. On 8 September 1948, Pawsey wrote to Appleton with numerous comments on the draft section II, "Solar Radio Noise". The draft text gave "precedence for discovery that solar noise associated with sunspots originated in small areas on the sun" to Ryle and Vonberg, not the RPL group. Pawsey pointed out that the two groups had

has discussed some interesting aspects of this paper based on RPR 24, a preprint of 16 June 1946.[39] This publication was to have a far reaching impact on the development of solar physics and radio astronomy techniques.

The MPP paper concluded with an attempt to unify the known properties of solar noise and cosmic noise, the radio emission observed since 1933 by Jansky and Reber, associated with the Milky Way. (See above for Payne-Scott's thorough discussion of this problem in the December 1945b summary paper.) The problem with the comparison which Payne-Scott had made in the December 1945b summary paper was now partially recognised:

---

worked independently and that the RPL work preceded the Cambridge work by almost 6 months (also submitted for publication exactly a month prior to the other group). The 13 month delay in publication of the RPL paper meant that the Cambridge data was published much earlier. In the end Pawsey wrote to Appleton: "I may leave the decision as to whether the wording should be changed, and if so the detailed wording, to you in this case." Even though Appleton wrote the next day that he probably would accept all the changes proposed by Pawsey, he did not. The final publication gave the preference to the Cambridge result. Pawsey may have been aware of the irony of the discussion since he was well aware of the suspicion that the real delay had come about because Appleton himself had delayed the publication in the *Proceedings of the Royal Society.*

[39]The published paper was RPP 12. The preprint, RPR 24 (found by Goss and Barnaby Norris in the RPL archives in 2007), had a different order of authors – Pawsey, Payne-Scott and McCready. B.Y. Mills (September 2008) has informed us that the Royal Society insisted on alphabetical ordering of authors. A few changes were made in the published version as compared to the preprint. These mainly consisted of modifications to the preprint concerning the dates and sources of information the RPL group had received about early overseas solar work which preceded the Sydney observations in 1945 and early 1946. As an example, the initial manuscript refers to the restricted reports from (1) the New Zealand group and (2) the AORG British group (reference to the 1942 observations of Hey, see chap. 6). The preprint stated: "In a prior letter, not available here until our initial work was completed, Appleton (1945) described reports of radio noise received on the short-wave communications band at times of marked solar activity and concluded that the noise originated in active areas on the sun. The observed intensities were, like Reber's, greatly in excess of the expected blackbody radiation." The published paper stated that the order of relevant information was initially the Hey paper from 1946 followed by the New Zealand data (unpublished data). The published paper then continued: "Radiation on still longer wave-lengths was reported by Appleton (1945), who described reports on radio noise received in the short-wave communication band at times of marked solar activity during the sunspot maximum about 1936. The observed intensities were greatly in excess of 6,000 K black-body radiation, and he concluded that the noise originated in active areas of the sun." As indicated in the previous footnote, these minor modifications indicate that Appleton may have been the referee. In the PC meeting of 30 June 1949, there was a complaint about the delays in publication in the *Proceedings of the Royal Society*; only one additional paper was published by the RPL group in this journal in 1952. No other major changes were made to the manuscript, except for a major typographical error in (7), introduced in type setting. This error changed the exponent of the wavelength in the Rayleigh-Jeans approximation of the Planck formula from two to three; the submitted paper had the correct value of two. A year later while Pawsey was in the UK, McCready wrote to him (NAA: C4659, 8) concerning a submission of an "Erratum" to the *Proceedings of the Royal Society* with a correction for (7). The erratum never appeared; apparently a letter was in fact sent to the editors of this journal. McCready mentioned in a letter to Pawsey that most reprints had been corrected; individual letters were posted to a number of colleagues who had been sent reprints with the typographical error. The RPL solar noise group was certainly embarrassed by this error.

> Cosmic noise was originally attributed to radiation from interstellar matter, rather than from stars, at a time when similar radiation from the sun had not been detected. The discovery of solar noise raises the question as to whether the cosmic noise is due to similar processes in stars. The basic difficulty remains that the intensity of cosmic noise is vastly greater that it should be if the stars emitted the same ratio of radio-frequency energy to light as does the sun. Nevertheless, the great variability of solar noise suggests the possibility of vastly greater output from stars differing from the sun and it seems that data at present available leave the question completely open.

The major problem was the low radio luminosity of the sun; if the sun were at a typical distance some tens of parsecs (comparable to the distance of a nearby star), the flux density would be about 10^{-8} Jy. This value would provide a meagre contribution to the background of radio emission in the galaxy. [40] (See earlier in this chapter for the resolution of this problem in the late 1950s.)

In order to interpret the data with the sea-cliff interferometer, a number of complex technical issues had to be solved. Appendix 1 of MPP was entitled "Derivation of intensity in terms of instrumental constants". This was clearly written by Payne-Scott; much of the text was taken verbatim from the December 1945b summary paper.

Appendix 2 (in MPP), with five subsections followed: "Factors complicating the analysis of the interference pattern". These sections dealt with complex issues that had to be solved by radio astronomers, mostly for the first time. Since delay compensation was impossible with a sea-cliff interferometer, the effects of the receiver bandwidth were calculated. With the small bandwidths used (2 MHz), the effective decorrelation was small. The major uncertainty was the issue of refraction due to the low lying atmosphere, the troposphere. In a conventional Michelson interferometer, refraction can be essentially ignored in producing any change in the interferometer delay (Thompson et al. 2001; the delay increments are almost the same for both paths of the interferometer). In this case, the angle of refraction must only be used in order to point the individual elements correctly.

But for the single antenna of the "sea-cliff" interferometer, the situation was much more complex. Two problems had to be solved for the first time. First, radio astronomers had to deal with *radio refraction*[41] and secondly, in contrast to a vertical interferometer, refraction changed the phase of the sea-cliff interferometer since the reflected ray from the sea experienced an additional path through the

[40] After the publication was submitted to the Royal Society in late July 1946, Pawsey and Payne-Scott attended the ANZAAS (Australia and New Zealand Association for the Advancement of Science) meeting of 21–28 August 1946 in Adelaide. Both gave 15 min presentations. Payne-Scott's was titled "Discovery of Cosmic and Static [sic, solar?] Noise" and Pawsey's, "Interpretation of Observations". The presentations were summaries of the exciting new data in the Royal Society paper. Pawsey wrote: "This work is a new branch of astronomy ... the discovery of this [radio] radiation will come to be recognised as one of the fundamental advances of astrophysics. The first stages [after this discovery] are those of general exploration of the phenomenon. To these our work belongs." (NAA: C3830, A1/1/1, Part 1).

[41] Refraction is the bending of the radiation due to the atmosphere of variable density above the optical or radio telescope. The radiation is bent towards the zenith. For a plane parallel atmosphere the angle goes as the tangent of the zenith angle (for modest angles). As an example for a zenith angle of 45°, the optical refraction is about 1 arc min. See Appendix L for more details.

lower atmosphere, compared to the direct ray from the radio source. In Appendix L, we present a new analysis of the nature and calibration of the radio refraction for the sea-cliff interferometer, clarifying the obscure descriptions presented by MPP.

In addition we have reanalysed the Bolton (1982) claims about the inaccuracies of the MPP refraction discussion. Bolton's harsh criticism implied that the large errors in the celestial coordinates of Cygnus A (at the level of 1°, published by Bolton and Stanley in 1948a and 1948b) were caused by following the procedures proposed by MPP. Bolton wrote:

> Unfortunately we used the formula devised by T. Pearcey to account for the apparent mean refraction deduced by Pawsey, Payne-Scott and McCready from their observations of sunspot radiation. The Pearcey formula contained a substantial ionospheric term which accounted for their erroneous assumption that radio and optical sunspot positions were coincident![42]

As we show in Appendix L, this claim has no foundation. The formulation had no ionospheric terms and the assumption of equality of radio and optical positions was never asserted.

The refraction determinations for the early solar observations by MPP were in fact reliable with a typical precision of about 3 arc min for elevations above 2°.[43]

Later in this chapter we will discuss an additional project carried out by Payne-Scott and McCready to investigate the effect of the ionosphere on dual frequency solar observations and a misguided attempt to interpret the difference in positions at 60 and 200 MHz in terms of simple ionospheric refraction. Certainly a major disadvantage of the sea-cliff interferometer was the restricted range of low elevations; not only was the observing time limited to an hour or less per day, but the low elevations also implied that large uncertain refraction corrections were required. Within a period of less than 8 years, the RPL group had abandoned the sea-cliff interferometer technique in favour of the conventional Michelson interferometer.[44]

---

[42]Bolton ("Australian Work on Radio Stars", 1955) also complained about the refraction corrections, but in a milder form. As he discussed positional determinations of radio stars he wrote: "The measurements were severely affected by atmospheric refraction, and failure to apply true corrections for this resulted in the initial positions given by the writer (1948), having errors of several degrees." The early Bolton and Stanley positions of Cygnus A had errors of about 1 degree in declination and 0.2° in right ascension. See Sullivan (2009) for a masterful discussion of the discrepancies in the determination of the coordinates of Cygnus A in the years 1948–1952 by two groups at RPL, the Cavendish Laboratory and Jodrell Bank. The optical identification was made after Smith (1951) determined the position of Cygnus A with a precision of better than 1 arc min.

[43]The minutes of the 19 March 1946 PC meeting stated: "an attempt is being made to determine the position of the source of the energy, from a study of the lobe patterns obtained ... The theoretical treatment of refraction right through the earth's atmosphere has been worked out with some approximations. A calculation in one case gave a result in which the source of energy lay somewhat on a horizontal band passing through a large sunspot. Little activity was noticed during the second appearance of the large spot in March [the giant spot of early February which had occurred in one solar rotation – 27 days – earlier.]".

[44]Surprisingly this type of instrument (Michelson interferometer) was called a "vertical interferometer" by Stanley and Slee (1950), in contrast to the sea-cliff interferometer.

Payne-Scott as Solo Scientist at Dover Height, July–August 1946; Payne-Scott's Absence from RPL, Late 1946

In the period after the major solar events of February 1946, the solar emission at 200 MHz had decreased to modest levels (based on data obtained from Dover Heights up to early April and data obtained by Allen from Mount Stromlo starting on 1 April 1946 to early May[45]). The typical 200 MHz intensities were less than 10^6 Jy (less than 10% of the peaks of early February) by the next solar rotation in early March. The RPL solar group began at this time to concentrate on multi-frequency observations of the solar emission. Elucidation of the metre wave spectra of the short term bursts (i.e., the determination of the radio energy as a function of frequency or wavelength) had already been proposed at the PC meeting of the previous November. At the PC meeting of 11 June 1946, Pawsey and Payne-Scott reported: "Future work will concentrate on an exploration of the spectrum, both of the mean level and of the bursts." A month later, on 8 July 1946, it was clear that work had begun in earnest. At 60 and 200 MHz, Payne-Scott had found that bursts did not correlate and simultaneous recording at 60 and 75 MHz had begun.[46]

By mid-1946, Payne-Scott was planning a new campaign at Dover Heights to follow up the successes associated with the prominent solar activity of February 1946. A new prominent sun spot appeared in July, with an associated large increase in solar noise. Newton (1955) has classified this sunspot as the fourth largest single spot in the period 1874–1952, a rank still maintained in 2008. The maximum area

---

[45] Allen made observations at CSO (Mount Stromlo) continuously from 1 April 1946 to the end March 1947.

[46] At the March to June monthly 1946 PC meetings, discussions of the determination of spectra of solar bursts continued. At the 8 July 1946 meeting, it was reported that 'General opinion however favoured a spectrum analyser style of approach, with all frequencies being studied at one place'. This was the first mention of the swept frequency type of instrument that was to become so decisive in the classification of Type I, II and III bursts by Wild and McCready (1950). Finally at this meeting there was the surprising statement that a letter had arrived from Appleton describing the results of solar noise coinciding with a sudden ionospheric disturbance – the controversial paper of 3 November 1945 described above. See also footnote 27, Chap. 6. Bowen knew about this paper when he wrote to Appleton in early 1946 (above). Is it possible that this publication was not known to the RPL scientific staff as late as July 1946? If that is the case, the internal communication at RPL was indeed poor. Also at the 13 May 1946 PC meeting, a discussion was held about a letter that Pearcey had received from a colleague in Cambridge describing Hey's spectra of solar bursts from 1.5 to 12 m wavelength (25–200 MHz). The result was published in July, with almost no delay, in the *Philosophical Magazine* (Appleton and Hey 1946a); large increases in the radio emission were correlated with short-wave radio communication fadeouts (at frequencies of 12–20 MHz) suggesting enhanced ionisation in the D layer of the ionosphere. The instantaneous radio spectra of a few bursts were determined, with some evidence of a low frequency cut-off due to the earth's ionosphere. Based on the observed sharp maximum of radio noise within 1 or 2 days of central meridian passage, the authors suggested that the sunspot radio noise was beamed in a sharp beam perpendicular to the sunspot. This letter with the report of the new Appleton–Hey data may well have spurred on the RPL group. Appleton and Hey proposed that a comparison with optical data suggested that the most intense optical flares were associated with radio bursts with delays of several minutes (after the flare onset).

was 4,720 millionths. A number of solar groups observed the radio emission from this event in addition to Payne-Scott: Allen (1947) from CSO (Mount Stromlo), Hey et al. (1948)[47] from AORG, Lovell and Banwell (1946) from Jodrell Bank[48] and Ryle and Vonberg (1946)[49] from Cambridge.

It was becoming apparent in the period 1946–1948 that various distinct types of solar bursts existed and that it was important to determine their spectra. The elucidation of the complex temporal, frequency and polarisation state of the solar bursts and outbursts was a major challenge; several years were to pass before the taxonomy of the radio sun was sorted out. This period has been described by Wild (1985):

> The situation ... was one characterised by mystery, incredulity and intense interest. A whole new field of research lay ahead with obvious objectives: to disentangle the confused conglomeration of phenomena; to interpret and understand them; and to put results to use in the mainstream of research for solar physics, astronomy and physics.[50]

As a test bed of the multi-frequency approach to the solution to this problem, Payne-Scott observed at Dover Heights for an intensive 2 week period during the transit of the large sunspot in July–August 1946. A total of about 110 h of observing was accumulated. These pilot observations (to be the basis for future multi-frequency programs) were carried out in order to lead to "several specific phenomena that

---

[47]The latter paper by Hey et al. probably represented the break of the collaboration between Appleton and the Army Operations Research Group, as described by Hey (1973). The publication (Hey et al. 1948) provided a thorough and detailed account of solar radio noise at 73 MHz from the later period, March 1946 to September 1947, compared to the limited scope of the Appleton and Hey publication of 1946a.

[48]Lovell and Banwell were observing with the meteor radar antenna during the day time; thus they were involved with the accidental discovery of the huge bursts on 25 July 1946. In *Astronomer by Chance* (1990), Lovell has provided an amusing description of this day. Peaks of emission in excess of 10^8 Jy were observed.

[49]See footnote 38, this chapter. The data represented Ryle's first interferometer observation from mid July 1946; the spacing of the interferometer ranged from 17 to 200 m. Also circular polarisation was determined by crossing the polarity of the two elements of the interferometer. The 100% circular polarisation indicated that strong magnetic fields in the sunspot group (already known from optical data) could affect the radio emission process. Two additional groups published results on the circular polarisation of these Type I bursts at about the same time in *Nature*. The first was Martyn's in the *Nature* of the previous week (31 August 1946, Martyn 1946a) based on CSO data at 200 MHz on 26 July 1946 (using the Yagi provided by RPL); the second was Appleton and Hey's (1946a) in the same issue of *Nature* (7 September 1946, just preceding the Ryle and Vonberg publication) from data at 60 MHz obtained on 27–28 July 1946 with the AORG (Army Operation Research Group) antenna. In October 1946 Bowen wrote to Pawsey from London, reporting on radio astronomy activities at Cambridge; he reported that Ryle was quite critical of the Martyn claim about the change of polarisation state at meridian transit of the sunspot (from right hand to left hand), in disagreement with the Cambridge data. Their data changed polarisation state at random times depending on whether the source of radiation arose on one side or the other of the sunspot.

[50]Thus a classification of the distinct types of bursts was required for the physics of the emission mechanisms to be understood. The association of the radio emission with well known optical phenomena on the sun (e.g., sunspots, flares, prominences, etc.) was also a major question for these novice solar physicists.

seemed particularly worth[y of] study". The observations were carried out at 60, 75, 200 MHz using tiltable Yagis and dawn observations at 30 and 200 MHz using the ShD broadside array.[51] Some limited data were obtained on the remarkably low frequency of 30 MHz, 10 m. A major problem for the recording of the data was vandalism (see footnote 51 this chapter) and a burglary at the beginning of the campaign in which a number of recorder pens were stolen! The result was that the timing and recorder pens were not optimal ("the timing and recorder pens were slightly out of line"), causing additional errors in the timing of the various frequencies. Perhaps the vandalism and the thefts were related. The method of reduction of the dawn records in the determination of the positions and widths of the sources was improved compared to the earlier observations; with these improvements both the source positions and source sizes could be determined with a precision of a few arc min. This represented a substantial improvement in comparison to the February campaign. A detailed treatment of the issue of the interferometer response in the presence of multiple sources on the sun was now included. This issue had not been discussed in MPP. Also the confusion of the determination of the properties of weak "enhanced radiation" sources in the presence of the larger scale (about 40 arc min) quiet sun was discussed. With this simple interferometer, the properties of weak burst sources could not be determined with any certainty. Payne-Scott wrote:[52]

> At 200 Mc/s with our experimental arrangement the sun's diameter corresponds to a phase difference of almost exactly 2 pi, so that a very shallow pattern would be produced by a uniformly radiating disc of the sun's dimensions . . . the minimum to maximum ratio being 0.76 [as expected for a source with radius about 20 arc min].

She also derived in detail the effects of sea-waves on the sea-cliff interferometer fringes; the effective height of the cliff varied with the wave action. These 8 s ripples in the radio data arising from the sea-waves could easily be distinguished in the

[51]These observations are possibly the last observations with this antenna (Fig. 7.3). Payne-Scott did not use it for her campaign in 1948, which made use of the Yagis at Hornsby (Chap. 8). Bolton (National Library of Australia, MS 9063, "Papers of John Gatenby Bolton") sent a letter to Sullivan on 13 December 1986, saying that by November 1946 vandals had almost destroyed the antenna, with only the basic steelwork remaining. "As this was largely rust by this time Stanley and I cut it up with an oxy torch and dropped the bits over the cliff in February 1947". Briton (RPL Chief) had corresponded with Lt. Colonel P. M. Moore, [Army] Commander, Eastern Command Fixed Defences at Watsons Bay (Sydney) on 15 May 1946, suggesting that they had heard that the army was considering plans "to scrap the equipment and dismantle the aerial". He hoped that RPL could borrow or even take possession of the army's CA [Coastal Artillery?] No. 1 aerial for experimental purposes; RPL planned to use this aerial for one additional year. Moore responded a month later that the army was willing to leave this equipment in place for an additional 6 months. Apparently, in the course of 1946, the army coastal defence group forgot about their CA No. 1 Aerial! (NAA; C3830, A1/1/1, Part 1). Vandalism continued in April–May 1947. On 8 April 1947, the minutes of the Propagation Committee reported : "The group is still worried about the use of the Dover site. More damage was done by vandals last weekend." The following month the PC minutes reported on 5 May 1947 that polarisation solar noise observations by Bolton were delayed by vandalism and that guards were then in place!

[52]RPL 9, "A Study of Solar Radio Frequency Radiation on Several Frequencies During the Sunspot of July–August, 1946", unpublished report, August 1947.

presence of the slower (about 2.5 min of time) fringes from the sea-cliff interferometer; also accurate tide tables were required for the analysis in order to calibrate the precise height of the reflecting surface on the sea. In an acknowledgement at the end of the introduction (publication date on an unspecified day in August 1947), she wrote that the work was done under the direction of Pawsey, design of the equipment being by McCready and Don Yabsley. The daytime observations (with the Yagi antennas) were done by Yabsley and Payne-Scott while the dawn observations (with sea-cliff antenna) were carried out by Urquart and McCready. The analysis of the data was done by Payne-Scott. The observations were from sunrise for an hour (up to about 8:00) and then from 9:00 to 16:00; at 60 and 75 MHz simple low gain (10–15.5 dB) Yagis were used, while the ShD antenna (only movable in azimuth) was used at 30 and 200 MHz. A simple 6.5 dB gain 200 MHz Yagi was also used for daytime observations. Apparently the Yagi antennas were pointed manually. The intensity scale had an uncertainty of almost 50%; radio telescope calibration has indeed come a long way in the last 60 years.

The detailed comparison indicated that, in general, there was an absence of correlated behaviour at different frequencies. Bursts tended to be more intense and variable at the lower frequencies. For much of the period when the intensity was most intense (between 27 and 29 July, during the meridian transit of the great sunspot), the ratio of intensities at 75 MHz compared to 200 MHz was almost a factor of 10. However, for much of the time the 200 MHz intensity was comparable to, or even in excess of, the lower frequency data, at a time when the total emission level was much lower. The quiet sun provided a minimum for the 200 MHz (about 6×10^4 Jy); at the lower frequencies, the sensitivity was not high enough to detect the quiet sun, expected to be a few times 10^4 Jy. For many of the bursts, the bandwidth of the radiation was inferred to be small since the data was not well correlated at the different frequencies (as now expected for Type I bursts). The sunrise observations at the three lower frequencies (Fig. 7.10) are impressively variable. The observed solar radiation fadeouts at 30 and 60 MHz (partial disappearance of the signal) suggested that the ionosphere was the likely cause.[53] The detection of solar emission at the lowest frequency (30 MHz) by the RPL group was also reported.

The determination of the positions of the 200 MHz emitting strips showed results comparable to those of the MPP publication; the strips included the majority of the sunspots suggesting that, as before, the radio emission arose from regions near the centres of solar activity (Fig. 7.11). On 27 July 1946 the strip was quite narrow and agreed with the optical position of this prominent sunspot. On the other 16 days, the strips included more than one spot; towards the end of the period, the strips were quite broad (20 arc min or so), arising from the thermal radiation from the entire corona.

---

[53]These fadeouts are probably related to severe ionospheric scintillation at these low elevations. Traveling ionospheric disturbances were quite common while the ionosphere was disturbed and the low elevation led to an increased path length compared to the zenith. Pawsey (1950a) also refers to the fact that at 60 MHz the observed changes in position were so large (more than a solar diameter) that the changes in position at this low frequency were quite implausible.

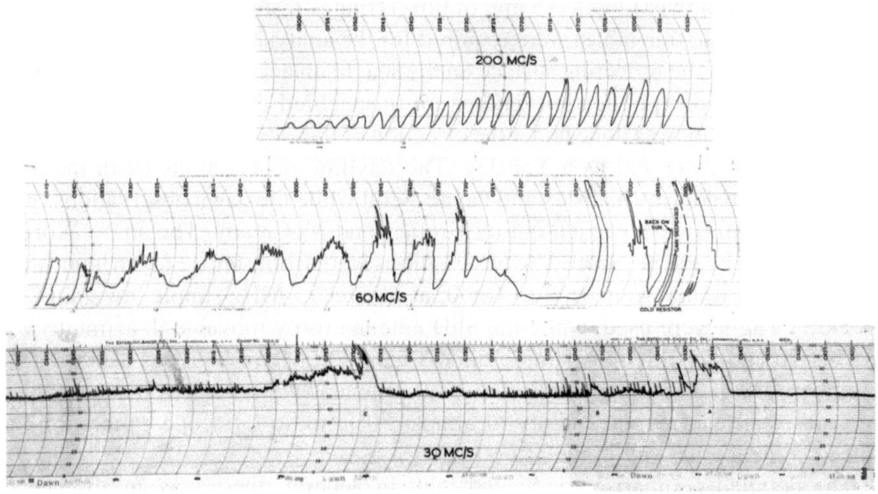

Fig. 7.10 Radio noise data from the major sunspot group of July–August 1946, from the unpublished report RPL 9 by Payne-Scott (published August 1947) (Figure 4 from RPL 9). For the first time solar radio data was observed at a number of different frequencies: 30, 60 and 200 MHz. The ShD (shore defence) radar antenna was used (perhaps the last data obtained with this instrument) as well as a number of simple Yagi aerials. A 200 MHz broadside array was also used with a polar mount for continual observations throughout the day, not just at sunrise. The dawn records from 27 July 1946 showed severe fading (intensity of the solar noise attenuated) at 30 and even 60 MHz, likely due to ionospheric effects at these low elevations. Time increases to the left; optical sunrise is about 0655 EAST. The 200 MHz fringes are attenuated about 45 min after sunrise as the sun moves out of the primary beam of the ShD antenna. ATNF historical photographic archive: B1285-10

Some of the bursts (typical increase of a factor of a thousand and in a few cases a factor of a million) were isolated with double humped shapes,[54] probably Type III bursts (Fig. 7.12). Type III bursts are now known to be broadband (about 100 MHz) fast drift bursts that can be associated with electron streams moving at velocities of a fraction of the speed of light. The frequency drift rate occurs at about −20 MHz/s (at around 65 MHz-observed first at the high frequencies and later at the lower); these events are often associated with the impulsive phase of solar flares. Payne-Scott spent much of her time in the following years working on these Type III bursts.

For these bursts, the time delays were of the order of seconds, expected for Type III bursts. The delay histogram is shown in Fig. 7.13. A major result of the July–August 1946 campaign was that some bursts showed some level of correlation at 60 and 75 MHz, with a definite delay of about 2 s as the higher frequency preceded the lower. For some bursts, the 200 MHz radiation could be associated

[54]Payne-Scott (Chap. 8) later studied these double-humped bursts at 65 and 85 MHz. They may well be the "F–H" Type III bursts (fundamental-harmonic), first detected some years later by Wild et al. (1954).

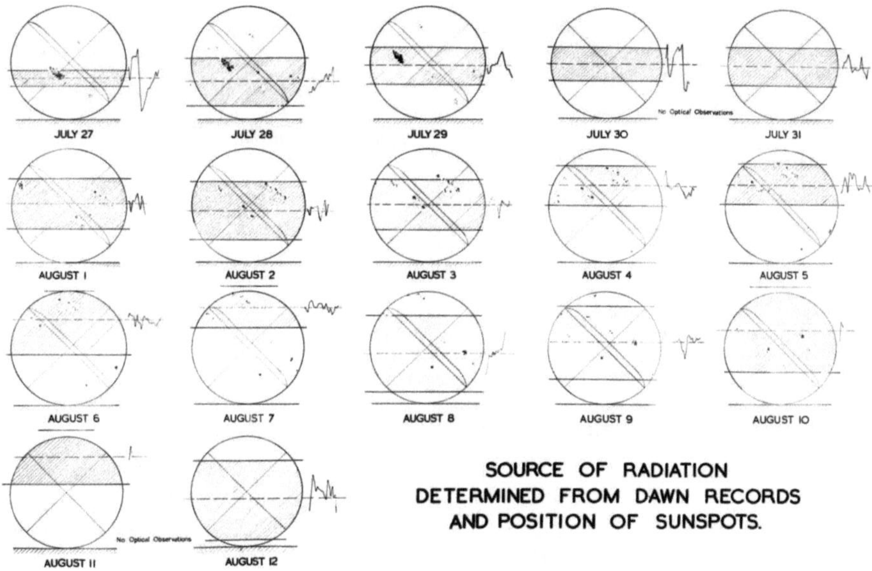

Fig. 7.11 Series of sketches of the sun on 17 days in July and August 1946, showing the width and positions of the radio solar noise sources on each day. The use of the sea-cliff interferometer had been considerably refined compared to the earlier use in February 1946; the precision of the determinations was improved. Similar format as Fig. 7.9. The small graph on the right-hand side gives the actual elevation calculated from each individual minimum in the interference pattern. The lobes were spaced by about 3 min of time (about 45 arc min of arc). On some days (e.g., 2 August) the equivalent radio emission strip is about 20 arc min; the sunspot groups were dispersed over the sun. The very wide strips towards the end of the period owe their width to thermal radiation from the whole disk of the sun. (Figure 6 from RPL 9) ATNF historical photographic archive: B1285-4

with the lower frequency events and this frequency also arrived earlier by a few seconds. However, this result was not published until the events of early March 1947 (see below). A few possible Type II bursts (rarer events which were called "outbursts" by Allen) may also have been observed; the intensities were even more impressive than the Type III events. Near sunspot maximum, Type II events occur about once per 50 hours, compared to about three bursts per hour for Type III bursts during sunspot maximum (Kundu, 1965). However, the recognition of the delays (from 200 to 60 MHz) at the level of 5–6 min for Type II events was not all certain in mid-1946. The major result was the recognition of the seconds of time delays for the Type III bursts, a controversial result that was not believed for some years (see below).

Finally, a suggestion of "fading" was observed at 60 and 75 MHz, and occasionally at 200 MHz. The correlated flux density changes consisted of an oscillation (up and down) with a fairly regular period from 0.5 to 2 min. This effect was not a simple interference phenomenon since the pattern was quite similar at the various frequencies. (See below for Pawsey's final word on fading.) The two noteworthy

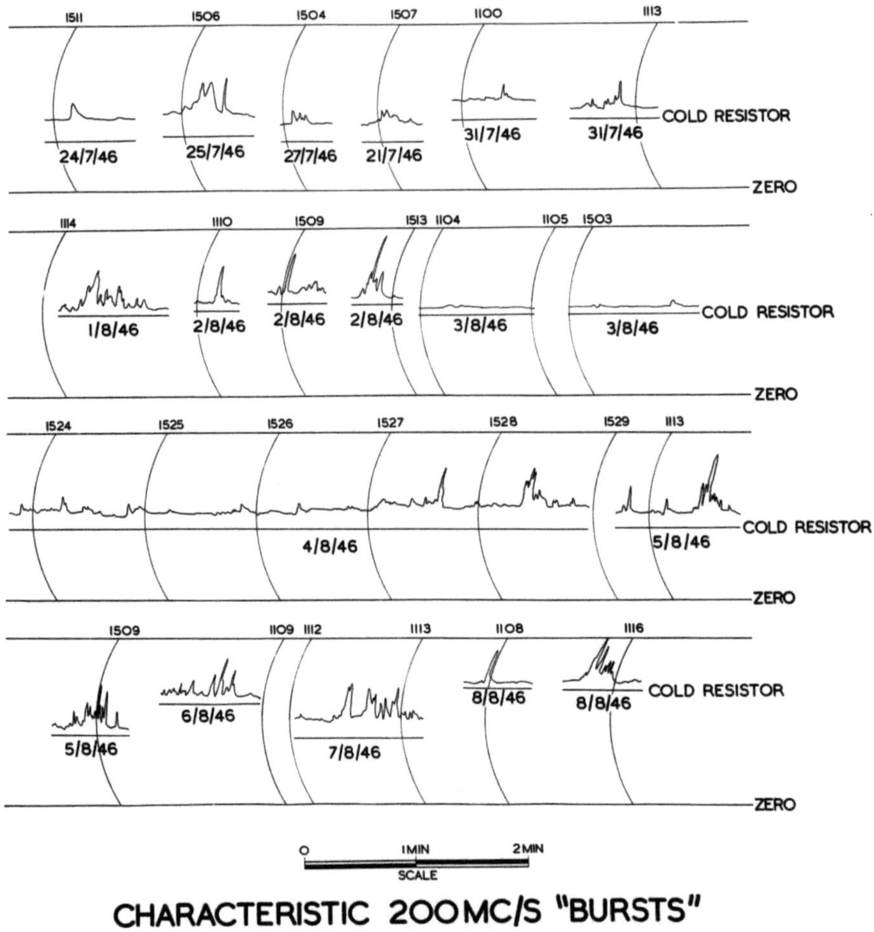

Fig. 7.12 Characteristic shapes of the 200 MHz bursts as observed with high time resolution over a number of days in July and August 1946; for the first time details of the shapes of the bursts in time were determined. In retrospect both Type I and Type III bursts were observed in these observations, i.e., isolated events (Type III) and noise storms (Type I). (Figure 7a from RPL 9) ATNF historical photographic archive: B12385-5

phenomena that were worth following up were this fading and the time delays between bursts.[55]

[55] At the PC meeting of 12 August 1946, the multi-frequency observations of the two events of late July were described by McCready. Payne-Scott was unable to attend this meeting; she was probably at Dover Heights from around 23 July to mid-August, observing during the day (the observations may have ended on 12 August). The 30 MHz data was given special emphasis in the report as well as the tentative detection of delays for the low frequencies (60 and 75 MHz). The detection of circular polarisation by Martyn (see footnote 49, this chapter) was summarised as well as eclipse plans for Brazil in May 1947. The purpose of the eclipse proposal was to be an improved high angular

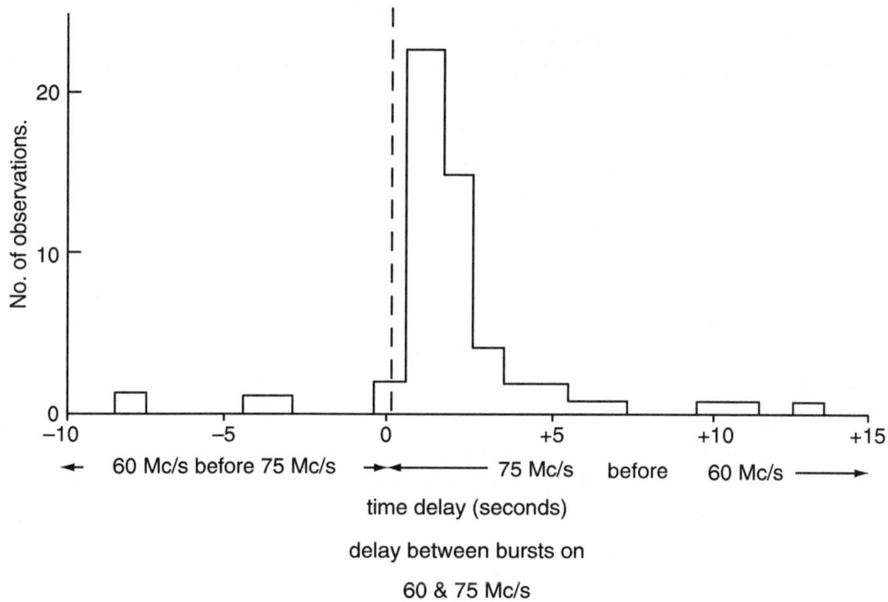

Fig. 7.13 From the times of fast recording every day (twice in 15 min) with simultaneous data at the three frequencies, delays were observed for some events. The 200 MHz bursts arrived first followed by bursts at 75 and 60 MHz. The time delays were about 1 s. The distribution of these delays is shown here where the delays between peaks as observed at 75 and 60 MHz are plotted. Clearly the data suggests that the 75 MHz isolate bursts precede the 60 MHz events by about 1.5 s. This result formed the basis for the many month campaign at Hornsby in 1948 (Chap. 8). (Figure 11 from RPL 9) (From CSIRO, ATNF)

There is a distinct possibility that Payne-Scott was not active at RPL after about October 1946. We can attempt to judge whether she was present from the set of PC minutes starting 11 June 1946–10 March 1947. She was not present at any of the

determination of the distribution of radio emission as the moon's limb covered the corona (see Hey 1955). Observations were to be carried out at 100, 200, 600 and 1,200 MHz. All work of the group with analysis and current observations was stopped at this date in August 1946 while preparations for the South American adventure were under way. It is ironic that in February 1947 the expedition was cancelled due to shipping problems from Sydney to South America. But the equipment was to be made available for use in Sydney! At the PC meeting of the following month (10 September 1946), John Bolton had been hired as an ARO (assistant research officer) to work in the receiver group under the direction of McCready. Two recent results from the 17 August and 7 September 1946 issues of *Nature* were reported. It is possible that a colleague in the UK had written by airmail, describing these results, both of which would have a lasting affect on the RPL group. The first was the detection of Cygnus A by Hey et al. ("Hey has observed bursts in cosmic noise in the direction of Cygnus.") and the second, the Ryle and Vonberg detection of circular polarisation with the use of a two element interferometer (see above, this chapter, also footnote 49 this chapter). Sullivan (2009) has reported that the issues of *Nature* arrived in Sydney from London with delays of 5–11 weeks until 1954, after which airmail was more common with much reduced delay.

meetings starting from 8 July 1946; she appeared again after seven meetings, or about 8 months, on 10 March 1947.[56] Except for a brief mention of her work by McCready at the 12 August 1946 meeting and the report from mid January 1947 described in footnote 56 (this chapter), there were no reported activities of her work during the eight meetings in the intervening period.

At the extensive meeting of the "Solar Noise Group"[57] on 6 June 1947 (where projects by Yabsley, Payne-Scott, Bolton, McCready, Piddington, Minnett, Pawsey, Smerd and a joint project by Bolton and Payne-Scott were described), plans for the next few years were presented. At this meeting, a proposal for the Payne-Scott report (RPL 9, August 1947) was outlined to summarise the solar work of the previous year, July–August 1946. "This report is hoped to provide a basis for a paper for publication jointly with other observers". On 14 October 1946, McCready had reported that observations at Dover Heights would begin again when Slee arrived at the end of October, implying that Payne-Scott was not available. From her personnel file (Chap. 4), we know that she went to the eye specialist for a medical examination in October 1946 (shortly after 16 October 1946, during the superannuation qualification process). Since we know that Payne-Scott had a miscarriage in the period 1945–1948 (see Chap. 11), is it possible that this event occurred in late 1946? If that were the case, it would explain her apparent absence from the scene from late 1946 to early 1947, a possible period of a few months. By June 1947, however, she was in full action again with a number of projects. The long delay in writing up RPL 9 (a year after the observations), and the fact that no publication was ever completed, may also be related to this possible loss of work of a few months. There is a strong likelihood that her colleagues, especially Pawsey, helped in "covering" for her absence. There is no discussion of any problems (such as medical leave) in her personnel file. By early to mid-1947, she was active and ready for a new phase of her research career.

Although the RPL 9 (internal report, "A Study of Solar Radio Frequency Radiation on Several Frequencies During the Sunspot of July–August 1946") was never published in a journal, it does contain much valuable material about the spectral indices of the bursts and especially the short time scale delays for the Type III bursts.[58] The final August 1947 report has 17 pages of main text with 15 pages of appendices dealing with various technical issues (e.g., Appendix 2 in RPL 9 provided a clear and detailed description of the interpretation of the dawn records from the sea-cliff interferometer, providing more detail than MPP) and descriptions

[56]The frequency of meetings was about once per month from mid-1946 to mid-1947. Payne-Scott attended about 65% of the 63 PC (plus Solar Noise Group) meetings from September 1945 to the date of her retirement in July 1951. She was, however, active in mid January 1947; on 14 January 1947 she wrote a short summary of the July–August 1946 observations at 60, 75 and 200 MHz with a four page table of times of occurrence and intensities at the three frequencies. The details of six of the bursts are shown, similar to the figures in RPL 9 (below). NAA: C3830, A1/1/5 Part 2.

[57]A preparation for the absence of Pawsey as he and his wife were about to leave for year's visit to the US and then Europe. See footnote 70, this chapter, and below.

[58]We are indebted to W.T. Sullivan III, for providing us with a full set of figures for RPL 9. The copy obtained from the RPL archives has only three of the 15 figures.

of each day's observations from 22 July to 12 August 1946.[59] This report laid the basis for the multi-frequency campaigns of March 1947 and especially for Payne-Scott's long range campaign of 1948 at Hornsby (mainly Type I and III bursts, Chap. 8). The introduction of RPL 9 provided a valuable summary of existing solar noise knowledge in mid-1947 while the "Conclusion and Future Programme" section of the report provided clues to the next major issues that were to be tackled as the next research project – fading and time delays between bursts.

The Behemoth Type II Burst of March 1947: Payne-Scott, Yabsley and Bolton Observe an Amazing Event

In the meantime, towards the end of 1946 the twin 100 MHz Yagi replaced the 75 MHz Yagi at Dover Heights (see Fig. 7.14 of the blockhouse with John Bolton, the newly arrived colleague). The largest sunspot of the modern era appeared in March and April 1947 (Newton 1955). The discovery of this behemoth solar outburst on 7 March 1947 has been described in an exaggerated fashion by Bolton (1982). The March 10 spot is number five in the Newton ranking (area 4,554 millionths), followed by its next appearance 27 days later on April 7 with an area of 6,132 millionths. This latter sunspot remains the record breaking sunspot since modern records have been maintained.

At this time the spectacular outburst of 8 March 1947, which became the prototypical Type II burst,[60] was detected. The data are shown in Fig. 7.15. The claim was made that the 60 MHz peak was about 10^{11} Jy, the strongest extragalactic signal ever received.[61] However, Paul Wild told Sullivan (2009) in 1987 that both he and Payne-Scott regarded the calibration of this large value to be quite uncertain.

The paper ("Relative Times of Arrival of Bursts of Solar Noise on Different Radio Frequencies", by Payne-Scott, Yabsley and Bolton) describing this event was

[59]The choice of the end date of 12 August 1946 is puzzling. The solar noise activity associated with this prominent sunspot was far from over. Payne-Scott in RPL 9 mentions a burst that is "off scale" on 12 August at 60 and 75 MHz. Allen (1947) indicated large 200 MHz bursts up to early September 1946. Is this abrupt cessation of the campaign in mid August 1946 related to her absence later on in 1946, perhaps because of problems with a pregnancy?

[60]Type II bursts are the strongest of the solar events, occasionally reaching 10^9 Jy at 100 MHz. These are rare slow drift events (−0.25 MHz/s that occur in association with flares. They have lifetimes of 15–20 min. Narrow band emission is often the occasion for harmonic lanes, i.e., radiation is observed at the fundamental plasma frequency and at twice this value (first detected by Wild et al. 1953, 1954). The emission mechanism is plasma oscillation, stimulated by outward moving shocks.

[61]However, Dale Gary and collaborators at the Owens Valley Solar Array (OVSA) did report a decimetric (30 cm) burst of about 10^{10} Jy on 6 December 2006. It is possible that at metre wavelengths this burst exceeded the March 1947 event (http://www.physorg.com, 15 December 2006).

Fig. 7.14 The blockhouse at Dover in 1947, about 2 months after the discovery of the giant Type II burst on 8 March 1947 by Payne-Scott, Yabsley and Bolton. Here John Bolton is shown on 1 May 1947. The 200 ShD antenna is no longer present. The 100 MHz Yagis are to the left while the 60 MHz Yagi is to the right (both crossed Yagi for the determination of circular polarisation of the solar radiation). The 200 MHz Yagi was on the far corner and not visible; for "radio star" observations the Yagis were used in a parallel configuration. ATNF historical photographic archive: B1031-6

submitted to *Nature* within 2 months,[62] on 21 May 1947, and was published on 23 August 1947; it had an important impact on solar research for many years. The delay in publication was about 12 weeks, a few weeks less than the delayed

[62] Details of Payne-Scott's participation in writing the paper remain unclear. Certainly her data were used in describing the seconds duration delays associated with the Type III bursts. Possibly Pawsey and Payne-Scott wrote the paper.

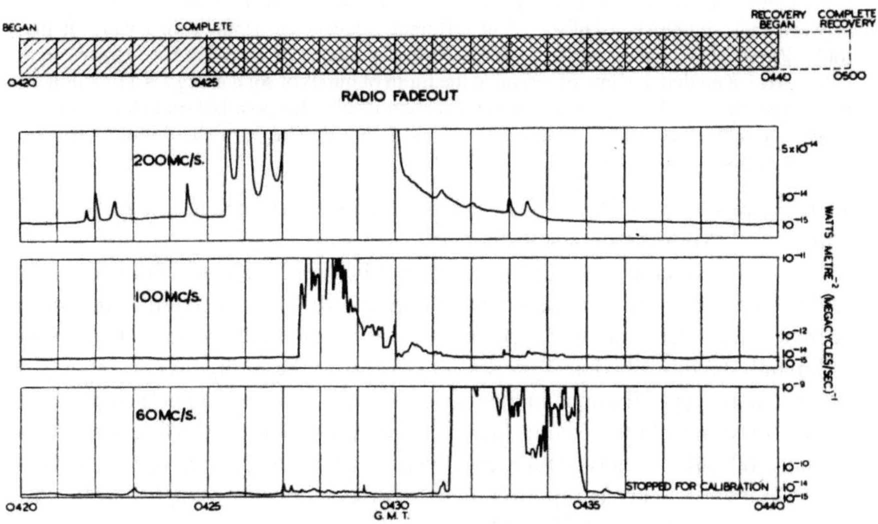

LARGE OUTBURST OF MARCH 8, 1947

Fig. 7.15 The giant Type II solar burst of 8 March 1947 as observed at Mount Stromlo and Dover Heights. The figure from the *Nature* publication of 23 August 1947. The 200 MHz data was obtained at Mount Stromlo and the 100 and 60 MHz data was observed at Dover Heights. The 60 MHz peak may have been a hundred billion Jy, but is uncertain. The minutes of time delay between the various frequencies is obvious. The radio fade out at a short wave frequency close to 20 MHz was used as a proxy for the time of the major flare. About a day later, a major aurora was observed in Australia on 9 March. In the publication the delays of several seconds for Type III bursts are also described. Reprinted by permission from Macmillan Publishers Ltd: *Nature*, vol. 160, p. 256, 1947. From "Relative Times of Arrival of Bursts of Solar Noise on Different Radio Frequencies", by Payne-Scott, Yabsley and Bolton

publication of the February 1946 paper of Pawsey, Payne-Scott and McCready, describing the initial Sydney detections. The paper was RPP 11.[63]

This *Nature* paper began with an unusual citation. Bowen had been on an overseas trip to Europe and the US in 1946. He gave a colloquium on 9 December 1946 at the RCA Laboratory in Princeton, N.J. In the February 1947 issue of *Observatory*, Reber and Greenstein wrote:

[63] Allen (1947) also includes the same 200 MHz data in his description of the campaign at Mount Stromlo from 1 April 1946 to 30 March 1947, just after the giant event of 8 March 1947. "... [A] rush of noise was heard in the loud speaker attached to the noise recorder". The peak intensity at 200 MHz was 3×10^7 Jy. These two publications (RPL and Mount Stromlo) were submitted at similar times in May 1947; neither referred to the other publication. Hey et al. (1948) did compare their solar noise data with Payne-Scott et al. (1947) in terms of time lags of radio bursts with respect to optical flares. Comparable delays had been inferred earlier by Hey et al. (1947) based on 25 MHz observations of a 25 MHz galactic plane fadeout (increase of D layer ionisation produced by a flare causing a sudden increase of the critical frequency to values above 25 MHz), followed by a sudden increase of radio noise from the burst itself 3 min later.

> If the coronal absorption depends on frequency, time lags of several seconds might exist between radio outbursts as observed at different frequencies. Unpublished work at the Radio Physics (sic) Laboratory at Sydney, Australia was reported on by E.G. Bowen in a talk . . . He stated that the time of arrival at the Earth of bursts of solar energy is a function of frequency. Bowen also reported that the direction of circular polarisation behaves rather erratically with time during solar outbursts and is not directly correlated with meridian passage of a sunspot group.

Thus the results from Payne-Scott's campaign were quoted in this rather circuitous manner, in some sense a self-citation.

Three main results were described in this publication. (1) Most of the bursts were not correlated at the three frequencies, suggesting that the emission at the various frequencies arose from widely separated levels in the solar corona. These bursts would mainly be described as Type I storm events in the future. (2) For events that would be called Type III bursts in the modern epoch, there was a good correlation in shape and ordering of the arrival times of the emission with 200 MHz first, then 75 and finally 60 MHz. These data were from the July–August campaign of 1946, obtained by Payne-Scott as reported in RPL 9 (see above), of August 1947. Of 60 cases of associated bursts at 75 and 60 MHz, the most common delay was found to be 2 s (Fig. 7.13). The determination of the delays between 200 and 75 MHz was comparable, but quite uncertain, since the form of the bursts was found to be disparate. In a few cases, delays were obtained with the 30 MHz emission arising a few seconds after similar events at 60 MHz. The explanation provided was that

> This time delay in the occurrence of bursts on different frequencies may be due to selective group retardation of the radiation in its passage through the ionised solar atmosphere . . .[64]

(3) The most significant result from the 8 March event was the discovery of what would today be called Type II bursts. The 1947 *Nature* paper indicated that some of these Type II bursts (with delays of minutes in the progression from high frequency to low) had probably been detected in the 1946 data; however, the 8 March 1947 event was the most prominent event observed up to that time. The 200 MHz data (see above) was provided by the Mount Stromlo instrument. There was a delay of 2 min between the onset of the burst at 200 MHz and the 100 MHz event; then, after a delay of a further 4 min, the 60 MHz outburst occurred. The shortwave fadeout (at an unspecified shortwave frequency, perhaps 20 MHz due to excess ionisation of the D layer of the ionosphere at a height of about 90 km) could be used as a proxy for the onset of a major flare on the surface of the sun; Radio Australia at 20 MHz disappeared and then reappeared about 20 min later. When optical observations began at Mount Stromlo on this day at 0440 UT (1440 EAST in Australia), a major flare was already in progress and the optical flare ceased at 0450 UT. The interpretation was that "... the outburst was related to some physical agency passing from high-frequency to lower-frequency levels. If we assume, following ideas suggested

[64] Clearly influenced by Jaeger and Westfold (see Chap. 8).

by Martyn, that radiation at any frequency originates near the level where the coronal density reduces the refractive index to zero, we can derive a rough estimate of these heights from electron-density data . . .". The derived velocities were in the range 500–750 km/s, comparable to the 1,600 km/s transit speed of auroral particles from the sun to the earth's magnetosphere. The displacement would correspond to about 0.3 solar radii in 6 min or an angular displacement of 5 arc min in this interval. In fact a prominent aurora was observed on 9 March, over a day later, in Australia, a rare event indeed in Eastern Australia. The inferred velocity of the particles causing the aurora was consistent with this time delay.

Payne-Scott et al. did not attempt the same calculation for the Type III burst data. The implied relativistic speeds of 25,000 km/s (almost 0.1 the speed of light) seemed too unreasonable at the time. It was only after Wild's 1950 paper (Wild 1950b) that this speed was proposed; it took some years before the implied relativistic speeds of advance of the agency giving rise to Type III bursts was generally accepted.

On 9 February 1948 (about a year later), Payne-Scott gave a talk[65] to the Propagation Committee on her progress with the Hornsby observations (see Chap. 8) of time delays for the "unpolarised bursts" (Type III). Her presentation was "Differences Between Times of Arrival of Bursts of Solar Radiation Emitted at Different Frequencies – Preliminary Measurements".[66] These observations are described below. Delays had been determined from the 18.3, 60, 65, and 85 MHz observations and she reported:

> Using Smerd's computations on radiating levels, based on Martyn's theory, the measured time differences would give a velocity of about 0.5×10^{10} cm/s for 60–18.3 Mc/s and 0.7×10^{10} cm/s for 65–60 Mc/s – in very good agreement, considering the errors and scatter in the measurements and the empirical values used in the calculations, but an apparently unlikely velocity. However, if bursts do originate at the borders of prominence materials, it is quite possible that the much higher density gradient there, may lead to a crowding of levels by as much as 100 times, reducing the above velocities to the order of prominence velocities.

The inferred velocities were about 0.2 times the velocity of light; but these values were dismissed as being unlikely at this time and no mention of this revolutionary explanation was included in the *Nature* paper of 1947. This presentation is probably

---

[65]The presentation was announced by Frank Kerr on 4 February 1948 (NAA: C3830, D4). Her presentation was item 2 on the program after (1) progress reports and preceding (3) "infra-low" (< 1 kHz, audio frequencies) observations of solar noise by McCready, based on the suggestion of Menzel and Salisbury, 1948 (see Sullivan 2009). By 31 March 1948, Bowen had decided that RPL would drop this project, apparently after McCready did some more work on it (NAA:C3830, A1/1/1, Part 3). Bowen then tried to interest Frank Wood of the Watheroo Magnetic Observatory in Western Australia. Apparently nothing was done. But by July 1949, Gordon Newstead of the University of Tasmania summarised some tentative results in a report; the authors were probably uncertain about the data as no publication ever appeared. The upper limits were far below the predictions of Menzel and Salisbury (NAA: C3830, A1/1/1, Part 4).

[66]File A1/3/1b from the Sullivan archive provides a comprehensive summary of her talk.

the first time the possibility was discussed of Type III bursts being caused by electron streams moving at a substantial fraction of the speed of light.[67]

There are at least two versions of the events that led to the *Nature* paper of August 1947. Bolton (1982) has written a dramatic and at least partially inaccurate version of the discovery. He had been invited to give a lecture at the annual meeting of the Astronomical Society of Australia at Noosa Heads, Queensland, in 1982. When the lecture was published, Don Yabsley took McGee (then the Editor of the *Proceedings of the Astronomical Society of Australia*) to task. McGee had not discussed the story with Yabsley before publication; Yabsley thought that the story had been misrepresented. He wrote:[68]

> This session (the July–August 1946 Dover Heights campaign described above) provided confirmation of something that I had already noticed, namely that when isolated, intense bursts were observed on two or more radio frequencies, the burst at the higher frequency generally preceded that at the lower frequency by a few seconds. Hence my inclusion as one of the authors of the "Relative Times of Arrival..." letter to *Nature* in 1947. In actual fact I don't remember playing any active part in the preparation of this letter – by early 1947 I had ceased participation in observations at Dover Heights, and was preparing new receivers for simultaneous observations at 200, 600 and 1,200 Mc/s at Georges Heights using a 16 ft × 18 ft paraboloid antenna. [See Orchiston et al. 2006, for a description of this research and for a photo of Yabsley and Pawsey in front of this antenna.] I think it is probable that Ruby and Joe Pawsey were responsible for the greater part of the text. John Bolton was the observer (at 100 and 60 Mc/s) on March 8, 1947. The 200 Mc/s receiver at Dover was out of action, and the 200 Mc/s record was provided from Mount Stromlo. My personal feeling is that Pawsey's name should have been on the letter as an author – perhaps he felt that three names on a Letter to *Nature* were quite enough.

Sullivan (2009) had a similar interview with Yabsley in 1986. "With regard to the paper by Payne-Scott, Yabsley and Bolton (1947), Pawsey contributed much to the project and to the paper and that originally he was intending to be a coauthor. But in the end he withdrew his name because he felt that three authors were quite enough for a *Nature* letter". The initial order of authors found in the RPL publications notebook is surprisingly *Bolton, Payne-Scott and Yabsley*, in contrast to the published version of Payne-Scott, Yabsley and Bolton (certainly not alphabetical)! We have no idea of the reason but as we shall see in Chap. 8, a lingering irritation may have continued, involving Bolton and Payne-Scott. Pawsey was thanked for "his interest and helpful discussions" and the Commonwealth Astronomer (Woolley) for the 200 MHz data.

The Bolton version in the 1982 paper had a different emphasis. He described the initial construction of a broadcast band receiver in late February 1947 to listen to

[67] As Suzuki and Dulk (1985) have discussed, "these streams move out through the corona along open field lines at a speed of about c/3, and their passage sets up plasma oscillations – Langmuir waves – which then radiate at their characteristic frequency". The energy range of the burst of electrons is typically in the range 10–100 kev. The electrons may be generated by magneto-hydrodynamic instabilities located where opposing lines of magnetic force are in close proximity (Wild 1974).

[68] Letter to Goss, 22 September 1997.

the cricket Test Match between England and Australia before the solar receivers were installed by Bolton and Stanley! Bolton wrote:[69]

> Finally on a Saturday afternoon [8 March 1947], as I unlocked the door of the blockhouse on my return from lunch, I heard the pen of one of the recorders hit the stop at the end of its travel. It was the 200 MHz recorder. I switched all three recorders from inches-per-hour to inches-per-minute and reduced the gain settings on all receivers to a minimum. Shortly afterwards the 100 MHz recorder hit its stop and the activity at 200 MHz decreased and 3 min later the 60 MHz recorder went off scale. Activity at all three frequencies ceased after about 15 min.

As we have seen, the 200 MHz receiver at Dover Heights was not working on this day and the time interval for this remarkable Type II burst was about 10 min. The 200 MHz data were provided by Allen and colleagues at Mount Stromlo. In fact, Bolton was involved in this period with an extensive series of high quality circular polarisation observations at both 60 and 100 MHz with Martyn. This data was never published (Bolton complained about this fact in the 1982 retrospective); however, Pawsey in the 1950 publication (Appendix E) made extensive use of this data and presented some 100 MHz polarisation data from 4 March 1947 (the previous Tuesday before the giant event). During the next passage of the sunspot, Bolton made an impressive series of observations at 60 and 100 MHz from 4 to 10 April (the meridian passage of the giant sunspot was close to 7 April). The effect discussed by Ryle (footnote 49, this chapter) was obvious as the reversal of the polarisation sense changed due to the relative dominance of different active regions of Type I storm emission.

Payne-Scott Winds Down at Dover Heights: Late 1947

A wide ranging discussion for the "Solar Noise Group" was held at the PC meeting on 6 June 1947. The rationale was Pawsey's impending departure (25 September 1947)[70] for a year's visit to science institutes in both the US and then Europe. The summaries given for each group were discussed in order to plan the next year's research during Pawsey's absence.[71] (We deal in detail with this visit in Chap. 9). Payne-Scott's major project was to finish the analysis of the July–August 1946

[69] *Proceedings of the Astronomical Society of Australia*, 1982, vol. 4, p. 349.

[70] The suggested date of departure of Pawsey and his wife Lenore in May 1947 by Haynes et al. in *Explorers of the Southern Sky* (1996, p. 211) is inaccurate, based on letters to and from Pawsey (NAA: C4659, 8). They arrived in San Francisco on the *Marine Phoenix* (a former troop ship of the US Navy, run by the Moore-McCormack Lines) on 14 October 1947, after departing from Sydney on 25 September 1947. They arrived back in Sydney from the UK on 29 October 1948, after stops in Aden, Perth, Adelaide and Melbourne. On Monday 25 October, Pawsey met Bolton in Melbourne as his young colleague was traveling to the 1 November 1948 solar eclipse in Tasmania. NAA:C3830, A1/3/18.

[71] A even more thorough discussion of these plans for each of the groups was held on 23 September 1947, 2 days before Pawsey's departure.

Dover Heights campaign, leading to a publication, possibly with Pawsey. As discussed previously, this never occurred.

Bolton and his group of Slee and Stanley were to continue solar work by making simultaneous observations of the sun with total and polarised intensity at 60, 85, 100 and 200 MHz. There was no explicit mention of "radio star" research!; however, item 3 was "plot of galactic noise at four frequencies in periods of no solar activity. Possible search for variations".[72] The statement was made that many interesting observations had been carried out since March 1947 but many were interrupted by "accidents" (probably broken equipment or even vandalism. (see footnote 51 this chapter). The quiet sun observations with the sea-cliff interferometer were planned to occur during periods of reduced solar burst activity.

A final fascinating "minor investigation" was planned as a joint project for Payne-Scott and Bolton – "Fading. To endeavor to obtain simultaneous records of this unusual type of noise variation from separated receivers at 60 Mc/s". This was a description of a "spaced receiver" experiment to investigate whether the fading at this low frequency was intrinsic or due to the ionosphere. It was known at the time that 200 MHz bursts provided identical records at sites spaced by many tens of km, implying that the influence of the atmosphere and ionosphere on the nature of the observed bursts was slight. In Pawsey's review paper of 1950a (Appendix E), a summary of unpublished data on this fading effect[73] was given; this "non-selective" (same behaviour at different frequencies, not at all like real solar phenomena, see above) had been observed at two positions displaced by 24 km (Dover Heights and Hornsby) at 60 MHz. The type of record was similar but the fading was not well correlated, suggesting that "... it may be due to modifications to solar radiation that are imposed near the earth, perhaps in the ionosphere".[74] Probably Bolton played only a small role in these observations.

In the period towards the end of 1947, Payne-Scott's activities started to wind down as the preparations were being made for the transition to the Hornsby Valley site (Chap. 8). At the PC meeting of 16 October (Pawsey was overseas in the US and Europe), she reported that "timing equipment is now in operation and the few bursts [presumably Type IIIs] that have occurred recently have been timed. Waiting for further solar activity and meanwhile planning photographic equipment [to film the bursts from the CRT tubes]". At the next PC meeting on 14 November 1947, a number of reports of solar work at Dover Heights were presented. Bolton, Stanley and Slee were busy with sunrise observations at 100 and 200 MHz and gave the encouraging report that the thermal level of the quiet sun was being detected at

[72]This program was in fact carried out in the following year by Payne-Scott at Hornsby (Chap. 9).

[73]The variation had a cyclic variation with a period of some minutes, similar to typical fading of an ionosphere-reflected signal.

[74]In a letter dated 11 June 1948 to Pawsey (to the Australian Scientific Research Liaison Office in London), Payne-Scott wrote that observations from Dover to Hornsby at 60 MHz showed "fair correlation" in contrast to the excellent correlation of 60 and 65 MHz at Hornsby. This slightly ambiguous result was interpreted by Pawsey to mean that fading was not an intrinsic property of solar emission, a result accepted in the modern era (NAA: C4659, 8).

60 MHz (about 1.5×10^4 Jy). Payne-Scott reported that no useful data had been taken on the relative times of arrival of bursts since the sun had not been active. She was anticipating new data from Hornsby,[75] the implication being that the Dover Heights site was becoming less useful due to 100 MHz daytime interference. None of this late 1947 data from Dover Heights was apparently published; probably the improvement of the data quality obtained in early 1948 at Hornsby meant that the older data was superseded.

Ionospheric Refraction from Dover Heights - Payne-Scott and McCready 1946

The presence of the ionosphere (heights ranging from 80 to 100 km in the daytime D layer and the more dense F layer at heights of 150–300 km, with the refraction dominated by the F2 layer) led to a number of disturbances affecting the incoming radio radiation. Since many of the effects depended on wavelength squared, any effects that would be trivial at 20 cm wavelength would be 625 times more prominent at 5 m wavelength (60 MHz). The sunrise data from 27 July to 7 August 1946 using the sea-cliff interferometer was utilised by Payne-Scott and McCready at 60 and 200 MHz (Fig. 7.10) to infer the properties of the ionosphere. By determining the positions of active regions on the sun, the positions at 200 and 60 MHz were compared: "... the true elevation of the source and atmospheric refraction are the same for both. The difference between apparent elevations at the same time then gave the difference between ionospheric refraction at the two frequencies."[76] However, both these assumptions were probably incorrect.[77] Thus the analysis did not lead to a straight forward conclusion.

There were two misconceptions: (1). Already in 1948, solar radio astronomers realised that that the source heights of the solar emission were dependent on observing frequency, due to the variation of the plasma frequency in the corona. For plausible coronal models, we would expect the height of the 60 MHz emission to be in the range 1.25–1.8 solar radii displaced from the centre of the sun. The data of Wild, Sheridan and Neylan (1959) suggested values of about 1.8 solar radii. The 200 MHz radiation would then have arisen at lower levels, 1.03–1.25 solar radii. Thus the difference in heights for emission from the solar limb would lie in the

[75]This is the first time that Payne-Scott's participation in the Hornsby site was mentioned at a PC meeting.

[76]The data was presented based on observations on August 1, 3 ,4, 5, 6 and 7, 1946. The range of apparent elevations ranged from 0.5, 2.5, 4.5, 6.5 and 8.5°. As an example, the ionospheric refraction on August 6 was 55, 37, 29, 26, and 14 arc min at elevations of 0.3, 2.5, 4.5, 6.5, and 8.5°, respectively; the calculated values using the Bailey (1948) model were 13, 12, 11, 9.5 and 8 arc min. The observed values were typically about three times larger than the predicted. However, the observed data at a constant elevation from day to day showed a large scatter, with a typical rms error of 5–6 arc min.

[77]Based on discussions with A.R. Thompson and Barry Clark, January 2007.

range 3–8 arc min. The data used by Payne-Scott and McCready was obtained when the active areas were 5–9 days from the meridian of the sun. Thus at this time (at least 7–8 days after central passage of the sunspot), the radiation at the two frequencies should have been displaced by about 6 arc min; at central meridian passage, the radiation at the two frequencies would have been observed along the line of sight with no displacement. Since the precision of the determination was about 6 arc min, this misconception was not decisive; however a systematic error was introduced.

(2) An additional major problem was that, although the wavefront at 60 MHz showed a large elevation change in the angle of incidence (up to 80 arc min of arc,[78] see Thompson et al. 2001), the excess path length for the *sea-cliff* interferometer would have been quite small. The ionosphere is so far away from the surface of the earth (height well in excess of 100 km) that the direct path to the antenna and the reflected path from the sea are nearly identical, as they pass through the ionosphere. Thus the delays in the sea-cliff interferometer for the two paths must be quite similar. A calculation[79] suggests that the excess path length would only be about 6° of phase for the 100 m cliff. Thus the change in inferred angle with the interferometer would only be a few arc min, much less than the values observed at 60 MHz. However, the data obtained by Payne-Scott and McCready showed that the ionosphere was probably the cause of these large displacements, possibly due to ionospheric wedges at low elevations. As A.R. Thompson wrote (January 2007):

> My general conclusion is that the progressive decrease of values [difference between 60 and 200 MHz positions] in [Payne-Scott, and McCready's] Table 1 with increasing elevation indicates that a substantial part of the deviations is due to the ionosphere. However, the situation is so complex that one cannot say anything very precise about the ionosphere from the observations.

A publication of these low frequency data represented the final research project based on Dover Heights observations by Payne-Scott.[80]

The publication ended with a prescient prediction; the authors suggested that the newly discovered point source discovered by Hey, Parsons and Phillips (1946) in Cygnus would be more suitable as a background source to investigate the

[78]The direction of the ionospheric refraction is the same as for tropospheric refraction (Thompson, Moran and Swenson , 2001, page 559), towards the zenith. The bending is due to the curvature of the ionosphere; if this structure were plane-parallel, then the bending angle would be zero. The ionospheric refraction scales as wavelength squared.

[79]Provided by Barry Clark, January 2007.

[80]Payne-Scott and L.L. McCready, "Ionospheric Effects Noted During Dawn Observations on Solar Noise", 1948. This publication was described in the PC meeting of 12 April 1948. Also Pawsey referred to the results twice in the 1950 review paper, once in reply to a question from G. Millington about the use of solar radio emission to study the properties of the earth's ionospheric F layer (at 150–300 km, too high to be studied easily from the earth by conventional echo studies). Pawsey wrote: "The observed effects were ... considerably greater than can be accounted for by current ideas concerning the ionosphere." The paper was RPP 51 with a first draft on 24 June 1948 and submission on 16 July 1948. The original order of authors was McCready and Payne-Scott.

ionosphere, than the very variable solar emission. In succeeding years, this prediction was fulfilled. The sun was quite an unsuitable probe of the ionosphere, since major solar activity was required to produce strong interferometer fringes while at the same time also producing a disturbed ionosphere (e.g., low frequency radio cutoff at low frequencies near 20 MHz).

Pawsey's Assessment of the Impact of the Solar Noise Group, 1946–1947

When Pawsey left Australia at the end of September 1947 for his trip to the US and Europe, he was probably quite proud of the remarkable achievements of the RPL solar group in the previous 2 years. The success of the RPL Solar Noise Group was already recognised through out the world; a major goal of this trip was to publicise its successes. In this short period, the group had made a prominent name for itself. An example of the recognition of the group can be gleaned from the following exchange with J.A. Ratcliffe (1902–1987; Pawsey had done a Ph.D. with Ratcliffe at Cambridge in 1934, "Intensity Variations of Downcoming Wireless Waves") in 1946. On 2 August 1946, Pawsey wrote:

> I have been principally interested over the last 6 months in the problem of radio frequency noise from the sun ... At the moment [Payne-Scott was observing the major solar emission associated with the July sunspot – see above] we are doing a bit of exploring, taking measurements of intensities at a number of different frequencies, some during the day and others at dawn [the latter is the sea-cliff interferometer]. We have found that the variation of solar noise on different frequencies is dissimilar and that that the dawn effect on 60 Mcs. is much more complicated that it is on 200 Mcs.

On 17 September 1946, Ratcliffe replied with a complimentary letter:

> First let me say how admirable I thought your paper on solar noise was [the *Nature* letter of February 1946]. There have been so many people scratching at this subject and making a few half-hearted measurements, that it is nice to see someone who has done it so thoroughly as you ... I think you will agree with me that this question of the emission of radio wavelengths from the sun is such a big one that it needs several workers on it. I do not view with any more dismay the possibility of overlap here than I do in the case of ionospheric research, for example. Now that that the Air Mail works so quickly (Ryle showed me your reply yesterday to his letter) we will make a special attempt to keep you fully in touch with what we are doing.[81]

The RPL Solar Noise Group had come of age; it was already one of the major research groups in the world. Its leadership in this new field was established and would remain. The unlikely dominance of Australia (together with the United Kingdom) in the first decade of post War radio astronomy had been established.

[81] Both letters NAA: C3830, A1/1/1, Part 1.

Chapter 8
Hornsby 1948: Type III Bursts Revealed; Conflicts with Bolton

The period of Pawsey's absence from RPL during a trip to the US and Europe (25 September 1947 to late October 1948[1]) was both frustrating and encouraging for Payne-Scott. After her own absence in late 1946, possibly due to the miscarriage (Chap. 7), her experiences during Pawsy's absence must have been trying. In this chapter we will describe these series of events and the likely impact that this had on Payne-Scott. Fortunately for her, the final result of the Hornsby campaign was spectacularly successful, with a lasting impact on solar radio research.

The events of this period were: (1) frustration with the progress of the Hornsby campaign and even the possibility that these observations would end prematurely, (2) anxiety that she would miss out on the new exciting project concerning a "vertical interferometer" (conventional three element Michelson interferometer) at Potts Hill to study the motions of solar outbursts (see Chap. 9), (3) growing conflicts with Bolton and even Stanley, (4) the lack of strong leadership of the solar noise group during Pawsey's absence, (5) pressure resulting from the fact that her results concerning the short term time delays (the frequency/time relation as a tool to probe motions in the sun) for Type III bursts were not accepted by the Cambridge solar researchers and even doubted for some period by Pawsey,[2] (6) Pawsey's return to Sydney in late October 1948 leading to a renewed leadership of the solar group, (7) publication of the Hornsby campaign in mid to late 1948, and (8)

[1]The purpose of this trip was to give Pawsey an opportunity to meet astronomers and physicists in the US, Canada and the UK; the meetings with the astronomers were his first contacts with astronomy outside Australia. He and his wife arrived in San Francisco on 14 October 1947 and visited the University of California and Stanford. He left the US for the UK on 27 March 1948 after numerous visits to US institutes in California, the mid-West and on the East Coast. He met Reber during this period, as well as Struve, Minkowski , and Seth Nicholson. In addition Pawsey met many others including Kuiper, Menzel, Spitzer, Goldberg, Shapley, and even Oort and Strömgren who were visiting the US at this time. After arrival in the UK on 1 April 1948, he was based at his old college, Sidney Sussex, at Cambridge with his host, J.A. Ratcliffe (his former Ph.D. thesis advisor). He was able to reestablish contacts with British colleagues for the first time since 1939.
[2]By mid-1948, Pawsey was a strong believer and proponent of these short time delays, probably based on the more thorough data obtained by Payne-Scott at Hornsby in 1948.

W.M Goss and R. McGee, *Under the Radar*, Astrophysics and Space Science Library 363,
DOI 10.1007/978-3-642-03141-0_8, © Springer-Verlag Berlin Heidelberg 2010

progress with the Potts Hill[3] 100 MHz interferometer project under the scientific leadership of Payne-Scott.

The 9 month period that Payne-Scott spent at the newly established Hornsby Valley field station was quite productive; she laid the foundation for unravelling the mysteries of the "unpolarised" bursts or "isolated bursts", which would later be called Type III bursts.[4] These bursts are probably the most studied burst phenomenon in astronomy; during periods of activity on the sun about three of these events per hour can be observed. The theory of the bursts (plasma emission from electron streams in the corona) is well understood today. The compelling rationale was to follow up the controversial claims of the *Nature* paper of August 1947; many colleagues were not convinced of the reality of the "seconds" delay and Payne-Scott had become very uncertain about the reality of the "minutes" delays for the Type II[5] outbursts associated with flares. Her plans for the 1948 campaign were formulated in a decisive meeting of the PC (Propagation Committee) on 23 September 1947 (the minutes refer to *Dr. Pawsey's Solar Noise Group*); seven major research campaigns were discussed. Before Pawsey's departure, the leadership issues in his absence were also discussed. No deputy was to be appointed, with the overall direction left in the hands of the Chief of RPL, Bowen. The details of the research programs were to be supervised by Lehany and McCready.[6]

As one of the major research campaigns, the Payne-Scott and Marie Clark[7] plan was summarised; the major undertaking was to be a confirmation of the reality of

---

[3]This instrument was initially under construction at Bankstown. In April 1948 (or earlier) the site was moved to the Potts Hill reservoir (Chap. 9).

[4]This term was introduced by Wild and McCready (1950) to describe the bursts which last for a few seconds, have bandwidths of 10 s MHz and drift to lower frequencies at a rate of about 20 MHz/s.

[5]The term was proposed by Wild (1950a) to describe the "outbursts" discovered by Appleton and Hey (1946a). Allen (1947) used the term "outbursts" for these minute time scale events.

[6]The text read: "Dr. Pawsey told members that during his absence he believed and expected that the research could carry on without appointing a deputy. Dr. Bowen would be in charge and would keep in fairly close touch with the work. To avoid burdening him with unnecessary details, Messrs. Lehany and McCready, as senior officers of the Group, were requested to keep Dr. Bowen informed of events by meeting at suitable intervals and co-opting other Group members whenever considered necessary." The project to investigate the spectral analysis of solar noise was discussed. The personnel mentioned were McCready and Medhurst (part time) "until he leaves, then new man". This new man was probably Paul Wild, who joined this project in 1948. There is a note that this new instrument represented a "common interest with Ruby Payne-Scott's time delay experiments. This represents a more ambitious presentation. Should be available to extend Ruby's results".

[7]Marie Coutts Clark's (1908–1991) career has been summarised by Hooker (2004); a contemporary of Payne-Scott, she graduated in physics at the University of Sydney in 1932. As Payne-Scott had done, Clark worked for a short period at AWA after teaching at a school. Clark functioned as an assistant, helping Payne-Scott at Dover Heights and later at Hornsby. Clark left RPL in July 1950. John Murray (letter, 26 January 2004) has provided us with additional details of Clark's role at RPL.

the delayed bursts as a function of frequency. The goal was "To *make sure* [underlined in original] if there is a systematic delay and sequence for bursts on different frequencies. Delay due either to time of travel ... or to selective wave retardation. Result to be lined up with magneto-ionic theory to decide latter. This is most important detail to fit into any explanation".

The first definite suggestion that Payne-Scott should transfer to Hornsby was made at the 14 November 1947 PC meeting. An additional problem was mentioned – 100 MHz day time observing at Dover Heights was affected by serious radio frequency interference in this suburb of Sydney. Plans for the move were developing rapidly; only 10 days later Payne-Scott wrote to Pawsey,[8] who was by then in Washington, DC:

> I am in the process of moving myself to Hornsby taking a trailer and the 65 MHz gear. The sun has been quiet lately, but I suppose may spring a surprise on us anytime.

Payne-Scott was free of the inhibiting atmosphere of Dover Heights (Chap. 7) and she could concentrate on her own work. An additional reason for the move from Dover Heights was the continued conflict between Payne-Scott and the group of Bolton, Stanley and Slee. These conflicts became more acute in the course of late 1947.[9] Within 2 months of Pawsey's departure, on 18 November 1947, Lindsay McCready wrote by hand to Pawsey, describing many of the events of the preceding 2 months:

> The letter would be incomplete without some gossip on personalities. To cut a long story short, Bolton and Ruby have had a "bust up" at Dover [Heights] partly due to technicalities (e.g., Ruby's local oscillator and his 100 Mc/s) and partly due to, I fear, her personality and, last but not least, both parties wanting to use the same gear for different experiments at the same time! Anyway, after careful examinations of the rights of all and of [the] facts we decided it be [sic] better if Ruby moved to Hornsby. No-one objects and both ok and Frank Kerr [who was well established at Hornsby with a low frequency lunar radar experiment] . . . now are quite happy about it she says she can get to Hornsby [by public transport] and arrive at the site before 9:45 a.m. and that she can't do that except on

[8] NAA: C4659, 8.

[9] McCready wrote to Pawsey about 6 months later on 9 June 1948 (NAA: C4659, 8): "... the 'feud' between Bolton and Ruby is still on, and leading to stupid and undesirable secrecy.' This open infighting probably continued until Pawsey returned in late October 1948. McCready showed weak leadership in handling this major disruption: "If you are wondering why we can't solve our own problems I would point out that everyone rightly regards you as leader and my impression is that Taffy has too many other worries at the moment to be bothered with things I should attend to ... far better to handle it the way we are attempting than getting Taffy to give orders ... Please don't imagine that people have gone completely haywire in your absence. As I said before I think I could patch it up and would have, but for many reasons decided to take no direct action as the matter is by no means urgent now that John [Bolton] is occupied in N.Z. [New Zealand RP Expedition from June to 16 August 1948] . . . and it has one advantage viz. Ruby is greatly spurred on!" In this letter McCready was also quite critical of Bolton for the first time: "There is evidence that he is attempting to go beyond his 'sphere of influence' and terms of reference as laid down before your departure . . . it must be done without the leader's approval and the knowledge of other members." Clearly McCready was out of his depth in dealing with these two strong personalities!

> rare occasions at Dover.[10] We are fitting her up in a separate trailer. I did not tell Taffy the full details – I mainly concentrated on technical difficulties they were naturally experiencing.

Pawsey's reaction to McCready is not known. However, he did write to Bowen on 8 December 1947 from the Australian Embassy in Washington, making a surprisingly forthright choice on behalf of Bolton:[11]

> The second point concerns solar noise in R.P. I had a letter from Lindsay [McCready] in which he mentioned that there had been some sort of a showdown between Ruby and John Bolton. This is not unexpected to me as Ruby seemed to me to be getting in the way at Dover. As I understand it, Lindsay has the situation well under control, at any rate when he wrote the letter, having arranged a transfer of Ruby's work to Hornsby. I don't think Lindsay had mentioned this to you, and I am only doing so because there might be some future complications in which Lindsay might require your backing. My feeling on the matter is that Lindsay's actions are likely to be above reproach. I also think that Bolton has, through his hard work and effective results, earned the right to take control of Dover, so that anyone working there, shall be doing so at his invitation.[12]

The move to Hornsby removed the day-to-day conflicts, although the clash of personalities continued.[13] Also the sunrise solar observations (with the sea-cliff interferometer) were less important at this time as the new swept lobe interferometer, which would provide high angular resolution throughout the day, was already under construction initially at Bankstown and later at Potts Hill (Chap. 9).

---

[10]This probably refers to the fact that the train service was more direct from Payne-Scott's home in Ashfield to Hornsby (all by train) than from her home to Dover Heights, which would have involved a more complex travel schedule by train and one or two buses. On some occasions, Payne-Scott would take the train to RPL at the University of Sydney, and then go by a service van, driven by other RPL personnel, to the Hornsby field station. The van went to the field station at irregular intervals. She was heard by John Murray in late 1948 arguing for 20 min with a clerk in the RPL administration after such an occasion; she had only bought a one way ticket from Hornsby to Ashfield at the end of the day. The bureaucratic colleague told her that CSIR regulations required "return tickets", not singles. He suggested that she should have left early in the day when the van returned to RPL!

[11]NAA: C3830, F1/4/PAW /1, Part 1.

[12]Bowen knew all about this imbroglio. He wrote to Pawsey (NAA: C3830, F1/4/PAW/1, Part 1) on 24 December 1947 (to the Australian Scientific Research Liaison Office at the Australian Embassy in Washington): "On the plans of Ruby, naturally I was fully aware of the trouble at Dover, although McCready in his usual gentlemanly manner smoothed it over very effectively. The new arrangements seem to be working well."

[13]An additional fact that may have reduced the day-to-day tension was that Bolton and Bowen had worked out a deal in early 1948 in which Bolton was to stop future solar work: "It does not appear practical to continue with any observatory work [continue to carry out routine solar observations to compare with data from Hornsby and Mount Stromlo and also probably implying that the group would occasionally observe the sun for testing purposes] on solar radiation." On 1 February 1948 Bolton wrote a long letter to Pawsey at the Embassy in Washington: "Bowen has given me permission to give up on the sun – at any rate just regard it as another source – and concentrate on the galactic source problem. I intend to go into solar evolution and atmospherics and see if I can find a home for my 100 Mc stars." (NAA: C4659, 8). Pawsey wrote back on 18 February 1948 and agreed with the fait accompli: "... am entirely in favour of your concentrating on the galactic work [radio stars]. The astronomers in the U.S. are waiting in a body on your results – so go to it."

Payne-Scott's assistant, Marie Clark (footnote 7, this chapter), contributed to the day-to-day observational chores in this period of continued solar activity; solar cycle number 18 reached an end in 1952–1954. Clark did not participate in the scientific research; her name never appeared in an acknowledgement in any of the Payne-Scott publications.[14]

Sadly, McCready wrote to Pawsey[15] in the UK on 6 June 1948 that there were conflicts between Clark and Payne-Scott. McCready had heard Payne-Scott speaking to Clark on the telephone in an impatient manner; he also thought that Clark was given too much responsibility "with observing work and maintenance at Hornsby". McCready seemed to side with Clark: "I am satisfied that the girl is willing and doing her best. I think Ruby's main concern at the moment is to get [well?] ahead of John Bolton in papers (my opinion only but not without circumstantial evidence) . . ." Later in the letter, McCready says, "I will do my best and of course watch relations and the group's welfare and have a chat with Miss Clarke [sic]." Since the observations were carried out continuously through September 1948, we can assume that the personal friction was to some extent reduced.

The Hornsby Valley field station was an important RPL outpost from 1946 to 1955 (see Orchiston and Slee 2005); the focus of the science at this site was *low frequency* radio astronomy. The location was a northerly suburb, about 25 km from the centre of Sydney. Frank Kerr et al. (1949) performed some of the early lunar radar experiments at 18 and 22 MHz using Radio Australia as a transmitter; as a by product, it was possible to study the earth's ionosphere (see Fig. 8.1a,b). Later on Shain and Higgins developed low frequency arrays at 9 and 18 MHz (Shain and Higgins 1954; Higgins and Shain 1954). Early continuum maps of the Galaxy with absorption due to intervening HII regions were important results from this instrument.[16]

---

[14]In early August 1948, a two part report (probably written by McCready to send to Pawsey in the UK) appeared at RPL: (1) "Status of Other Experimental Work" and (2) "Proposed Publications of Radio Astronomy Group (Dr. Pawsey's Group)." The latter listed 17 publications including the Pawsey review paper (Appendix E) and the Payne-Scott publication on the Hornsby data of 1948. Paper No. 11, "Correlation of Solar Noise Intensity with Sun's Rotation," by Payne-Scott and Clark, remains a mystery. This proposed publication for *Nature* never appeared; the claim was made that "analysis should be complete in 6 weeks and first draft ready in 8 weeks". Perhaps the conflicts between the two colleagues (see below) prevented the completion of this project.

[15]NAA: C4659, 8.

[16]The origin of the Hornsby Valley field station was described in a PC committee meeting on 11 June 1946 (rubric – High Power Vertical Radar) where Kerr outlined the requirements of a new site for 10–43 MHz systems for the study of cosmic and solar noise. A striking fact was that he required a non-cliff site (in contrast to Dover Heights), a deep valley or "hollow" to avoid ground echoes due to the Lloyd's mirror effect from sea reflections. The first discussions of radar observations concerned radar reflections from the sun! At meetings in July and October 1946, sites at Hornsby were discussed with the "Old Mans Valley" site chosen after another site was found to be unavailable. By mid-November 1946, work had started on this site. This site was used for some years for low frequency radio astronomy; the station was closed in 1955. John Murray has told us (letter January 2004) that the most of the equipment used by Payne-Scott in late 1947 and 1948 had been transferred from Hornsby to Potts Hill, probably towards the end of 1948, while the remainder of low frequency equipment was destroyed in a bush fire near Hornsby later in the 1950s.

a

b

Fig. 8.1 (**a**) The low frequency arrays used by Payne-Scott during 1948 to determine the properties of solar bursts. The field station at Hornsby (Old Mans Valley, north of Sydney, see Fig. 1.4) was used initially for Moon bounce radar experiments carried out by Kerr et al. (1949). The transmission was at 17.84 and 21.54 MHz from the Radio Australia station in Shepperton, Victoria; the reception of the Moon bounce radar signals was at the Hornsby site. Figures 8.1a,b and 8.2: all photos from 11 February 1948. During 1948 Payne-Scott observed solar bursts at 18.4 MHz (plane polarized broadside array), 19.8 MHz (fixed rhombic with a bearing 60° east of north, elevation 15°), 60, 65 and 85 MHz. ATNF historical photographic archive: B1266-15. (**b**) Close up view of the broadside array, transmission lines and instrument huts where the receivers were located. ATNF historical photographic archive: B1266-5

Payne-Scott made excellent use of two important properties of this site – improved radio frequency isolation from man-made interference (in 1948 Hornsby was close to the outer suburbs of Sydney) and the presence of the low frequency instruments. She used an 18.3 MHz broadside array (Fig. 8.1a,b) and a 19.8 MHz rhombic antenna for these 15 m wavelength ground breaking solar observations; these data were to be compared with 60, 65 and 85 MHz antennas being installed at Hornsby. The low frequency instruments near 19 MHz added a new dimension in the determination of the delays compared to the higher frequency which had been used at Dover Heights (e.g., 60 MHz); the time of arrival delays (highest frequencies arriving first in time) involving the low frequency 19 MHz instrumentation were many seconds compared to values of about 1 s for the observations at the frequencies near 60 and 75 MHz (Fig. 7.13). Thus Payne-Scott's transfer to Hornsby involved at least two considerations – technical arguments for an improved site and personality clashes.

Work at Hornsby began at a rapid pace towards the end of 1947 and early 1948. The publication, submitted exactly a year later in early 1949, stated that the total observing campaign ran from January 1948 to 23 September 1948. The PC meeting of 6 January 1948 implied, however, that the observations may have begun before Christmas 1947. At this meeting, Payne-Scott already announced that data had been collected at 60 and 75 [sic, probably 65] MHz with generally little solar activity; however, several events had been observed to have been associated with short-wave fade-outs (caused by increased ionisation due to the radiation from a solar flare) at 14.4 MHz (Radio Australia, VLQ3). The exciting news was that large delays in the unpolarised bursts of 5–10 s between 60 and later 18 MHz had already been discovered. The 85 MHz system was not yet operating.

As the campaign got underway in early 1948, the variable component of solar radiation at 85, 65, 60 MHz was observed (separate simple Yagis with a single polarisation were available at all frequencies; at 85 MHz crossed Yagis were available and thus circular polarisation could be determined. The sun could be followed for many hours per day with these higher frequency Yagi antennas). In Fig. 8.2, the Yagis at 60 and 65 MHz are shown, amidst the 19 MHz broadside array. These three frequencies and 19 MHz were used on an almost daily basis. In contrast to the earlier observations at Dover Heights, these systematic observations enabled Payne-Scott to characterise both the detailed time behaviour and polarisation properties of these meter wave bursts and outbursts. The continuous nature of the observations in 1948 was a major component of the success; Payne-Scott wrote an internal memo in December 1948 (Payne-Scott, 1948b, RPL 30) illustrating the coverage of all observations made at RPL from 20 MHz to 24 GHz[17] from 1947 to the end of 1948. The only long term continuous coverage before this had been provided by Yabsley at 200, 600 and 1200 MHz from August to November 1947 (Lehany and Yabsley 1949) and then by Payne-Scott for the whole of 1948 at 85 and 60 MHz. She noted that all other data "are very scrappy" for this 2 year period.

[17] These high frequency data from RPL were published by Piddington and Minnett (1949).

Fig. 8.2 The two Yagis used by Payne-Scott for solar work at the frequencies of 60 and 65 MHz at Hornsby Valley in 1948. The additional Yagi at 85 MHz is not shown; this consisted of a pair of crossed Yagis for reception of circular polarisation. ATNF historical photographic archive: B1266-2

The low frequency antennas could not track and thus considerable corrections had to be made to the intensities as the sun was observed off-axis.

Ms. Merle Watman[18] of the SBW (Sydney Bush Walkers) has provided a humorous comment concerning Payne-Scott at this time. "[We] ... heard that Ruby was one of the most brilliant science students to come out of Adelaide (sic) University." Ruby often left the train at Hornsby station after a weekend bushwalk to "take some readings somewhere" and the party would joke that "Ruby had gone to take her echoes off the moon". In fact as stated above, Kerr, Shain and Higgins were engaged in lunar radar experiments at 18 and 22 MHz, but Payne-Scott was not involved in these.

Clark and Payne-Scott were certainly busy in this early period of 1948. The 20 February 1948 PC minutes state that they had already succeeded in creating histograms of burst timing between 65 and 60 MHz (only 0.3 s) and 85–60 MHz (0.7 s). The more reliably determined difference between 60 and 19 MHz was about 9 s. These results had been presented by Payne-Scott at the controversial talk she had given a week earlier at the PC meeting (Chap. 7) at which the high velocities (0.1–0.3 times the speed of light) had been discussed and rejected as being implausible.

On 12 April 1948 the Propagation Committee met with Bowen in the Chair; Pawsey was to return 6 months later. Payne-Scott reported that fast cathode ray tube

[18]From an interview with Dr. Elizabeth Hall in 1999.

photos were obtained to determine the rise times of bursts at two frequencies with the Hornsby antennas. Three weeks later the circular polarised antennas at 85 MHz were in operation; the bursts showing the delays showed no polarisation in contrast to the storm bursts associated with the enhanced emission (storm bursts) (Type I continuum). This is how the name favoured by Payne-Scott originated – *unpolarised bursts*,[19] later Type III. It was suggested that the Hornsby work should cease soon but continue in a limited fashion at Potts Hill; this did not happen due to delays in the completion of the Potts Hill solar instruments (Chap. 9).

Payne-Scott has Doubts at Hornsby: Advice from Pawsey

After a few months of observing at Hornsby towards the end of 1947 and early 1948, Payne-Scott must have become quite uncertain concerning the value of the new multi-frequency data, with doubts about the reality of the frequency delays of the unpolarised bursts (Type III). At this time, her fellow RPL colleagues, Wild and McCready (1950), were within a year of using the swept-frequency spectrograph at Penrith (70–130 MHz); this data produced the ground-breaking classification of the Type I, II and III solar events, based on observations from March to June 1949. Also, as reported above, a number of colleagues in both Australia and the UK (also Appendix D) had already expressed doubts about the reliability of the seconds delays in the unpolarised bursts. She and Bowen were both becoming convinced in early 1948 that the Hornsby research was of little value. McCready wrote[20] to Pawsey on 6 June 1948: "... it appeared that her [Payne-Scott's] work at Hornsby was getting nowhere. Taffy himself suggested she fold it up and get into something more profitable ... I got her to give a talk at the Propagation Committee [the 9 February 1948 colloquium described in Chap. 7] mainly for Taffy's benefit ..." Another major factor contributing to Payne-Scott's anxiety was the planning and construction of the 100 MHz swept-lobe interferometer (Chap. 9). Clearly, because of its engineering and scientific challenges, she felt more attracted to this promising new project.

Probably at McCready's instigation, on 18 May 1948 Payne-Scott wrote a four page handwritten summary of her work at Hornsby to Pawsey (at the ASRLO, Australian Scientific Research Liaison Office, Africa House, London).[21]

The letter began:

> Lindsay McCready is on holiday and has forwarded to us your last letter concerning time delays. I thought I should write outlining the present situation and asking your opinion,

---

[19]In 1948–1950, Pawsey used the term "isolated bursts" to describe these events; both terms were descriptive of properties of Type III bursts. Apparently, Payne-Scott's passionate use of her term was based on her conviction that the lack of polarisation was the most distinguishing characteristic of these bursts.

[20]NAA: C4659, 8.

[21]NAA: C4659, 8. The letter is written on thin airmail paper with some "print-through", as well as severe fading. We are indebted to H. Wendt for considerable assistance in deciphering this valuable letter.

> particularly as I do not think the problem will be settled by the spectroscopic analysis equipment [the Penrith swept-frequency spectrograph] you suggest.

Next she provided a thorough description of the results of the first 5–6 months of observations at Hornsby. She described the circular polarisation of the Type I bursts ("enhanced radiation") and the lack of polarisation of the Type III bursts. She then pointed out that the Type I bursts showed little correlation at adjacent frequencies. The Type III did show correlation at adjacent frequencies; this result had been one of the major conclusions (Chap. 7). She then discussed the delays from the three frequency Hornsby data (much of this text about the delays is quite illegible in the NAA document). The letter also has a discussion of the time resolved Type III bursts recorded at high speed on an Easterline Angus recorder at the high speed of 12 in. a minute. The exponential form of the decay of the burst was fitted with a decay constant at the various frequencies; the constant at 85 MHz was "rather higher" than at 60 MHz. There was also a hand drawing of the burst form, showing a sudden rise and then exponential decay. "A characteristic unpolarised burst shows a finite rise time, rounded top, and shows decay, reminiscent of the transient response of a medium with a natural resonant frequency."[22] Finally it was clear that Payne-Scott's major purpose in writing this letter was a request for advice: "... I would like your opinion on whether to leave it there or to try anything more ... I am convinced of the reality of the delays [?]..."[23]

An important aspect of the letter was the expression of her frustration at not being able to begin her expected participation in the planning of the 100 MHz instrument (as described above, based on McCready's letter to Pawsey). She wrote:

> You have probably heard that I am taking on 100 Mc/s interferometry and hence would like to get clear of all the [responsibilities at Hornsby?] ...

Eleven days later, on 29 May 1948, Pawsey wrote a remarkable response.[24] This letter was handwritten on the train (Pawsey's handwriting is fairly illegible even when written on a stationary desk!) and contained even-handed advice; we can suppose that this communication represented some comfort to Payne-Scott after all the controversy which had occurred since Pawsey had left Sydney, exactly 8 months previously. He began:

---

[22]Quote from the later publication of these results by Payne-Scott (1949a) since the text of the letter is unreadable.

[23]There are some other interesting pieces of news in the letter. She thanked Pawsey for some contribution to her postcard collection. She told Pawsey that Owen Emery (from Mount Stromlo and the Radio Research Board – RRB) had tuberculosis. She said she had seen Bernard Mills who was making good progress, having also had tuberculosis. She asked if Pawsey had seen Burgess in the US, responding to a probable mention of Burgess in a previous letter (footnote 25, this chapter). At this time, Pawsey had been in the UK for about 5 weeks. It is striking that neither Payne-Scott nor Bolton mentioned their mutual conflicts of 1947–1948 in any of their letters to Pawsey. Bolton wrote many more letters and did complain to Pawsey about numerous other problems, but never about the conflicts with Payne-Scott.

[24]NAA: C4659, 8.

> I think the proposed subject [Type III bursts] would make an excellent paper subject to your being able to give sound evidence for the points you mention. You have to chose [sic] between doing work on bursts or starting interferometry. I think you should realise on what you have done – Do only what is necessary to make it a good plan. In the interferometry field you start behind Ryle and will probably take some time to catch up. Therefore go ahead with it in second place – complete the burst story first.[25]

Eleven days later Pawsey wrote to Bowen with a detailed summary of his letter to Payne-Scott. He emphasised that Payne-Scott should publish her results "only . . . [if] the 'facts' she mentioned could be considered to be reasonably established".[26]

On 22 July 1948 the Propagation Committee met with Piddington, not Bowen, in the Chair. Payne-Scott gave a report. In contrast to the plan to cease solar observations at Hornsby as previously discussed, the research was to continue for some months. As before, the main goal was the determination of burst delays using a fast recorder, in order to compare with Jaeger and Westfold's[27] theoretical predictions (see below for details). A description of the continuous monitoring of the short-wave station VLQ3, a 10 kilowatt station at 14.4 MHz, was given; ionospheric fade-outs of this station could be used as a proxy for the onsets of major optical solar flares. At the next meeting of the Propagation Committee on 6 September 1948, Bowen was in the Chair. Payne-Scott stated that she was busy analysing the Hornsby data. These series of observations were coming to an end; a number of the existing Yagis at other field stations were being moved to Potts Hill, including the 97 MHz system from Dover Heights and the 62 MHz (probably the 60 MHz system) from Hornsby. These aerials were used for continuous monitoring of the sun for a "total power" determination of the solar intensity during the day. An additional incentive to complete the move from Hornsby was the upcoming

---

[25]NAA: C4659, 8. Pawsey discussed the waveform of the bursts, i.e., the time resolved fast recording of the Type III bursts as described by Payne-Scott. Also he described to her his detailed advice for carrying out the "Burgess experiment" (see below, end of this chapter). Finally he told Payne-Scott that he had, in fact, met Burgess (Ronald E. Burgess, 1917–1977, National Physical Laboratory, UK, Teddington, Middlesex) at the ASRLO in Washington in late 1947 or early 1948; Burgess was apparently either visiting the US or working at the ASRLO for some period. Burgess had been involved in the controversy with E.G. Appleton (footnote 27, Chap. 6). Burgess had been instrumental in quantifying concepts of the properties of radio frequency noise in two influential publications, "Noise in Receiving Aerial Systems" (1941), and "Fluctuation Noise in a Receiving Aerial" (1946). See also Payne-Scott's contribution to these issues of the thermodynamics of radio frequency noise in Appendix G.

[26]NAA: C4659, 8. On 9 June 1948 McCready wrote Pawsey (NAA: C 4659, 8) in London with an enthusiastic report about Payne-Scott's reaction on receipt of the 29 May 1948 handwritten letter. "Her reaction was as I anticipated viz she still wants to keep interferometry. Nevertheless, I sowed a few seeds but I doubt whether they will be fertile unless you add [some fertilizer ?] in a subsequent letter." Payne-Scott had started the Burgess experiment at Hornsby (see below).

[27]At this meeting, Wesfold gave a report on "Solar Noise (Theory)"; he described "... what happens to the radiation ... after a burst has been produced . . . The magneto-ionic theory has been worked out for the case of a non-uniform distribution of ion density with height, and the results applied to solar bursts. By measuring the rate of decay of bursts and knowing the [radio-frequency] of the observations, it is possible to calculate the collision frequency at a given electron density."

1 November 1948 solar eclipse, which lead to a consolidation of solar noise work at Potts Hill (Orhiston, Slee and Burman, 2006). These total intensity records were then compared with the high resolution Michelson interferometer (also called a spaced interferometer at RPL) at Potts Hill from 1949 to 1951.

The Publication of the Hornsby Observations

"Bursts of Solar Radiation at Metre Wavelengths" was published in 1949a in the *Australian Journal of Scientific Research*, the manuscript having been submitted on 5 January 1949, only 3 months after the Hornsby observations drew to a close.[28] Pawsey wrote to J.A. Ratcliffe, at the Cavendish Laboratory of the University of Cambridge on 17 December 1948,[29] "I have been delaying writing to you on this subject because I thought that Miss Payne-Scott's paper would have been available by now."[30] Thus it is clear that Payne-Scott was in a hurry to complete this paper before the end of 1948.

The manuscript provided a discussion of the variable component of solar radiation at these four low frequencies over a period of 9 months. Two types of variable high intensity radiation were distinguished. One was the "enhanced level" and "storm bursts" which were circularly polarised; these were the Type I events also called "noise storms" (i.e., Type I continuum plus Type I bursts) in the early literature. The other, a particular type of short duration, was the unpolarised burst.

Based on the Hornsby data from 1948, Payne-Scott found that the unpolarised bursts tended to occur nearly simultaneously over a range of frequencies. In many cases the higher frequencies arrived first, followed by a delay of order seconds for the lower frequencies. No cases of minute-scale delays (such as the famous March 1947 event) were observed. An outburst is shown in our Fig. 8.3, with time scales of some minutes; any Type II behaviour, 85 MHz radiation arriving some minutes before 60 MHz, was quite uncertain. As Sullivan (2009) points out, "The complex frequency–time structure now seen in the Type IIs made it clear why Payne-Scott had not been able to find them despite long term observations during 1948 with separate receivers at several frequencies". The Penrith swept-frequency instrument

[28]This paper was designated RPP 62, with a first draft prepared on 18 May 1948, well before the observations were completed in September. The manuscript was sent to the *Australian Journal of Scientific Research*, Series A, on 24 December 1948. The original title was "Characteristics of the Variable Components of Solar Radiation at Metre Wavelengths".

[29]NAA: C3830, A1/1/1, Part 3. See footnote 46, this chapter. In early August, a report had been prepared for Pawsey, who was still in Europe, about the status of this manuscript as well as other RPL ones. A complete summary was presented with the explanation, "Diagrams still to be drawn. Should be completed in 4 weeks". NAA: C3830, A1/1/1, Part 3.

[30]A poignant note was added by Pawsey: "I sent you a food parcel a couple of weeks ago and I hope you receive it by Christmas. Wishing you and your family the compliments of the season." All correspondence between them was on a familiar basis, beginning either "Dear Jack" or "Dear Joe", in contrast to most other correspondence, such as that between Ryle and Pawsey in this era, which began "Dear Pawsey" or "Dear Ryle".

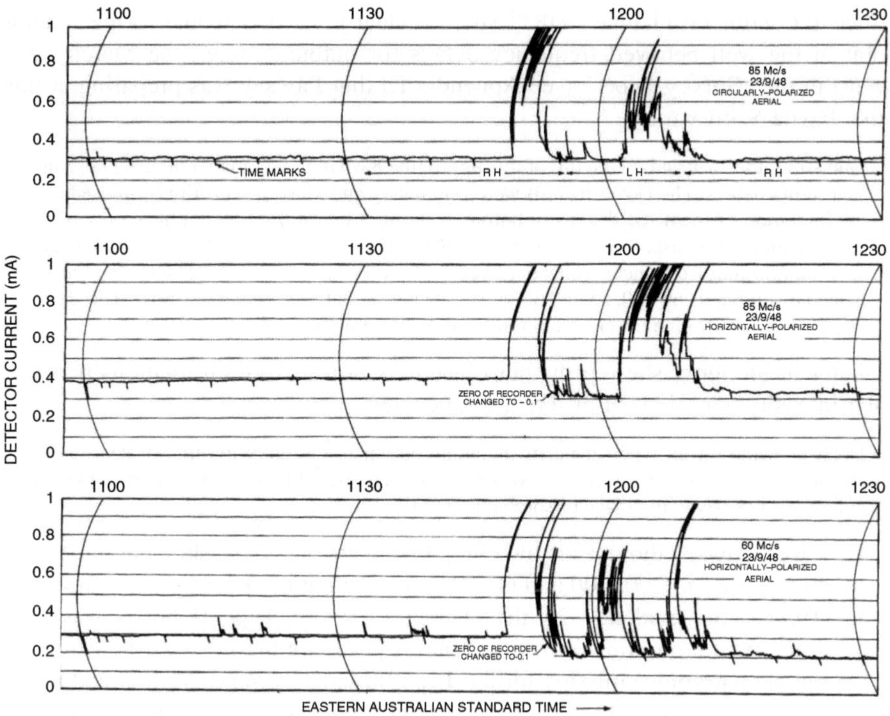

Fig. 8.3 An example of an outburst observed by Payne-Scott at Hornsby, with the Yagis at 60 and 85 MHz. A radio fade out began at 1151EAST, an approximate time for a solar flare. This event is not a clear-cut Type II burst. The 85 MHz data is shown in the top two panels (different polarisations as indicated) and the 60 MHz in the bottom panel. Data from 23 September 1948. Figure 2 of Payne-Scott (1949a), *Australian Journal of Scientific Research (A),* vol. 2, p. 214. Figure taken from ATNF historical photographic archive: B1640-3. See http://www.publish.csiro.au/nid/51/paper/CH9490214.htm

(Wild and McCready 1950, which came into operation only a year later) which displayed the spectrum of the solar burst from 70 to 130 MHz once per 0.1 s, was well matched to the complex spectra of the Type II bursts (as well as Type I and Type III). In fact Payne-Scott began to doubt the reality of the March 1947 Type II event in the 1949a publication: "However on the question of the longer delays, the present author has never since, in the recording of hundreds of bursts, obtained any evidence for delays of the order of minutes. Either the case reported earlier [March 1947] was unusual, or the record was misinterpreted; as the relative amplitude of different portions of a complex burst may be very different on different frequencies, such a misinterpretation of a single case is quite possible."

Payne-Scott's nervousness about the ordered delays on the minute–time scales was also expressed in additional correspondence with Pawsey in an official letter (i.e., typed), written a month after the handwritten letter of 18 May 1948 discussed above. Payne-Scott wrote to him again at the Australian Scientific Research Liaison

Office in London on 11 June 1948.[31] She was clearly worried about making a major point of the well behaved frequency delays for outbursts (e.g., the March 1947 event) in the "Survey Paper" (see Appendix E) that Pawsey was preparing at this time. Payne-Scott wrote:

> I am still rather worried by the use of this figure [the 8 March 1947 giant outburst] in so far as it shows delays. The Dover records were very messy and Martyn at the time criticised our interpretation of them. Of dozens of outbursts during the last 6 months, I have only once seen an analogous case, in which outbursts appeared to have a markedly different appearance on different frequencies. This occurred on May 14 last, when a fade-out was accompanied by outbursts on 200, 85 and 60 Mc/s . . . when an outburst began on 200 and 60 *there was nothing at all on 85* [her underlining] for over a minute . . .

Later in the letter she repeats her objections in a section on relations between solar intensity variations at different frequencies:

> Delays – same for all types of bursts including outbursts – usually in order of decreasing frequency, occasionally reversed – of order of seconds.[32] [Thus there was no obvious evidence of bursts with delays of some minutes.]

However, Pawsey did use his own Fig. 4 (see our Fig. 7.15) showing the March 1947 observations at 200, 100 and 60 MHz. Following the Penrith observations of Wild (1950a), the evidence for the characteristics of Type II bursts was well established; the major clincher for the role of plasma emission was the detection of the second harmonic emission for the Type II bursts by Wild, Murray and Rowe (1953). The major contribution of the 1949a Payne-Scott paper was a detailed description of the unpolarised bursts; few details are given of the enhanced radiation (Type I) bursts (Fig. 8.4).[33] She did remark that the Type I bursts were seldom observed at the low frequency of 19 MHz, in contrast to the Type III bursts. She then summarised the spectral index of the Type III bursts by comparing the relative

---

[31] NAA: C3830, F1/4/PAW/1, Part 2.

[32] In the letter she also says that the paper has "too much figure 4 [the figure from the August 1947 paper] for my liking". She again pointed out the need for local optical observations to be certain of the onset of the flares instead of relying on shortwave fadeouts as a proxy. She even admitted that the swept-frequency spectrograph might be required to sort this out! (This turned out to be true; Wild (1950a) published his definitive paper on Type II outbursts a year later.) This same message of doubt had also appeared in the unofficial (handwritten) letter of 19 May 1948: "... never seen anything like this [the 4 min delays for various frequencies for outbursts] since. There is the possibility that it has a bad interpretation."

[33] Martyn (NAA: C3830, D5/4/62 from Sullivan archive) was most impressed by the circular polarisation data at 85 MHz as shown in a letter of 4 January 1949 to Bowen in Sydney. This new data "cuts across the line that Bolton and I [Martyn] had planned to do, notably on *the building up of polarisation on the various frequencies as a spot develops* with a view to the special study of the *origin of bursts and general level*". (Martyn's emphasis) Martyn was convinced that polarisation was the key to solving this problem. He wrote: "Miss Payne-Scott has made a very good job of doing it ... the main thing is that someone has done it, and we [Bolton and Martyn] can get on now to something else. I presume that Bolton too is quite happy. We shall certainly have to revise our ideas about what we should do now."

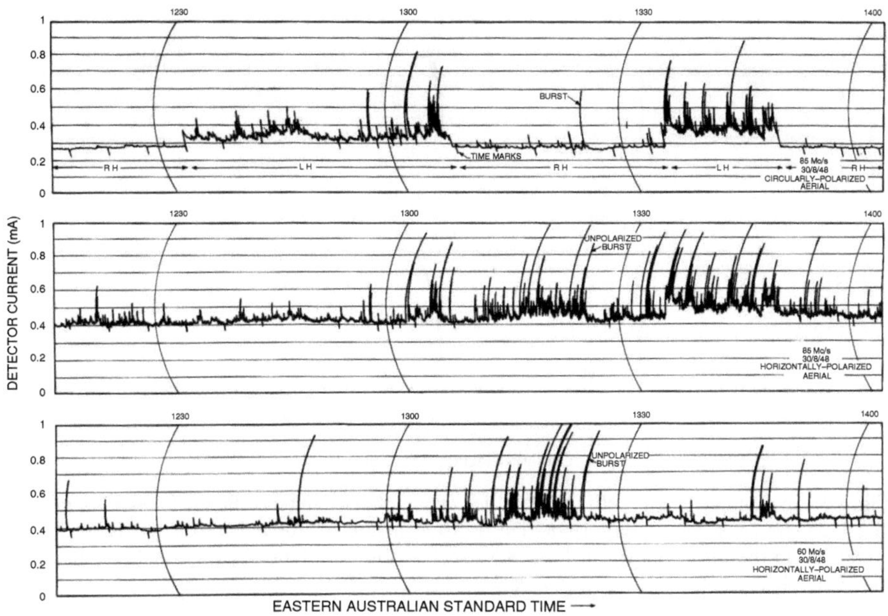

Fig. 8.4 An example of a Type I event (called enhanced radiation by Payne-Scott) observed at Hornsby at 60 and 85 MHz. The 85 MHz data is shown in the top two panels; the polarized nature of the 85 MHz radiation is evident in the top panel as the observations are switched between right hand and left hand polarisation. Data from 30 August 1948. Figure 3 of Payne-Scott (1949a). ATNF historical photographic archive: B1640-2

intensity of the 60 and 85 MHz data and found that the lower frequency intensity was usually twice that at the higher frequency.

The bursts appeared to show an exponential decay; double humped bursts were common, the second part being a possible "echo" of the original burst. Their characteristics were shown to conform broadly with the hypothesis that the bursts originated in localised transitory disturbances in the high corona and radiated over a wide frequency range (at least over the range 60–85 MHz. See Fig. 8.5 for a Type III burst of 19 July 1948).

The decay constant of the bursts could be predicted on the assumption of an excited medium at the region of origin; the double peaks were explained on the assumption that the second peak was an "echo" of the disturbance after reflection at the appropriate lower level in the corona.

The occurrence of time delays between the arrival of the "corresponding" unpolarised bursts on different frequencies was confirmed, the higher frequency commonly arriving earlier, with delays of about 0.7 s between 85 and 60 MHz and 9 s between 60 and 19 MHz. There was good correlation between major radio fade-outs and large bursts on these frequencies. Payne-Scott interpreted this extensive data set (certainly the most thorough collection of burst data in late 1948) in terms of the theory proposed by Jaeger and Westfold (1949). These colleagues had

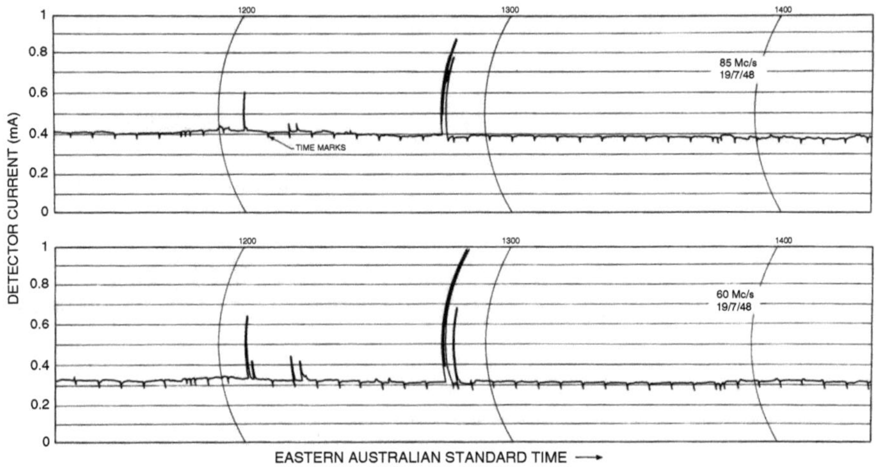

Fig. 8.5 An example of a Type III burst, called an unpolarised burst by Payne-Scott. Observations at 85 and 60 MHz on 19 July 1948. The isolated nature of this burst structure is obvious as compared to the bursts shown in Figs. 8.3 and 8.4. Figure 8.1 of Payne-Scott (1949a). ATNF historical photographic archive: B1640-4

suggested that the unpolarised bursts arose well out in the solar corona, with a common level of origin. They also proposed that the decay constant could be identified with the local collision frequency, with bursts at 85 MHz arising at heights in the corona of 5×10^5km (Westfold 1949). The production of the double hump bursts arose from:

> The production of bursts outside the echoing region [the point where the radiation would be turned back, or the observing frequency is equal to the critical frequency of the plasma] offers an explanation for the common occurrence of double-humped bursts, the second part being an echo of the original burst. The observed faster decay rate of single bursts will then be expected, as they will arise inside the echoing region where the collision frequency is higher than it is further out.

As a final conclusion, Payne-Scott carried out a quantitative comparison of the measured delays and the predictions of the Jaeger and Westfold theory. The measured delays between 85 and 60 MHz were 0.7 s compared to a prediction of 0.5 s. However, the measured delay between 60 and 19 MHz was 9 s while the theoretical prediction was only 1 s. Clearly there were problems with this latter comparison; however, Payne-Scott suggested that the theory of different frequencies propagating at different speeds was an adequate explanation of the second time scale delays (the selected group retardation explanation).

Sullivan (2009) has provided a spirited description of the resolution of this problem only a year later by Paul Wild in his classic paper on the Type III bursts (Wild 1950b). Wild had continuous data from 70 to 130 Mc/s. Although the predicted intensity frequency spectra were in rough agreement with the Jaeger–Westfold theory, the predicted time-frequency spectrum was in disagreement with the slope of the dynamic spectra. Wild was thus forced to make the audacious suggestion that there were outward moving disturbances moving at relativistic velocities of 20,000–1,000,000 km/s, much faster than the fastest moving disturbances of about 1,000 km/s known at this time. Sullivan has provided an anecdote of the interaction between Wild and Jaeger regarding the derivation of the equations governing the arrival times as a function of frequency from an ionised medium. Wild had derived a multiple page mathematical treatment of the major result. A few days later Jaeger produced a few lines of equations, that were incorporated (with credit) in the Wild paper. Ironically, Payne-Scott had given her colloquium on 9 February 1948 (Chap. 7) with a discussion of unpolarised burst velocities of about 50,000 km/s, a value so large as to be rejected!

At about this time, a major event occurred with the creation of a new name for the Propagation Committee. At the meeting of 7 March 1949, Pawsey was again the Chair, having arrived back in Australia in October 1948. A discussion was held about the future purpose of the committee, including its name, content and frequency of meetings.

On 5 April 1949, the first meeting of the "Radio Astronomy Discussion Group" took place with a 1 hour talk by Piddington on the derivation of the properties of the solar chromosphere. Thus "Cosmic Noise" and "Solar Noise" became the new "Radio Astronomy". The following week, on 11 April 1949, a general discussion of all projects for the RPL radio astronomy group took place, probably a planning session following Pawsey's return.

Cambridge: Collaboration and Controversy

Through out the first decades of the post World War II renewed activity in radio astronomy, interaction between RPL and the three British groups consisted of an interplay of collaboration and competition with occasional spurts of controversy. The year 1948 was a turning point in the contact between RPL and the Cavendish Laboratory, just as the groups were beginning their long tradition of radio astronomy research. The leader of this British group was initially J.A. Ratcliffe (1902–1987), followed a few years later by M. Ryle (1918–1984).[34] Additional details of the complex interactions with the Cambridge group are presented in Appendix D. A remarkable exchange occurred in March 1950; Ryle claimed in a letter to Pawsey that the majority of the solar bursts as observed in Cambridge were associated with the nearby passage of light aircraft![35]

---

[34] Sullivan (2009) provides a summary of the formation of this university research group starting in 1945, which was staffed initially mainly by ex TRE (Telecommunications Research Establishment) personnel returning to academia.

[35] This letter has been located in the National Archives in Australia and is shown in Appendix D (NAA: C3830, A1/1/1, Part 5).

Since Pawsey had been Ratcliffe's student, although he was only 6 years younger, it was natural that he was a guest of his former advisor when he visited the Cavendish Laboratory in May 1948. Pawsey spent quite some time discussing solar radio noise with his colleagues in Cambridge; these discussions led to numerous letters to and from staff at RPL. Since Pawsey had his foot in both camps, his role was decisive in guiding the nature of the interactions.

Just before Pawsey's departure from Sydney in late September 1947, there had already been an exchange of letters with Ryle in which Ryle expressed doubts about the frequency time dependence of the Type III bursts at the level of some seconds described in the *Nature* paper concerning the 8 March 1947 event (Chap. 7). Pawsey wrote in return on 3 July 1947:[36]

> With regard to the "seconds delay" cases, I am not entirely happy about the evidence. I believe that the tendency exists but do not know how often relative to zero delay, which is common [Type I], or to unco-ordinated [sic] effects.

Ryle elaborated on these doubts in his 60 page review of 1950 in *Reports on Progress in Physics*. In addition, Hey, in an impressive review of "Radio Astronomy"[37] in *Monthly Notices of the Royal Astronomical Society* in 1949, was ambivalent. He reported on the Hey et al. (1947) data of a delay of several minutes from the onset of a shortwave fadeout in the ionosphere (due to increased UV ionisation arising from a solar flare travelling at the speed of light impinging on the D layer of the ionosphere) and a subsequent large solar radio burst; the onset of the ionospheric absorption was inferred from a sudden decrease in galactic noise at a frequency of 25 MHz. But Hey also stated: "It may be added here that these workers (Hey et al. 1948) had not found any marked time differences between different radio wave-lengths."

In his review paper of 1950 (Appendix E), Pawsey (1950a) was again making a case for the "seconds delays", bolstered by Fig. 7.13 (Chap. 7). At the time of the presentation of this paper to the Radio Section of the Institute of Engineers in London in December 1949, Ryle made a number of criticisms. These questions and Pawsey's reply are included in the 1950 published version (with the 2 year delay of publication, some content of the review paper was quite out of date). Ryle wrote this text at the end of the main publication of the Pawsey review:

> Our observations at Cambridge have indicated that, whilst this is frequently observed [high frequency arrived first], quite a large number of outbursts (sic) occur in the opposite order . . . This, I think, suggests that the sequence is not fundamental to the process of excitation. (see also Appendix D)

Pawsey wrote in his reply that the recent work of McCready and Wild with the new radio-spectrum observations (with the new swept frequency instrument at Penrith) had given "evidence of a distinct spectral type associated with flares and

[36] NAA: C3830, A1/1/1, Part 1. The letter from Ryle to Pawsey has not been found.

[37] The title is noteworthy; by this date the new name (radio astronomy) for the discipline was becoming an acceptable term.

characterised by broad bands of excited frequencies which drift slowly towards lower frequencies [Type II bursts]". He did not mention the unequivocal evidence of the fast drift bursts described later on by Wild (1950b).

Within 3 weeks of arriving in the UK, Pawsey attended a meeting of the Royal Astronomical Society (RAS) in London. The meeting, with Appleton in the Chair, was held on Friday 23 April 1948, and lasted 2 hours; it was strikingly called "A Geophysical Discussion on Solar Radio Noise". (A summary of this fascinating discussion is published in *Observatory*, 1948, vol. 68, p. 178.) A number of the important players in the field were present: Appleton (Chair, Nobel Prize 1947), Hey, Ryle (Nobel Prize 1974), Hoyle, Chapman, Pawsey and Martyn, the latter two being from Australia.[38] (Stratton asked a major question which was answered by Hoyle.)

All seven presented short papers on two major topics, the quiet sun emission from the million degree corona and the burst emission from sunspot regions. On the latter topic, there was disagreement. Ryle argued that there was a connection between the hot corona and the radio bursts; he required a thermal origin with an effective temperature in the general corona of roughly 10^6K and about 10^{1}°K in the vicinity of sunspots. Hoyle also suggested a type of thermal emission. Martyn gave a nice comparison of the two problems: "This discussion has shown that there is agreement on the quiet Sun, so in this respect I propose to let sleeping dogs lie!"[39] Martyn was critical of the thermal emission explanations of the burst radiation; he was quite prescient as he suggested that plasma oscillations would be a more efficient radiation mechanism. It was only in the late 1950s that plasma oscillations achieved general acceptance as the cause of some types of solar bursts. But the details of the theories were complex and time consuming;[40] some years were to pass before a consensus was achieved.

---

[38] As Gillmor (1991) has pointed out, Martyn was recognised as the fourth most important founder of ionospheric physics in the twentieth century with the first three being Appleton, Chapman and Ratcliffe.

[39] Pawsey presented a summary of the research underway at RPL (with the collaboration of Yabsley) to characterise the thermal emission of the sun from cm to metre wavelengths. He made the significant point that the base emission at metre wavelengths was thermal: (1) the wave form of the emission was the same as fluctuation noise, (2) the emitting area was the entire sun, (3) the polarisation was random and (4) most importantly, the intensity as a function of frequency followed the Martyn prediction for thermal radiation. Pawsey also gave a longer review of both solar and cosmic noise (Bolton and Stanley's work on radio stars) in Session III, "Radio Noise", of a 2 day meeting (7–8 April 1948) on "Convention on Scientific Radio", sponsored by the Institution of Electrical Engineers in London. He ended his presentation: "Returning to Mr. Ryle's paper, I should like to say how much we admire the experimental techniques which he evolved. ... it is done in the best Cavendish tradition. The results which he described have been very largely repeated between us – in Australia and in Cambridge – by somewhat different methods, but I am happy to say that we agree on all conclusions." This latter point was to be contradicted by many events in the coming decade.

[40] See Sullivan (2009) for a more detailed description of the development of theories of solar bursts in the late 1940s and early 1950s.

Pawsey was anxious to ensure that the Hornsby experiment would provide a decisive result about the reality of the time frequency behaviour of both the Type II and III events. Two weeks after the RAS meeting he wrote by hand to McCready, "People [here in the UK] do not believe the high frequency precedes low."[41] A month later, he made a proposal to establish a collaboration with the Cambridge group and also to ensure the priority of the RPL claim to time delays. This complex plan was only partially successful as can be seen from his letter from Pawsey to Bowen of 24 June 1948 (footnote 31 this chapter).

> In my discussions at the Cavendish I found Smith [F. Graham-Smith, later Astronomer Royal, 1982–1990], one of Ratcliffe's men who works on solar noise, contemplating carrying out a series of observations on the form of bursts which is very similar to that done by Ruby. It is a clear case where the discouraging sort of duplication of work could occur if we do not get together before hand. I indicated the general lines of Ruby's program without giving much of the results and we agreed that I should write you with the following suggestions ... We should ask Ruby to prepare an outline of the paper she intends to write and send it to Ratcliffe for Smith's information in the near future, and when he has results to publish would of course expect to acknowledge any relevant work of Ruby's ... By this arrangement we gain freedom from fear of immediate prior publication by Smith and consequent undue haste in publication. His gain is obvious ... At the same time it emphasises that competition is likely and I should think strongly recommend Ruby to get on with the job promptly. I think it should take full precedence over any interference work [the 100 MHz interferometer project at Potts Hill] she may have on hand.

There is no record of the detailed response from RPL. Graham-Smith reported to Goss[42] in 2008 that he had had no idea of this surprising 1948 negotiation and had never heard of any of these discussions. Graham-Smith was:

> "Happily immersed in the early days of interferometry and position finding [his participation in solar research decreased in the course of 1948–1949]. I am surprised that Pawsey was more concerned with overlap in the research agenda ... This was totally unnecessary. If two groups were involved in solar burst research, this was jolly good! Pawsey seems to have thought that there was limited science in this field and that one group would mop it up ... with cut throat competition with the other group. This is ridiculous looking back from 60 years later on. Joe Pawsey was quite concerned ... about the Australians being upstaged ... I had always thought of Pawsey as an "internationalist" in world wide astronomy [later on Pawsey was a prominent leader in both in the International Astronomical Union (IAU) and in the International Union of Radio Science (URSI)]".

Already in 1948, there were growing signs of distrust between these two main groups working on solar radio noise.

A partial conclusion of this complex process was initiated later in 1948 following a letter from Ryle to Pawsey on 23 November 1948.[43] Apparently the contacts

---

[41] NAA: C4659, 8.

[42] Interview with Goss at the Jodrell Bank Observatory, 19 May 2008. Professor Sir Francis Graham-Smith was Director of the Royal Greenwich Observatory (1976–1981) and later Director of Jodrell Bank Observatory of the University of Manchester (1981–1988).

[43] NAA: C3830, A1/1/1, Part 3.

between RPL and the Cavendish had been minimal since Pawsey had left the Cavendish Laboratory about 4 months earlier. Ryle wrote:

> During your last visit to Cambridge we discussed our programme and among other things the question of our continuing work on solar bursts. You were writing to find out how your experiments on the correlation of bursts on different frequencies were going. We have done nothing further on the subject, but I think there is much more important work to be done and if it does not overlap with your programme I should like to start it up again with a new member of the team [Smith was then concentrating on the radio star position work]. There is no immediate hurry, as he will be fully occupied for the next 2 or 3 months, but I should like to get him thinking about his own line of experiment soon, and the analysis of solar bursts is the first choice at the moment.[44] Please remember us to your wife. We have often wondered if your family recognised you again![45]

A few weeks later, on 9 December 1948, Payne-Scott replied to Ryle, sending a copy of her draft paper on the Hornsby campaign (the paper was submitted a month later to the journal). She wrote:

> You will see that one of the main points is the distinction between variations (bursts) in the circularly polarised "enhanced radiation" and unpolarised bursts. It is the latter that show good correspondence on different frequencies and time-delays. Much of the past confusion originated because no distinction was drawn between the two kinds of short-period variation in solar noise.[46]

No record of a reply to her letter has been found in Australia or in Ryle's archives in Cambridge; even the letter from Payne-Scott which we have found in the National Archives of Australia was not found at Cambridge.[47]

Payne-Scott's letter provided the solution to the confusion about the frequency time behaviour that had plagued the discussions in the previous year. The seconds of time delays were *only* relevant for the Type III (unpolarised bursts), not the Type I bursts. The confusion would continue for a year or two; but the decisive observations of Wild and McCready (1950) put an end to the controversy. By this time,

[44]In pencil at the bottom is a note by Pawsey: "Reply when copy of Ruby's paper available."

[45]Pawsey and his wife Lenore had been away from the three young children (ages 11, 9 and 3 in 1948) for about a year. The two grandmothers (from Canada and Melbourne) had looked after Stuart, Margaret and Hastings in the Vaucluse (Sydney) home.

[46]NAA: C3830, A1/1/1, Part 3. Eight days later, on 17 December 1948, Pawsey wrote to Ratcliffe. (See above and footnote 30, this chapter). There was some confusion concerning who would answer whom! Pawsey wrote to Ratcliffe concerning the "investigation of bursts"; he had delayed writing until Payne-Scott's paper was available. The paper was to be ready in a few days and then he would send a copy to Ratcliffe. Probably Pawsey was not aware that Payne-Scott had previously sent a draft to Ryle. Pawsey also wrote to Ratcliffe that Payne-Scott had included some theoretical speculations concerning the origin of bursts in the new paper (the comparison with the Westfold theory).

[47]Sullivan has looked through the Ryle archives in Cambridge without finding this letter (circa 1980). See footnote 26, chapter 9 and Appendix D.

from 1949 to 1952, the Cavendish radio astronomers were beginning to move away from solar noise research; "radio star" research became the major emphasis as the group became more independent of Ratcliffe (Sullivan 2009).[48]

The Burgess Experiment: Payne-Scott at Hornsby

A final additional observation was carried out at the Hornsby site by Payne-Scott; she and Pawsey had hoped that the results might well be decisive in providing clues about the emission mechanism for Type I bursts. The origin of this test had been proposed by R.E. Burgess of the National Physical Laboratory (UK) (1917–1977) in early 1948 (see footnote 25, this chapter). The PC (Propagation Committee) minutes of 12 April 1948 read: "Burgess has suggested a study of the video spectrum of enhanced radiation to see whether it is the same as that of random noise. This will be tried when the sun is active."

Pawsey[49] had written a long report in May 1948 while he was in Europe; he was quite frustrated by the lack of any successful theory to explain the non-thermal radiation of solar bursts, presumably Type Is. He was apparently convinced that the statistics of the noise would provide vital clues as to the origin.[50] Various questions he posed were: (1) Was the waveform similar to random noise? (He even suggested how to determine the amplitude distribution.) (2) What was the waveform of the bursts? (3) Was there a standard form of the bursts? After visiting Ryle at Cambridge in April or early May 1948, he had been told that solar noise bursts simply consisted of single 0.5 s bursts of a standard shape (see footnote 53, this chapter). Thus a detailed study should show these "standard bursts". Pawsey concluded his frustrated letter: "[These] together would make a first class short paper: 'The waveform of solar noise'."

On 11 June 1948, Payne-Scott wrote to Pawsey[51] ("Dear Dr. Pawsey ...") that she was planning to do this test and that the main problem was the fluctuating nature of both the bursts and the enhanced level, as well as the rarity of these events. The observations were finally obtained on 5 and 6 August 1948 when a large sunspot

[48]Exceptions were the three Cambridge PhD theses on aperture synthesis observations of the sun by Stanier (at 500 MHz, 1950), Machin (at 81.5 MHz, 1951) and O'Brien (at 38, 81.5 and 210 MHz), determining the brightness profile of the quiet sun.

[49]Letter of 18 May 1948 to the radio astronomy group at RPL after his visit to the US, posted from the UK. NAA: C4659, 8.

[50]Others had also discussed the nature of the noise spectrum of solar bursts. The most imaginative and poetic description may well have been produced by Reber (1946), who described the bursts at 480 MHz as "like wind whistling through the trees when no leaves are on the limbs". He used the term "swishes" to describe the effect of radiation in excess of the slower variable background solar emission. In 1948, Reber (1948) presented a more quantitative definition of "bursts" (few minutes to hours) and "swishes" (a second or less). "Sometimes overlapping swishes would produce grinding noises".

[51]NAA: C3830, F1/4/PAW/1, Part 2.

was present. The Hornsby Yagis were used at 85 and 60 MHz. Type I bursts were observed (enhanced radiation). The publication in the *Australian Journal of Scientific Research* immediately followed the detailed paper describing the 9 month campaign of observations of solar bursts (see above). The paper was entitled: "The Noise-like Character of Solar Radiation at Metre Wavelengths," Payne-Scott (1949b).[52]

She reported that R.E. Burgess had suggested a comparison of the characteristics of solar noise with a noise diode in the receiver. A number of the early solar researchers had already noted the character of solar noise; the suggestion had even been made that the burst radiation might have the character of terrestrial lightning. Payne-Scott carried out a quantitative assessment of the nature of the noise statistics (an early version of a power spectrum, based on the distribution of amplitudes), probably for the first time for solar radio emission. The paper compared the enhanced radiation at 85 and 60 MHz with receiver noise signals via a comparison with a noise diode and with a single frequency signal (CW – continuous wave). This Type I noise storm was quite intense; the solar noise was 14 times more intense compared to the receiver plus sky noise. A quantitative comparison of the noise statistics was done using the theory of Rice (1945). The paper concluded:

> The distribution of amplitudes with this mixture of solar radiation and receiver noise coincides with those for receiver noise alone ... The results show that to within the limitations of this experimental technique there is no observable difference between solar noise and thermal noise over the small frequency range accepted by the receiver and that solar radiation certainly does not reach us as a series of discrete frequencies separated at intervals of more than, say, 2 Mc/s.[53]

The acknowledgements stated: "The suggestion that the distribution of solar noise could be investigated by using a biased detector was made to us by R.E. Burgess of the National Physical Laboratory, England." In the end this observational test was much less decisive than Pawsey and Payne-Scott had anticipated; no major clues as to the origin of the Type I bursts was in fact provided by the determination of the statistics of the solar noise. The data did prove, however, that the solar bursts did not have the characteristics of terrestrial lightning.[54]

---

[52]The paper was initially entitled "A Comparison Between Solar Noise and Thermal Noise" (RPP 65). The paper was submitted on 22 November 1948, although the RPL records indicate that the first draft had been typed 2 months earlier, on 8 September 1948.

[53]Possibly a reference to the proposal that Ryle made to Pawsey that most solar bursts consisted of a "0.5 s standard burst component".

[54]In early 1950 (NAA: C3830, A1/1/, 1 Part 5) there was additional correspondence between Burgess and Pawsey. After praising Pawsey for the "radio-astronomy" work of his group, he asked again about a point that they had discussed in Washington in 1948. Burgess was concerned about the imperfect reflection off the sea and the effect of the finite bandwidth on the determination of the angular size of sources using the sea-cliff interferometer. Pawsey replied to the letter of 24 January on 21 February 1950. He gave a detailed response with numerous equations, with obvious input from Steve Smerd and Payne-Scott. He also included the appendices to the Stanley and Slee publication of 1950 (see Appendix C, footnote 15). The nature of the approximate solutions was a tricky issue that could lead to large errors. A few months later, John Bolton (NAA: C3830, F1/4/

In summary, in spite of the controversy with Bolton, the 9 month period was very productive for Payne-Scott and provided a solid basis for the future of solar noise research at RPL. She was then poised to begin the final stage of her scientific career at Potts Hill with the 100 MHz interferometer. For the first time RPL scientists could observe the solar bursts and outbursts at any time of the day, not just at sunrise. Payne-Scott had clearly shown that Type III bursts (her unpolarised bursts) showed rapid changes in frequency drift, from high to low frequencies. Thus a fast exciter speed for Type III bursts was clearly required; within a few years this was the accepted explanation. Pawsey's 13 month absence had a major impact on the evolution of Payne-Scott's career at RPL.

BOL/1, letter from Bolton to Bowen) met Burgess at Slough at the National Physical Laboratory (of the UK); Burgess gave Bolton advice about "wide bandwidth systems and ground reflections". In addition, Bolton and Burgess discussed the issue of imperfect reflections off the sea and the finite bandwidth problem. Bolton wrote that the treatment of these problems in the publication by Stanley and Slee (1950) was in fact carried out by Westfold. Bolton asserted that "the estimate of errors [by Pawsey] was on the right side so I think the problem is only of academic interest".

Chapter 9
Payne-Scott at Potts Hill, 1949–1951: Movies of the Outward Motions of Solar Outbursts with the Swept-Lobe Interferometer

Payne-Scott's Career in Turmoil 1948; Choice of the Swept-Lobe Interferometer Site

The next major stage in Payne-Scott's career had been planned in September 1947 before Pawsey left for his year-long trip to the US and Europe in late September.[1] This involved a new instrument, constructed at RPL, and a new site, Potts Hill. The observations made there had a lasting impact on the progress of solar physics in the latter half of the twentieth century. The new instrument was the first "vertical interferometer"[2] (a Michelson) at RPL; it was also the first swept-lobe interferometer in radio astronomy.

The solar radio astronomers in Sydney had recognised the awkward nature of the sea-cliff interferometer: observations were only possible at dawn and the large errors produced by uncertain refraction corrections to the phase of the interferometer fringes made for substantial uncertainty. Also the use of eclipses (Hey 1955) of the sun by the moon to study small scale solar radio structures had proved to be an awkward technique (an example was the eclipse of 1 November 1948 observed in eastern Australia); the times of observation were infrequent and the results were often ambiguous, leading to uncertain interpretations. Thus the ability to make high resolution observations at most times during daytime represented a major step forward. As Wild (1968) has pointed out: "... another Pawsey-inspired experiment

[1]The PC meeting of 23 September 1947 (2 days before Pawsey's departure for San Francisco) contained a thorough discussion of a new solar interferometer for a new field station at Bankstown. R.F. Treharne and Alec Little were in charge of this 100 MHz instrument. The site location was moved to Potts Hill Reservoir in April 1948, at which time Payne-Scott was in charge of the project. Already on 6 June 1947 the PC minutes contained a brief reference to an "improved interferometer," to be designed by R. Treharne. No location was specified and the instrument was to be "capable of yielding interference patterns in a fraction of a second with a view to extending this technique to 'bursts'."

[2]See footnote 44 Chap. 7. As Barry Clark has suggested (private communication, April 2009), we might think that a "sea cliff interferometer" would have been called the "vertical interferometer"!

W.M. Goss and R. McGee, *Under the Radar*, Astrophysics and Space Science Library 363,
DOI 10.1007/978-3-642-03141-0_9,

was put in to operation and brilliantly performed by Payne-Scott and Little. The idea was to locate . . . the instantaneous position of the dominant source on the sun at any one time."

An ingenious feature of this new interferometer was the rapid "imaging"[3] capability of the instrument in making radio astronomical positional determinations on the sun at the rate of 25 times a second. In addition the polarisation state of the incoming radiation could be determined at the time scale of a second; this feature of the swept-lobe interferometer was probably also a first in radio astronomy. The project was undoubtedly inspired by Pawsey with some preliminary design by Ross Treharne.[4] A few months later, in September 1947, Alec G. Little[5] (1925–1985) joined the project.

Based on seven letters from Lindsay McCready and Don Yabsley to Pawsey[6] during the latter's trip in 1947–1948, additional facts about the planning of the swept-lobe interferometer, the choice of sites and the conflicts associated with Payne-Scott's participation can be established. McCready was optimistic about Treharne's design based on the letters of 7 October 1947 (just after Pawsey's departure for the US) and of 18 November 1947. However, by January 1948 (no day mentioned on the letter; footnote 6, this chapter), he was frustrated with progress due to Treharne's "absence" (no details given) and the fact that Little was on holiday. The initial tests of the system, which had shown possible interferometer fringes, were not valid; gain fluctuations due to variations in power supply voltage were the source of the "fringes" from the interferometer.[7]

Apparently the first site chosen was Bankstown (a nearby suburb a few km to the south of the final site at Potts Hill), but a number of problems developed. Possibly on 24 March 1948, McCready wrote to Pawsey (footnote 6 this chapter) (his date of 24 March 1947 cannot be correct). There is a scrawled note at the top of the letter:

---

[3]The "imaging" was indeed crude; the technique only involved rapid scans in a one dimensional manner of the fringes across the sun. With a single major solar event, it was then possible to determine this one dimensional position at the rate of 25 times per second. With modern high resolution radio solar telescopes, a true two dimensional image of the sun can be determined at a rate of many times per second.

[4]NAA: B2/2, Part 2. This new instrument (to become the Potts Hill interferometer) had been discussed initially at the 6 June 1947 meeting of the Solar Noise Group. R. Treharne had been given the job to design "...[100 MHz] equipment capable of yielding interference patterns in a fraction of a second with a view to extending this to bursts. Initial ideas are to use manual phase variation to aerials connected by a transmission line". At this time no specific site for the instrument was suggested.

[5]Freeman (1991) and Sullivan (2009) have provided short summaries of the remarkable career of Little, who became an Associate Professor of Physics at the University of Sydney; he played a major role in the development of the MOST (Molonglo Observatory Synthesis Telescope) in the 1980s, before his untimely death in 1985.

[6]NAA: C4659, 8.

[7]The PC Minutes of 14 November 1947 also reported on the initial successful tests (and the completion of the motor driven local oscillator phase shifter), followed by the minutes of 20 February 1948 reporting on problems due to the mains voltage stability (the commercial power at 240 V AC).

"Bad news: Robbery and vandalism at Bankstown. Have to find another site now for 100 Mc/s interferometer." (See footnote 51 Chap. 7 for severe vandalism problems at the Dover Heights field station of RPL.) In an earlier Solar Noise Group meeting of 16 October 1947[8] (the group was called "Dr. Pawsey's group" even he was absent), Treharne and Little had reported that the interferometer was almost ready for tests, but that the site problems were not resolved. The building to house the apparatus at Bankstown had been "Sold". With no explanation, Burgmann and Eagles (an administrator) were trying to find a replacement!

However, by 23 April 1948 (see footnote 6 this chapter), McCready was quite pleased to announce to Pawsey that a new permanent site had been found. This location was the Potts Hill Reservoir site, controlled by the Sydney Water Board.[9] Given the problems with security at the previous Bankstown site, the fact that this location was "fenced and guarded" was a major advantage. Also the contact person for RPL was a water board engineer, who was "most anxious to help us". He had taken a tour of the RPL; he knew "all about solar noise"! McCready pointed out that the location of the new site (between Lidcombe and Bankstown) could be reached by car from the RPL in the University of Sydney grounds in only 5 min extra time than it took to reach the field station (used earlier) at Georges Heights, north of the main Sydney Harbour.[10] In a later letter (footnote 6 this chapter) (6 June 1946), McCready was quite pleased with the continued level of cooperation from the Sydney Water Board, as the negotiations continued for the use of the new site. He suggested that a prototype of the interferometer should be constructed at the site; numerous engineering details would still need to be completed.

By mid-April 1948, Ruby Payne-Scott was associated with the project (see Chap. 8 for her communications with Pawsey in May 1948 about her activities). McCready wrote to Pawsey on 23 April 1948 that Payne-Scott was "tapering off at

[8]Pawsey had left Sydney for San Francisco about 3 weeks earlier.

[9]NAA: C4659, 8. The history of the site selection and some aspects of the scientific research programs at the Potts Hill site were described by Frank L. Kerr (1918–2000) in the January 1953 issue of the *Sydney Water Board Journal* (Sullivan archive). Although by this date the swept-lobe interferometer was no longer in use, the determination of solar noise positions from active regions was described in detail. The article began: "In 1948, the Chief of the Division of Radiophysics at C.S.I.R.O., Dr. E.G. Bowen, sought and the Board was pleased to grant, permission for the Division to install certain equipment on the Board's land at Potts Hill for the investigation of solar radio activity ... The Board has been very happy to have been able to provide the Division with the space and other activities needed." An important factor in the success of RPL in the utilisation of the site was the role played by H.A. Stowe. He was the Chief of Electrical Engineering of the Water Board and a keen radio "ham". In 1953, Kerr described the new 36 ft antenna being constructed for HI use (Wendt 2009).

[10]The "Solar Noise Research Report" (at a meeting of "Dr. Pawsey's group") on 30 April 1948 reported: "A search for a site to replace Bankstown has resulted in the choice of Pott's [sic] Hill reservoir (which should also provide a suitable stretch of water for K-band [must have meant L band or 20 cm; this was to be the solar grating array of Christiansen] interferometry; negotiations with the [Sydney] Water Board have begun. Meanwhile equipment is being made to replace that stolen, and some obvious faults in the system are being remedied." The acquisition of the new site was trying for McCready.

Hornsby and orienting to [Treharne's project]".[11] As she wound down her activities at the Hornsby Valley field station, the PC minutes indicated that she was taking over the scientific leadership of the project as well as the development of the ingenious, simple calibration scheme. Already at the meeting of 30 April, only Payne-Scott and Little were mentioned; at the next meeting on 22 July 1948 the announcement was made that Treharne had left the project.[12] Subsequently, throughout the period from mid-1948 to the time of her resignation in July 1951, the reports on this project were given by Payne-Scott. Active observations were carried out between May 1949 and August 1950.

The Potts Hill site was certainly more convenient for Payne-Scott. She and her husband, Bill Hall (Chap. 4), were thinking of building a house in the nearby suburb of Oatley (about 14 km from Potts Hill); this house was started in 1950–1951, with major construction done by Payne-Scott and her husband (Chap. 12). Bill and Ruby Hall (she started to use her married name when she retired in 1951) moved into the new house in August 1951, shortly after she resigned from RPL.

The layout of the Potts Hill field station (about 16 km from the centre of Sydney, near the suburbs of Regents Park, Chullora and Birrong,) is shown in Fig. 9.1, taken from Wendt (2009), based on the site layout in the early 1950s. In Fig. 9.2, an RPL map from the planning stages is shown.. The sketch, dated June 1948, gives the placement of aerials 3 and 4, which were never constructed. An aerial view from the 1950s is shown in Fig. 9.3. The site was closed in 1962, when other RPL sites and the Parkes radio telescope became prominent (the 64 m telescope opened on 31 October 1961); Wendt (2009) has provided a thorough history of the site in the period 1948–1962. On 16 August 1948, McCready wrote to Pawsey (footnote 6 this chapter) extolling the qualities of the Potts Hill location: "Potts Hill field station is [a] beautiful sight with all the wattle trees in full bloom." (It was early spring in Australia.) On this same date, Don Yabsley wrote to Pawsey (footnote 6 this chapter) with an optimistic assessment of the status of the new instrument. First lobes (or first fringes) had been obtained by Payne-Scott and Little; the group was beginning to worry about methods of improving the calibration of the instrument.[13]

[11] Also Alec Little was finishing off the X-ray experiments at RPL; this project then moved to Melbourne.

[12] Treharne resigned from RPL in the middle of 1948 to join Philips Electrical Industries of Australia. In a letter from Bowen to Pawsey of 5 May 1948 (containing a general report on activities of the laboratory since the previous year, during Pawsey's trip), he wrote: "[Treharne] unfortunately now has other interests and the work has not been going at all well recently. Ruby is there taking it over and, and due to difficulties at Bankstown, is acquiring a new site for the work at Potts Hill." In the same report, he is quite positive about the significance of Payne-Scott's determination of the time of arrival of bursts at Hornsby (Chap. 8). NAA: C3830, F1/4/PAW/1, Part 2.

[13] Also on this date, Yabsley reported that Bolton had just returned from the radio astronomy expedition to New Zealand. Bolton and Stanley were getting ready at this time to join John Murray and Don Yabsley in Tasmania for the 1 November 1948 solar eclipse (John Murray, letter, April 2009).

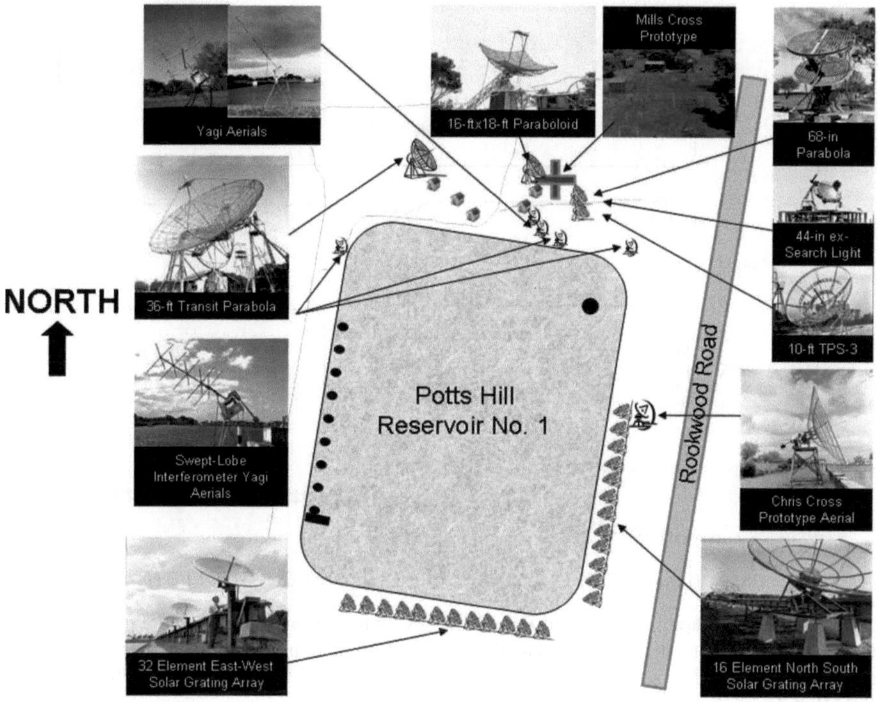

Fig. 9.1 Detailed site layout of the Potts Hill Reservoir RPL field station in the early 1950s. The instruments used by Payne-Scott and Little were the swept-lobe interferometer and the solar radio noise Yagis. The East West solar grating array used by Christiansen et al. was completed in about November 1951 and the North South array in mid-1953; the 36 ft transit telescope was completed in January 1953. The East West extent of Reservoir Number 1 is about 280 m. The insert photos are from the ATNF historical photographic archive. Figure provided by, permission from Harry Wendt

In the middle of 1948, Payne-Scott's concerns about her future plans continued. John Bolton was not a factor, as had been the case earlier in 1948 (Chap. 8); in fact he was away in New Zealand for some months in 1948. At this time Payne-Scott was highly motivated to start the 100 MHz interferometer project *before* the Hornsby research was finished; due to the uncertainties of this project, she thought there was a possibility that the new exciting Potts Hill work would begin without her being able to work full-time on the new interferometer (see Chap. 8). On 6 June 1948, McCready wrote a troubled letter to Pawsey (footnote 6 this chapter) about the Potts Hill project. On the one hand he was pleased with the negotiations with the Sydney Water Board: "the equipment losses thru [sic] vandalism not yet overcome but pretty close to it." On the other, he was worried as to whether Payne-Scott had the time to handle all the engineering planning; she would possibly need "an army of assistants". He thought the logical choice would be Don Yabsley since Chris Christiansen would have the automatic multi-frequency recording apparatus working at 200, 600 and 1,200 MHz at Georges Heights, thus freeing up Yabsley. And furthermore, Yabsley and Payne-Scott worked well together. But Yabsley was

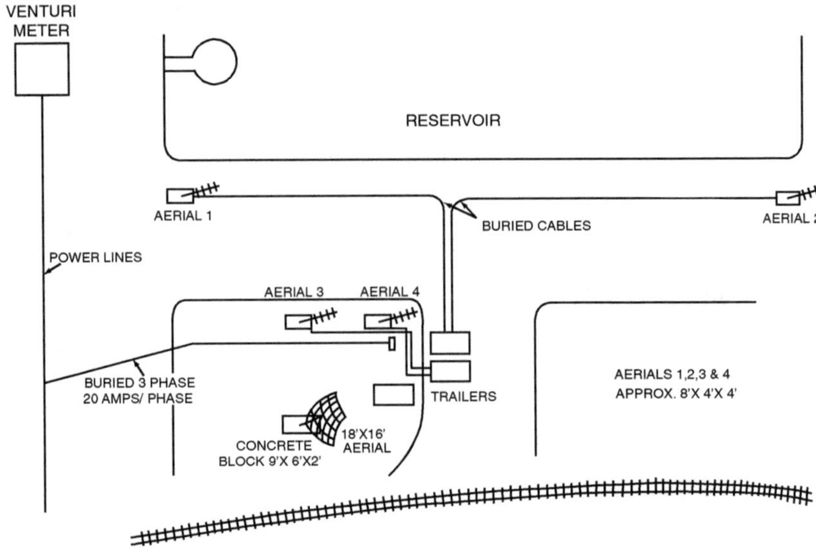

Fig. 9.2 A sketch map of the planned Potts Hill RPL field station. South is to the top of the map, East to the left. Note that in this plan there were four antennas for the planned swept-lobe interferometer at 97 MHz. In the end only three were built (Fig. 9.8). From June 1948 (NAA: A/1/1/1, Part 3)

quite worried about the "terrific amount of work" already on hand. McCready wrote that Payne-Scott would not react well to this suggestion of joining forces with Yabsley. This created a difficult situation for McCready:

> I have not discussed it with Ruby yet but I know her reaction will be violent hostility and complete assurances that Alec [Little] can do the work. I am not expecting you, 12,000 miles away, to make decisions or do anything. I will be my best and of course watch relations and the group's welfare . . . you suggest in a future letter that there is a better chance of Ruby getting one or two important papers out quickly by concentrating entirely on the loose ends of her present work [at Hornsby] and she should seriously consider letting someone else take on the interferometry at 100 Mc/s. I would prefer of course that you did not mention that I have written this.[14]

The group in Sydney was clearly floundering because of the lack of Pawsey's leadership. Within a week a welcome solution to the crisis arrived with Pawsey's letter of 29 May 1948 to Payne-Scott (footnote 6 this chapter, also described in Chap. 8). Both McCready and Pawsey saw that the Hornsby project could be

[14] In the 6 June 1948 letter, McCready also seemed to blame himself for the conflicts which he had been unable to resolve. The excuses for the imbroglio were (1) Payne-Scott was herself very keen to start the new Potts Hill project and (2) the work at Hornsby was going nowhere. Even Bowen had suggested that she "fold it up and get into something more profitable". But as we have seen, the pessimism about the Hornsby endeavours was not justified.

Fig. 9.3 Aerial photo from 19 March 1954 taken from the North looking South. The main portion of the RPL field station is shown in the foreground of reservoir no. 1. The 36 ft transit telescope is visible as well as the rail line in the foreground. The rail line was used to bring coal into the nearby power plant; the smoke stacks are visible in Fig. 9.5. ATNF historical photographic archive: B3253-1

completed on time, allowing for an active participation of Payne-Scott in the planning of the new Potts Hill interferometer project.

Thus in the end Payne-Scott was able to finish the Hornsby research and wrote the paper describing the properties of solar bursts (Chap. 8) before the end of 1948. More importantly, she did manage to maintain her control of the scientific leadership of the 100 MHz interferometer project, especially after Treharne left RPL. Alec Little and and Ruby Payne-Scott became an efficient team.

Engineering and Scientific Planning for the Potts Hill Swept-Lobe Interferometer

The detailed scientific motivation for the 100 MHz swept-lobe interferometer was presented at the 23 September 1947 PC meeting, just prior to Pawsey's departure for the trip to the US and Europe. Under the rubric "Solar Interferometer Research

Programme-Bankstown Treharne and Little", a detailed rationale was presented: (1) a determination of the sizes of the "bursts" (Type I storm bursts) as compared to the "enhanced radiation" (Type I background continuum). At that time, the expectation was that the bursts were smaller than the background during "noise storms", (2) the determination of the positions of the "bursts" and "enhanced radiation" sources in comparison with the positions of sunspots, (3) the determination of the correlation of optical flares with "outbursts" (probably Type II and Type IV, see below), (4) a study of the thermal emission of the quiet sun[15] in both total intensity and polarised intensity, (5) the determination of the properties of the source in Cygnus and other "circumpolar" cosmic sources, and (6) the extension of this technique with additional frequencies 65, 200, 1,200 and 3,000 MHz.[16]

The techniques to operate the swept-lobe interferometer were outlined at this meeting as was the new method of varying the phase of the interferometer by changing the phase of the local oscillator for the heterodyne system. An improved method of recording the data using a movie camera (cine) was also suggested. The system had no variable delay lines and thus the bandwidth was restricted to 150 KHz. For slower time scale observations (e.g., enhanced solar radiation or cosmic sources), a drift scan mode was constructed using the natural fringe rate due to the rotation of the earth. The initial system consisted of two antennas on an east–west line; thus only the right ascension of the sources could be obtained, since observations were carried out near transit. Possible extensions to additional antennas were discussed in order to obtain two dimensional positions or even distributions for extended sources (i.e., aperture synthesis).[17] Pawsey was quite impressed with the status of the project in late September 1947 (2 days before he left Australia); he wrote in the PC minutes: "This plan looks further ahead than the others.[18] Many aspects will not be touched for some time [this was indeed true], perhaps never."

Beginning with the period of Pawsey's visit to Cambridge in June 1948, a dialogue began between the Cavendish Laboratory and RPL concerning detailed position determinations (and errors) using Michelson interferometers (the Cambridge group described their radio telescope as a "spaced aerial interferometer"). Probably the initial motivation was the large 2° discrepancy in the Cygnus A

[15] With the final instrument, this observation was difficult due to limited sensitivity. The quiet sun has a total flux density of about 2×10^4 Jy at 100 MHz; the flux density sensitivity of the interferometer was a few thousand Jy for observations of a few hours and only a few hundred thousand Jy for the subsecond observations. In addition, the fringe size of about 40 arc min was comparable to the size of the radio quiet sun of 40–50 arc min.

[16] This was never carried out; the project was partially fulfilled by Wild, Sheridan and Trent (1959) and Wild, Sheridan and Nylan (1959) at the Dapto field station for the frequency range of 40–70 MHz (Type I, II and III bursts).

[17] This sophisticated use of the instrument was never fulfilled in practice.

[18] Other groups were: Bolton, Stanley, Slee (Dover Heights Program No. 1); Payne-Scott and Marie Clark (Dover Heights Program No. 2); Lehany, Yabsley and Harragon (Georges Heights at 200–3,000 MHz); McCready (spectrum analysis of solar noise with no mention of Wild!). Smerd and Westfold (theory) also presented long range plans.

positions from Cambridge and RPL. On 22 June 1948 Ryle wrote to Pawsey[19] with a copy of a four page report by F.G. Smith, "The Determination of the Positions of Discrete Galactic Sources". This report outlined the theory of the determination of the right ascension and declination of sources along with a succinct description of sources of error for the spaced interferometer case. Later Smith published a more detailed account in *Monthly Notices of the Royal Astronomical Society*, 1952; in this publication he discussed the position determinations with both spaced and sea-cliff interferometers. An example of a major source of error was the effect of misalignment of the aerials from a perfect east–west orientation. The report concluded that the accuracy of the Cambridge position of Cygnus A was about 15 arc min in declination and about 20 sec of time (about 4 arc min) in right ascension. Pawsey sent the report to Sydney; the document had a wide distribution among the RPL staff. Pawsey also had a few ideas about the Cygnus A discrepancy; in 1948, there was little resolution of the problem.[20]

The exchange of letters and the report from Smith of June 1948 set the stage for a follow-up discussion that ensued a year later. Clearly, the Cavendish report had had an impact on the RPL group, who were learning the details of spaced interferometry for the first time. By mid-1949, the Potts Hill interferometer was coming on line. Payne-Scott and Mills worked on the theory of position determinations and had carried out a detailed analysis of the possible errors. On 28 June 1949, Pawsey wrote to Ratcliffe:[21]

> Some of our people have been working over the past few months using Ryle's spaced aerial interferometer technique to observe both sun and discrete cosmic sources. They have been impressed with the rather tricky nature of the possible sources of error in the case of attempts at precise position finding ... might be a good thing to maintain rather close contact with Ryle and his group with the idea that each might be able to help the other. Ruby Payne-Scott and B.Y. Mills have prepared the enclosed notes which you might pass on to Ryle with a request from us for his collaboration. Incidentally the "dawn" [sea-cliff] technique is just as bad or worse but the errors are different.

[19]NAA: C4659, 8. As was often the case with letters from Ryle, he apologised to Pawsey for the late reply. The end of term at Cambridge and especially "four parties of visitors" were the reasons for the lateness. On the same day (NAA: C3830, A1/1/1, Part 3), Ryle wrote Bolton (the report by Smith was enclosed) with a summary of the discussions that had been held with Pawsey about the large discrepancies. A repeat of the observation was suggested. Motion of the source was not precluded: "... if you find it has moved – then the astrophysicists must think again!" A flurry of letters resulted to and from Bowen and Pawsey.

[20]Sullivan (2009) has a fascinating description of the time evolution of the determination of the Cygnus A position. Baade and Minkowski (1954) identified the object with a high redshift galaxy, based on the accurate position obtained by Smith (1951) with the Cambridge instrument. The accuracies of the radio coordinates determined by Smith were about 0.2 arc min in right ascension and 1 arc min in declination.

[21]NAA; C3830, A1/1/1, Part 4.

The "Notes on Interferometer Errors" contained an extensive discussion of errors due to both atmospheric and ionospheric refraction,[22] as well as alignment and timing errors and the effects of changes in cable lengths due to temperature fluctuations. Mills provided a preliminary position for Cygnus A obtained at Potts Hill; observations of this northern source were quite difficult from Australia since the source only reached an elevation of 15° above the northern horizon. Both Ratcliffe and Ryle answered,[23] the latter in detail. Ryle seemed to disagree with a number of the points in the Payne-Scott–Mills report. In particular he was dubious about the severity of the cable length stability problem described by the RPL scientists, suggesting that the Australians had misinterpreted their data: "I suppose you are quite sure that the stability of your signal generator is sufficiently good not to be responsible?" Also, Ryle suggested two additional sources of error that had not been considered in the RPL report. These were related to Ryle's questioning of the method of determining the aerial impedances as well as the effective frequency of the system.[24]

As construction began at Potts Hill, there were again delays before commercial power could be installed at the site of the new observatory. Finding a suitable movie camera for the real time recording was also a major problem. By the time of the 22 July 1948 PC meeting, Payne-Scott was clearly in charge of the new instrument. She again presented a summary of the purpose of the project and began to provide regular progress reports at the PC meetings. She urged other groups to consider the use of the Potts Hill field station for other projects; in the end a number of additional groups did move to Potts Hill (e.g., the Christiansen grating array, the prototype Mills Cross, the previously used Georges Heights antenna, numerous small high frequency antennas, and the new 36 ft transit antenna constructed in 1952–1953).[25] At the PC meeting on 6 September 1948, Payne-Scott gave a progress report on the initial tests of the new

---

[22]In this report, there seems to be no appreciation of the fact that atmospheric refraction can be neglected (to first order) for a spaced interferometer, based on the assumption of a plane parallel atmosphere. The almost equality of the phase change in both arms of the interferometer leads to a cancellation of the effect. As far as we can ascertain, the first published recognitions of this important fact in Australia were made first by Little and Payne-Scott (1951)" – "A uniform plane sheet of refractive material does not introduce any differential path difference between parallel rays, and hence the refraction due to a uniformly stratified region above a plane Earth can be ignored . . .", and then by Mills (1952) – "It can be shown that for an interferometer with a plane atmosphere the refraction correction is automatically applied if the free-space velocity of radio waves is used in the declination calculation".

[23]NAA: C3830, A1/1/,1 Part 4.

[24]Ryle concluded his letter to Pawsey by stating that a joint Jodrell Bank-Cambridge experiment had shown that radio star fluctuations were mainly due to a cause induced "by some relatively local effect". Thus the ionosphere was the likely culprit, not the intrinsic behaviour of the source. Ryle pointed out that their conclusion agreed with the results of the "spaced receiver" data obtained in 1948 from New Zealand to Australia. The Cambridge–Jodrell results (with no mention of the RPL results) appeared in back to back papers in *Nature* (1950) by Smith (Cambridge) and Little and Lovell (Jodrell Bank).

[25]Wendt (2009) has provided a detailed description of the various projects carried out at Potts Hill in the years 1948–1962.

interferometer.[26] As Pawsey returned at the end of October from his long trip overseas, three meetings of the PC were held in November. He was catching-up on the status of the various groups at RPL; in addition he gave several talks about his impressions of overseas radio astronomy groups. After February 1949, the meetings were again scheduled at monthly intervals. Pawsey brought a new sense of urgency to the activities of the radio astronomy group at RPL after his extended absence.

Early Testing of the 97 MHz Interferometer

After Pawsey's return, an increasing energy could be detected in the activities of the cosmic noise group, soon to be the radio astronomy group.[27] On 8 February 1949 Payne-Scott reported that 97 MHz (changed from 100 MHz to avoid external interference) interferometry was ready to start at Potts Hill; and the Yagis for continuous monitoring had been moved from Hornsby (62 MHz) and Dover Heights (97 MHz). By the time of the next meeting of the PC committee on 7 March 1949, Payne-Scott had exciting news; continuous recording with the two Yagis was in operation and some successful observations had been made with the interferometer. "On 1 day, while unpolarised radiation [Type III] was being received all day, an isolated burst [maybe a Type I?] appeared to come from a quite different place on the sun."[28] In Figs. 9.4 and 9.5 two photos of the swept-lobe interferometer are shown. The former shows the details of the western-most Yagi with vertical and horizontal polarisation; the latter shows the layout of the three elements next to the northern boundary of the reservoir. In addition there was a total power Yagi operating

---

[26] In her letter to Ryle on 9 December 1948, Payne-Scott described the end of the Hornsby campaign (Chap. 8) and her plans for the new interferometer. "I have now transferred my interests to the type of interferometer that you use, but provided with a phase-changing system so that we can sweep through the complete lobe in 1/25 s instead of relying on the motion of the source, and hence providing a means of measuring the position on the sun of short-period variations of either type [we assume she meant Type I and possibly Type III bursts]. I should like to hear your views on the stability of the calibration of this type of interferometer ..." She then described the new Wild and McCready swept-frequency spectrograph (called a "spectrum analyzer") to scan the solar spectrum from 50 to 100 MHz. "It should solve many problems that spot-frequency observation leaves open, particularly that of the distribution of intensity with frequency in a burst."

[27] The change of name has been described in Chap. 8. As discussed by Sullivan (2009), the first use of the term "Radio Astronomy" was by Pawsey in a letter in January 1948 and by Ryle in April 1948 at a meeting of the Royal Astronomical Society (he did, however, use the term inside inverted commas). Ryle wrote: "More refined observations in this new 'radio astronomy' should provide us with much new information on such processes." The new term caught on quickly.

[28] Also at this meeting, Wild gave his first report on the Penrith swept frequency instrument, after only three days of data collection! He attempted to compare the results of different types of bursts with this new instrument (70–120 MHz) with the fixed frequency instruments previously used by Payne-Scott. At this time, the clear spectral distinction of the various types of solar bursts was not yet evident.

Fig. 9.4 The Western most antenna of the three elements of the swept-lobe interferometer; this 97 MHz radio telescope determined the motions of solar bursts in a one dimensional scan many times per second. Used by Payne-Scott and Little in 1949–1950. With the crossed dipoles circular polarisation could be determined. From 28 July 1950. ATNF historical photographic archive: 2217

at 98 MHz, used to calibrate the data from the swept-lobe interferometer at 97 MHz. The polarisation state of this single Yagi was also switched at a rate of two times per minute to check the polarisation state of the solar radiation at radio frequencies.

Figure 9.6a shows one of the single total power Yagis at 62 MHz, located close to central hut; this antenna was used to determine the total emission from the sun, presumably while the interferometer was in use. A record from this instrument is shown in Fig. 9.6b, an example used by Wild in 1985 to illustrate the nature of fixed frequency solar noise observations in about 1950.

By the time of the 11 April 1949 meeting of the Radio Astronomy Committee, Payne-Scott was pleased with preliminary observations. Two construction projects were underway, adding both circular polarisations as well as a third aerial; the latter was required to sort out lobe ambiguities (to identify the correct lobe – without this feature the location of the solar radio emission source could arise from at least two possible positions). Mills was also busy with initial observations on three of the cosmic radio sources looking for polarisation with the 97 MHz interferometer; none showed any detectable signals. He already realised that the ability of the Potts Hill interferometer to detect weaker radio objects was not suitable for systematic observations of cosmic radio sources, due to the limited sensitivity.[29] A few weeks later, on 28 April 1949, Payne-Scott was complaining about the lack of solar activity even

[29] Already at the 8 February 1949 meeting he had expressed his doubts about the sensitivity: "It appears that a larger aerial is desirable" to detect more than a few cosmic radio sources. The swept-lobe interferometer had, after all, been planned to detect large solar bursts and outburst.

Fig. 9.5 Photo of the swept-lobe interferometer site on 12 November 1950. The Western antenna is clearly visible; the signals from the three antennas were combined in the "Hut." ATNF historical photographic archive: 2312

though the continuous recording (two single Yagis) continued. She was working on the scheme to achieve accurate phase and amplitude calibration for the system; by 16 June 1949 a calibration scheme was almost in place. The first main results were then available; a Type I noise storm (continuum plus noise bursts) was located 300,000 km above the visible photosphere of the sun.[30]

At the 15 August 1949 Radio Astronomy meeting, Payne-Scott announced that the third antenna was almost ready for use and that the recording mechanism using the film camera for strong bursts was in operation. The next report on the Potts Hill interferometer was given at the 17 November 1949 meeting; the complete system was working with three antennas as well as crossed polarised feeds to determine the complete state of polarisation. Payne-Scott reported that both Type I and Type III bursts were being observed.[31] After reporting a large outburst at the 7 February 1950 meeting (surprisingly with no prominent sunspots; perhaps the relevant active region was over the limb of the sun), a detailed report of a

[30]At this meeting, Mills and Thomas announced the results of their first Cygnus A position determinations. Discrepancies of about one degree were found with both Bolton's and Ryle's previously determined positions. At the next meeting on 7 July 1949, Mills announced that he was giving up on the Potts Hill interferometer for cosmic source positional determinations due to the limited sensitivity. The total flux density of Cygnus A at 100 MHZ was about 15,000 Jy with noise about 3,000 Jy. In 1951, he and Thomas published the Cygnus A position, followed by a refinement by Mills (1952) with an error 0.5 (RA) by 1.5 (Dec) arc min; even with this small error rectangle the optical object eluded detection. See footnote 20, this chapter, for the optical identification that followed in 1954 from Smith's (1951) improved radio position.

[31]At the 1 September 1949 meeting, preparations were summarised for a second series of RPL solar eclipse observations for the 22 October 1949 eclipse in Sydney, Tasmania and Victoria.

Fig. 9.6 (**a**) At Potts Hill a number of single Yagis were also operated to observe the total radio emission power from the sun. At 62 MHz, this single polarisation system was used. The photo was made on 14 July 1952. ATNF historical photographic archive: B2805-5. (**b**). Example of the type of data obtained at 62 MHz, taken by Wild (1985). Wild wrote : "Single frequency (62 MHz) record of a flare-associated outburst followed by a prolonged period of storm [Type I] on 4 March 1949". ATNF historical photographic archive: B3045

comparison of 97 and 600 MHz data (by Don Yabsley) was carried out for solar bursts on dates shortly preceding the meeting of 14 March 1950. At some periods there was good agreement, at other times poor. In late January 1950, Payne-Scott gave a thorough summary of the status of the results from the swept-lobe interferometer (as well as a summary of the Hornsby results, Chap. 8) at an Australian meeting of URSI, ANCORS (Australian National Committee for Radio Science). The details of this impressive contribution are summarised in Appendix M, including the now forgotten designations of solar bursts suggested by Wild in early 1950: Types α, β and γ. Within a few weeks these types became the well known Types III, II and I, respectively.[32]

A short report was given by Payne-Scott on the 97 MHz interferometer at an action packed meeting on 27 July 1950.[33] Wild gave a long summary of the first campaign from the Penrith instrument, the swept-frequency spectrograph (Chap. 8; Wild and McCready 1950). During a detailed discussion of the revolutionary results of the short campaign from February to June 1949 (249 h), reported later on by Wild and McCready (1950) and Wild (1950a,b, 1951), Christiansen (as secretary of the committee) reported:

> During the discussion on this report [by Wild] the use of the term "isolated burst" [meaning Type III] was responsible for an outburst from Miss Payne-Scott.[34]

Wild reported on his results concerning Type I, II and III bursts, making the strong claim for the latter that group retardation did not explain their spectral characteristics. He suggested instead (as Payne-Scott had in the 9 February 1948 colloquium) that outward moving disturbances at 10% of the speed of light were required. No record was given as to how this remarkable claim was received by Wild's colleagues. We can assume that the improved nature of Wild's determination of these speeds led to an increased credibility. (See footnote 105 Appendix M.)

At subsequent meetings of the Radio Astronomy Committee in early 1951, Payne-Scott gave detailed summaries of the results from the 97 MHz interferometer.[35] A cursory description of the publications (in preparation) was given on 23 January 1951, followed by a detailed description of all the results for Type I and Type IV (although at this time not recognised as a distinct category) on 6 February 1951. For the Type III bursts, Payne-Scott remarked that bursts appeared

---

[32]We do not know Payne-Scott's reaction as the name of the "unpolarised" bursts evolved to "isolated" bursts, then α bursts, and finally the accepted name, Type III bursts.

[33]A number of personnel-McCready, Yabsley, Marie Clark, and Norm Labrum-left the Radio Astronomy Group at this time.

[34]Her preferred term was "non-polarised".

[35]The frequency of the meetings of the full committee had decreased in the second half of 1950; Bolton gave a presentation on 12 December 1950, after his return from an overseas trip, reporting on US and European radio astronomy. In 1951 there were five meetings before Payne-Scott's resignation on 20 July, including a meeting on 18 July.

to originate around sunspots but with a large positional scatter; this result was never published (see below). Finally a parting shot was fired at David Martyn: "In no case [for Type I bursts] was a change of polarisation observed when a sunspot group crossed the meridian (as predicted by Martyn)". On 8 May 1951, the minutes reported that discussions had been held at Mount Stromlo with the view of working more closely with the group of astronomers there.[36] The minutes stated that "several people here went to a colloquium there. This might be reciprocated. Ruby Payne-Scott to talk early in July". It is not clear if she ever gave the talk. Ron Bracewell was the secretary of the Radio Astronomy Committee on 23 July 1951, 3 days after Payne-Scott's resignation from CSIRO. No mention was made of her departure in this or later minutes of the Radio Astronomy Committee.

Publication of the Results at 97 MHz: Paper I, the Instrument

The first of three publications provided a description of the instrument (Little and Payne-Scott, December 1951, "The Position and Movement on the Solar Disk of Sources of Radiation at a Frequency of 97 Mc/s. I. Equipment").[37] The paper was submitted on 28 June 1951, just before Payne-Scott's resignation. The second paper was submitted on the same date as Paper I. 1 July 1951. This swept-lobe instrument remained an unique instrument until the late 1950s when Wild, Sheridan and Trent (1959) and Wild, Sheridan and Nylan (1959) extended the Dapto swept-frequency interferometer to include the facility to determine positions of solar bursts in the EW direction with a precision of about 1 arc min over a frequency range of 40–70 MHz.

A restatement of the motivation for the construction of the new interferometer was described by Little and Payne-Scott (1951):

> In these existing interferometers [previous instruments] the position of the source is calculated from the interference pattern, which is produced by motion of the source through the lobes patterns of the aerials. But if, as in the case of bursts, the source has a life of only a few seconds, it will not have moved sufficiently during this time to produce the required interference pattern. Consequently, these methods are not directly applicable where the disturbance is of short duration or is rapidly changing its position, and this paper will describe a modified spaced-aerial [note that the term vertical interferometer was not used] interferometer which has been built to measure the position and polarisation of such a variable component.

---

[36] One reason was given for working more closely: "One interesting development is that there are some US astronomers working there. These groups previously at Johannesburg."

[37] RPP 135. The first draft was prepared on 10 August 1950 (close to the cessation of data taking) and the complete manuscript was sent to the *Australian Journal of Scientific Research*, Series A – Physical Sciences, on 28 June 1951. RPP 136 was Paper II in the series and prepared and submitted at the same time. The papers appeared back to back in the December 1951 issue on pages 489–525.

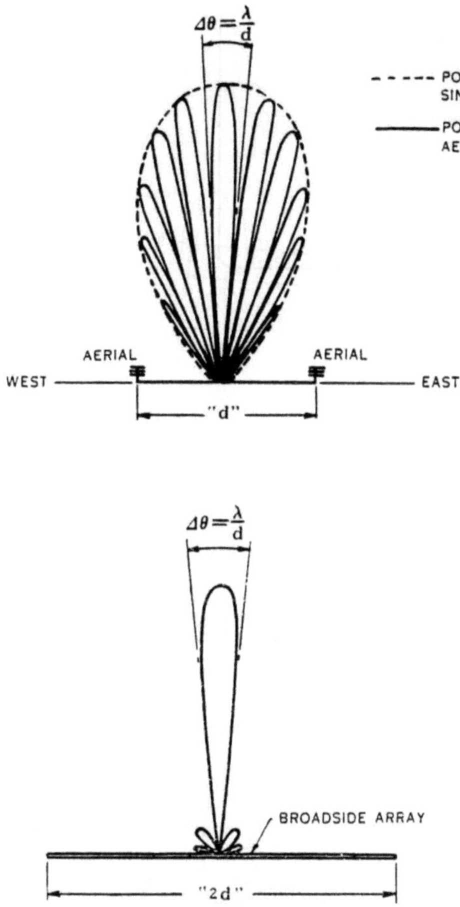

Fig. 9.7 Little and Payne-Scott (1951) provided a tutorial on interferometry in their publication on the equipment used for the swept-lobe interferometer Fig. 2 in their publication. *Australian Journal of Scientific Research* (A), vol. 4, p. 489. CSIRO Publishing, Copyright © CSIRO. See http://www.publish.csiro.au/nid/51/paper/CH9510489.htm

In this system the lobes were swept over the sun 25 times a second with two long baselines (240 and 280.5 m); the use of these two baselines with fringe spacing of about 44 and 38 arc min enabled lobe ambiguities to be resolved. The simple theory of the two element interferometer was sketched in Fig. 9.7. (See Fig. 9.8 for the layout and a simple block diagram of the interferometer.) Solar sources of sizes comparable to the size of the solar disk were not detected with this fringe spacing (footnote 15, this chapter). In addition the polarisation state of the radiation was determined at a rate of once per second. The EW position of the bursts and outbursts could be determined with a precision of about 2 arc min. The observing sequence was to observe on the 240 m baseline with parallel hand correlation (vertical vs.

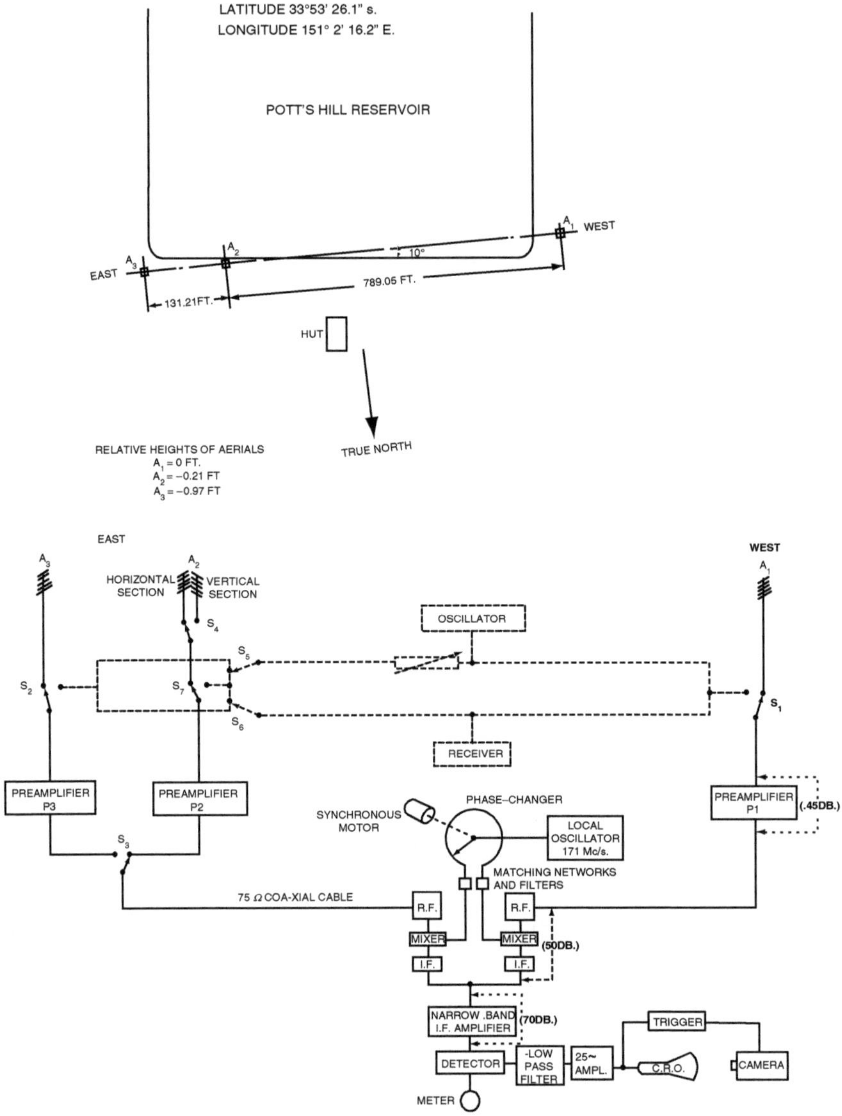

Fig. 9.8 Details of the interferometer used for the swept-lobe 97 MHz instrument at Potts Hill. Little and Payne-Scot (1951). The lobe spacing for the long baseline of 280.5 m was 38 arc min; the lobes were swept across the sun 25 times per second; the circular polarisation state was determined once per second. The novel calibration method was developed by Payne-Scott. Their Fig. 4. *Australian Journal of Scientific Research* (A), vol. 4, p. 489. CSIRO Publishing, Copyright © CSIRO. See http://www.publish.csiro.au/nid/51/paper/CH9510489.htm

vertical, etc.) followed by cross hand correlation (vertical vs. horizontal, etc.) in order to determine the polarisation state of the solar bursts. Finally, the parallel hand correlations were determined on the 280.5 m baseline. The total time to complete the sequence was about 1 s. A film recorder was used (a modified 16 mm cine-camera) which was triggered automatically when the intensity of the radiation reached a suitable high level. (Figure 9.9 shows a sample determination with the camera of circularly polarised Type I emission.) For circumstances when there was a steady level of solar bursts ("enhanced radiation"), the system could be used as a normal fixed lobe interferometer (Fig. 9.10) (the mode of observing used by Mills and Thomas for the Cygnus A observations). The flux density uncertainty in this case was reduced due to the longer integration times to values of a few thousand Jy (footnote 15 this chapter). Any state of the radiation polarisation could be determined – i.e., random, linear, elliptical or circular. In fact in 1948 Payne-Scott had worked out the equations for the determination of the polarisation state (modern usage would be "Stokes" parameters to describe the polarisation state of the radiation).[38]

The calibration of this instrument was quite straightforward; due to the low sensitivity, the use of known point sources for calibration was not possible, as is the case for modern interferometers. A large part of the Little and Payne-Scott paper consisted of a detailed description of the determinations of the positions, flux densities and sizes of the solar sources. Payne-Scott designed the simple calibration system, which relied on a set of precise calibration cables and the injection of test signals from portable Yagi systems (the dotted lines in Fig. 9.8). The observations occurred every week-day for about 2 h on each side of meridian transit of the sun; during periods of high solar activity, observations were continued during the weekend. Data was taken for 30 noise storms (Type I storm, Paper II), six outbursts (possibly Type IV events, Paper III) and 25 unpolarised bursts ("isolated" bursts, Type III). Later Little tried to analyse the Type III bursts; however the positions did not yield consistent results and these results were never published.[39] Thus the prediction in Paper II (see below) that "Future papers in this series will deal with the application of the interferometer to these phenomena [unpolarised bursts – Type III and outbursts – Type IV]" was only correct for the outbursts. The detailed description of the direct determination of the motions of the Type III bursts had to wait for the enhanced Dapto interferometer data presented by Wild, Sheridan and Nylan (1959).

---

[38] This derivation was probably one of the first derivations in radio astronomy of the response of an interferometer to a polarised signal. In a letter to Pawsey (in London) on 11 June 1948 (NAA: C4659, 8), Payne-Scott derived the response of linearly polarised antennas to a signal of arbitrary polarisation.

[39] Alec Little, as reported to Sullivan in 1978 (Sullivan 2009).

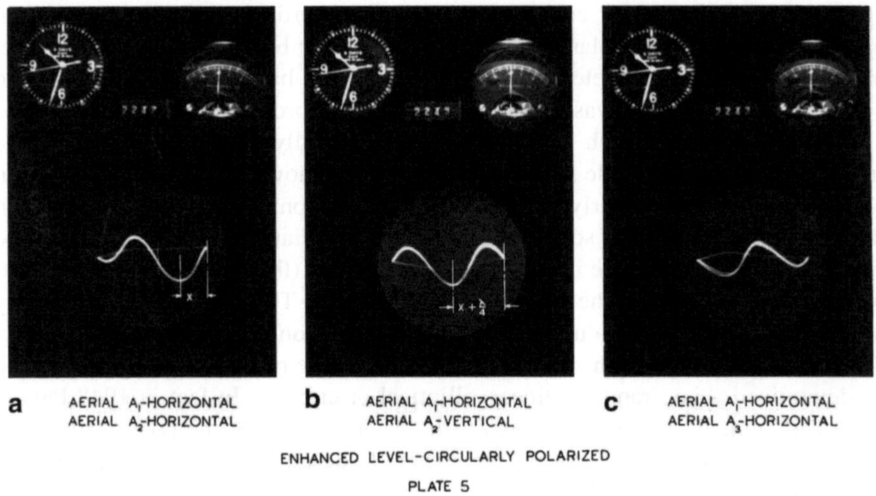

Fig. 9.9 At a swept rate of 25 times a second, the data recording with the swept-lobe interferometer required new techniques. The data were recorded on film with a 25 Hz time base with a modified 16 mm cine-camera, triggered automatically when the solar radio noise exceeded a predetermined threshold. The cathode ray screen showed the effect of the rotation of the phase-changer, which produced a sine wave on the display. The position of the minimum of the sine wave varied with the position of the radio source on the sun. The screen was photographed, together with a clock, a film frame counter and a meter registering the intensity of the solar noise. An example of the circular polarisation of a Type I burst is shown in Fig. 9.9. If the solar radio radiation was circularly polarized, the interference pattern would shift through a quarter lobe as shown in the figure, with no change in amplitude. The direction of the shift depended on whether the radiation was left handed or right handed polarized. As expected, this Type I noise storm was circularly polarized. Little and Payne-Scott (1951, their Plate 4 in the publication but Plate 5 here. Solar bursts less than about 5,000 Jy could not be observed with this method due to the limited sensitivity. From May 1949 to August 1950, 30 Type I events were observed in detail as well as six outbursts. The data on Type III bursts was inconclusive and never published. ATNF historical photographic archive: 2304-1

Paper II, Noise Storms (Type I)

In Paper II (Payne-Scott and Little, 1951, "II. Noise Storms" RPP 136, see footnote 37, this chapter), the authors described the results of their observations of 30 noise storms (Type I continuum), with detailed descriptions of the progress of six of the storms.[40] Payne-Scott and Little determined in an elegant fashion (the reduction of these records from the movie must have taken Payne-Scott many hours of tedious work) that the Type I continuum originated high in the corona over large sunspots. By locating the position of the source of radio emission for these six storms of long

[40] These results had been summarised in detail at the 27 July 1950 Radio Astronomy Committee meeting.

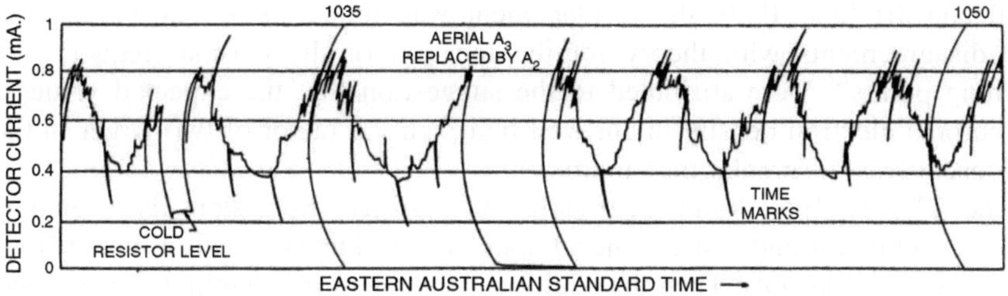

Fig. 9.10 The use of the Potts Hill swept-lobe interferometer in the fixed-lobe interferometer mode. Longer integration times made it possible to observe strong non -solar radio emission (e.g., Cygnus A) or semi-steady Type I storm emission. Little and Payne-Scott (1951). Their Fig. 9. *Australian Journal of Scientific Research* (A), vol. 4, p. 489. CSIRO Publishing, Copyright © CSIRO. See http://www.publish.csiro.au/nid/51/paper/CH9510489.htm

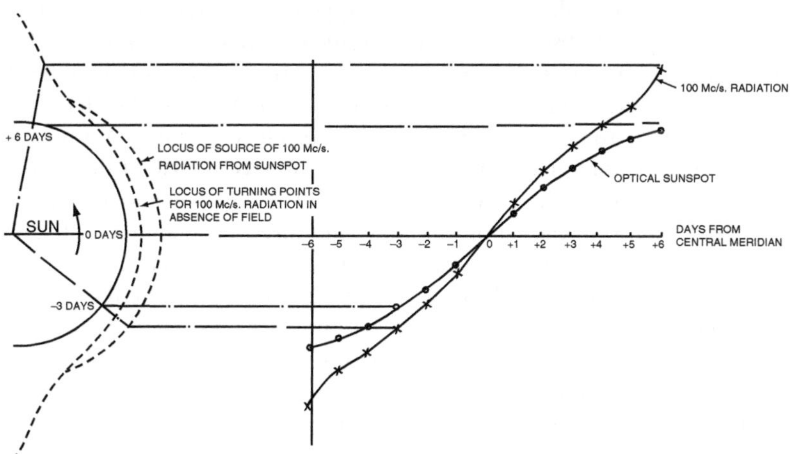

Fig. 9.11 The relation between the motions of the 97 MHz Type I solar storms and the associated sunspot group over a solar rotation proved that the radio emission arose high in the corona; the observed range was 0.3–1 times the photospheric radius above the sun's visible surface. This deduction is one of the more important conclusions of the collaboration of Payne-Scott and Little. From Payne-Scott and Little (1951).Their Fig. 8. *Australian Journal of Scientific Research* (A), vol. 4, p. 508. CSIRO Publishing, Copyright © CSIRO. See http://www.publish.csiro.au/nid/51/paper/CH9510508.htm

duration over periods of 10–16 days, she could show that the angular rate of change of position for the radio bursts was faster than that of the optical sunspots (Fig. 9.11). If the assumption was made that the source of the noise-storm lay at some height above the visible surface of the Sun (near the position where the plasma frequency was 97 MHz) and that the noise storm was radially displaced above the relevant spot group, it was then possible to calculate the radial displacement of the two. For most of the observations, this displacement was found to be 0.3–0.4 radii above the sun's visible surface. For the storms starting 10 February

1950 and 10 June 1950, the displacement was found to be 0.8–1 solar radii. The disagreements with theory for the location of the plasma frequency (the "turning points") were attributed to the suggestion that the expected values for the coronal electron density might well be roughly a factor of two larger in these periods, at least near solar maximum.

The 27 July 1950 Radio Astronomy Committee discussion also stated that Payne-Scott had found that the short bursts (Type I bursts or "storm bursts") were found close to sunspots. Then the additional statement in the minutes was made:

> Mr. Mills said that he had found the position of these bursts also to coincide with that of the "enhanced radiation" (Type I continuum)[41] ...

Payne-Scott and Little next described the detailed relation between sunspots and noise storms. They carried out a comparison of both the number of sunspots and the maximum area of the largest sunspot with the associated Type I bursts. The Mt. Wilson data was used for the comparison (in Appendix F, Payne-Scott's attempts to coordinate joint optical observations are described). The major result was that "... the size of the largest spot in the group is a more certain indication than the size of the whole group of the chance that a noise storm will be associated with the group". Thus a large spot group would probably not produce a noise storm unless it contained at least a single large spot with area of at least 0.04% of the Sun's disk. Since the close relationship between the size of a sunspot and the associated magnetic field was well known at this time, the direct association of the Type I bursts and the presence of a strong magnetic field was suggested.

The final result described in detail was the behaviour of the circular polarisation of Type I bursts. The radiation tended to be circularly polarised with linear and elliptical states occurring only on rare occasions. The radiation was often found to be nearly 100% polarised (i.e., RH or LH). The direction of the sense of rotation of the circular polarisation was found to depend on the magnetic polarity of the associated spot, with LH being present when the largest spot was a north seeking pole (positive polarity) and RH for a south seeking pole. (Positive indicated that the magnetic field was directed towards the observer.) The conclusion of the Payne-Scott and Little (1951) paper was: "The observations quoted here point to magnetic fields as the important factor in the production of solar noise storms at a frequency of 97 Mc/s." The paper then ended with a comparison of the data with theories of the emission mechanisms, prevalent in 1951. The continuation of numerous controversies with the Cavendish astronomers can be anticipated in the final sentences of the paper:

> Unfortunately most of the theories put forward to explain the origin of storm radiation do not lead to any definite predictions of region of origin, relation to sunspots, or polarisation, and hence are not capable of verification or refutation by the observations recorded here ... The exciting mechanism must then be non-thermal ... Ryle (1948) has suggested that storm radiation is due to local temperature increases, but the heights of origin observed seem

[41] Based on a discussion with Mills (2007), it seems likely that the attribution is incorrect. The report may well have originated with Little.

> much greater than his theory would indicate and the polarisation appears to be in the wrong direction.[42]

The importance of this publication concerning the detailed description of the angular and polarisation properties of the Type I bursts observed from May 1949 to August 1950 can be judged by two major reviews published two, and then 34, years later. In 1953, the influential review article by Pawsey and Smerd, "Solar Radio Emission", was published as Chap. 7 in *The Sun* (series, *The Solar System*), edited by G.P. Kuiper and B.M. Middlehurst. In 1985, K. Kai, D.B. Melrose and S. Suzuki wrote a summary article, "Storms", in *Solar Radiophysics*, edited by McLean and Labrum.

The 1953 article of 66 pages summarised the pioneering work of the RPL group from 1945 to 1952; the article, one of the most thorough of the early review articles concerning solar radio astronomy, was widely quoted and had a major impact in the astronomy world. Pawsey and Smerd summarised in detail the Payne-Scott and Little interferometer results. They pointed out that the 97 MHz data indicated that the sense of rotation of the circular polarisation corresponded to the ordinary mode of propagation, if the magnetic field of the sunspot extended into the corona. In addition the rarely observed linear polarisation state[43] was further evidence for the presence of ordinary, not the extraordinary, propagation mode. Pawsey and Smerd then discussed the significance of the high intensities observed by the interferometer. The true brightness temperatures were uncertain since the angular sizes were generally unknown; upper limits to the angular sizes of several arc min could be determined. Thus typical brightness temperature lower limits of 10^8 to 10^{10} K were common. Based on the burst duration of only seconds, they also concluded that a simple thermal origin was not likely and that "enhanced radiation probably originates as coherent radiation from groups of charged particles". Pawsey and Smerd also reiterated the importance of the threshold effect (sunspot size related to the presence of Type I noise storms) discovered by Payne-Scott and Little. Pawsey and Smerd emphasised that this was further evidence that if the threshold were surpassed, Type I emission would be emitted due to the presence of a high magnetic field.

After 30 years of detailed observations by numerous solar radio astronomers, Kai et al. wrote in 1985:

> Although Type I storms have been extensively studied for more than 30 years, it is only in recent times that any convincing theories have been advanced to explain their occurrence. It is now generally accepted that Type I bursts must be some form of fundamental frequency plasma emission ...

---

[42] Sullivan (2009) in his Chap. 13 ("The Radio Sun") discusses the *thermal* theory proposed by Ryle to explain solar bursts. The problem was that a fatal flaw was incorporated into the theory; he had assumed that the collision cross sections and rates of emission for the high temperatures of the solar corona were the same as the earth's ionosphere at a modest temperature of only 300 K.

[43] Modern evidence indicates that the detection of linear polarisation may not have been significant; the Faraday depth of the corona is large at 97 MHz (Bastian, private communication, April 2009) leading to large de-polarisation of linearly polarised radiation.

Kai et al. described the three major conclusions of the Payne-Scott and Little 1951 publication: (1) the determination of the source height above the visible sun, (2) the association of polarisation with sunspots indicating that the storms are polarised in the ordinary wave mode and (3) the association of the Type I bursts with the prominent sunspot in a group. Kai et al. wrote: “It is now generally accepted that large well-developed spot groups tend on the average to give rise to Type I storms more frequently than smaller spot groups.”

Paper III, Outbursts (Type IV)

The final paper in this series by Payne-Scott and Little is “III. Outbursts”, 1952.[44] The date of submission (1 July 1951) is noteworthy – only 19 days before the final day of Payne-Scott’s employment at CSIRO’s Division of Radiophysics.[45] At the time of writing the paper, the authors made the assumption that these six outbursts were Type II bursts. However, as Wild (1968) has pointed out, the observations reported were in fact Type IVM (moving Type IV events) bursts, the rarest of the burst and outburst types.[46] Based on Wild’s 1968 publication, the evidence from the Potts Hill 1949–1950 data was not consistent with Type II bursts: (1) the movement took place later in the lifetime of the disturbance than was expected for a Type II burst and (2) the motions observed at a fixed frequency were not expected for Type II bursts,[47] since only one level of the corona would radiate at this frequency, according to the plasma emission hypothesis. The Type IV event is now known to occur after a major solar flare (about 10% of Type II events are followed by an associated Type IV) and lasts for some tens of minutes, with some durations of up to 2 h. The radio spectrum of Type IVs was characterised by a smooth continuum possibly due to the synchrotron emission process. The typical angular size has been

---

[44]These results were presented initially at the 27 July 1950 Radio Astronomy meeting in which three outbursts were described. Two of the radio events started near an optical flare and the radio emission then moved outwards towards the solar limb.

[45]Paper III in the series was RPP 137 and was published in March 1952. A comparison of the results published in Papers II and III with the ANCORS publication of January 1950 (see Appendix M) shows that a thorough understanding of the behaviour of Type I bursts was achieved shortly after the observations began in May 1949; a clearer elucidation of the Type IVM outbursts was arrived at only after more extensive observations in the period February–August 1950.

[46]Stewart (1985) has pointed out that in the years 1966–1980, the Culgoora radioheliograph only recorded 56 Type IVM bursts, in comparison to 560 Type II events. The stationary component of Type IV events are associated with non-storm, flare-related continuum emission (Robinson 1985). Indeed the Type IV class is confusing!

[47]In at least one case studied by Stephen White, Tim Bastian and colleagues, motions of Type II bursts had in fact been detected near the solar limb with velocities of about 900 km/s (Bastian, email 19 April 2009; data from the Green Bank solar radio burst spectrometer and the Nançay Radioheliograph).

determined to be 8–12 arc min (at 169 MHz, Pick and Vilmer 2008). The solar source associated with the moving Type IV bursts travels outwards through the corona after a solar flare or an eruptive prominence mass ejection.

In the paper by Payne-Scott and Little, the velocities deduced were comparable to those of Type II outbursts, as is now known to be the case for Type IV moving bursts (in excess of some hundreds of km/s). The moving Type IV bursts (IVM) were only recognised a few years later based on data obtained with the 32 element grating interferometer at Nançay in France. Boishot and Denisse (1957) suggested that the emission mechanism was synchrotron radiation from electrons with energies in the range of 0.1 to one million electron volts.[48] Subsequent Sydney observations by Wild, Sheridan and Trent (1959) showed that the Type II bursts, as observed at *different* frequencies, did arise from different positions in the corona, as expected from the plasma hypothesis. For the moving Type IV events all frequencies arose from about the same position; the position did move outwards throughout the period of the event. Both the height and the motions of the Type IV events were thus inconsistent with the origin via plasma waves. It was soon found by later investigations in the 1950s that the radiation of the Type IV events was partially circularly polarised; this was also observed in the Potts Hill data from 1949 to 1950. Based on the relation to the magnetic field of the leading sunspot of the active region, the sense of polarisation was found to correspond to the extraordinary mode of propagation.

Despite the revisions in understanding of these Type IVM events which were based on subsequent work, the Potts-Hill determination of the velocities of the Type IVM events in 1949 and 1950 provided an important characterisation of the velocities, positions and polarisation of these events of Type II followed by Type IVM. Often these events were associated with terrestrial aurorae and magnetic storms about a day later. Six large events were observed with five being associated with large optical flares and short-wave fade outs (probably at 9.6 MHz) (5 September 1949, 17 and 22 February 1950 and 5 and 11 August 1950; the latter dates are close to the dates of the cessation of observations at Potts Hill). The additional event of 29 June 1950 had no optical data but short wave fade outs were observed.[49] For two of the events (5 August and 22 February), sudden commencement magnetic storms were observed by the Watheroo Observatory in Western Australia.

---

[48]The radiation had the characteristics of a smooth continuum, consistent with the gyro-synchrotron theory. However, a number of problems existed with the model. Stewart (1985) has summarised the pros and cons of the gyro-synchrotron model; a major problem was the presence of substantial circular polarisation. He suggested a hybrid model consisting of second-harmonic plasma emission followed by gyro-synchrotron emission. On the other hand, Smerd and Dulk (1971) made a strong case for gyro-synchrotron emission.

[49]None of these Type II/IVM events was associated with a major sunspot; for the 1949 event, and 17 February and 5 August 1950, the sunspot areas were only in the range 1,100 millionths, much less than the giant sunspots of 1946 and 1947.

Ron Bracewell had recently returned from the UK where he had obtained a Ph.D. from Cambridge in late 1949.[50] Based on his contacts in the UK, he was frustrated by the continued isolation of his RPL colleagues in Australia and began to write a series of five short articles in the UK journal, *Observatory*. His purpose was to provide publicity for the Australian results. Bracewell was convinced that the RPL group was suffering from isolation within the world astronomical community. The *Observatory* articles started in October 1950 and ran through to August 1954. The first article provided advance publicity for the swept-lobe interferometer at Potts Hill; the final publication of the Potts Hill results was more than a year and a half in the future, in the *Australian Journal of Scientific Research*, a publication with limited circulation in the scientific world. Bracewell's first article[51] was "An Instrumental Development in Radio Astronomy", a title that gave few clues to the subject discussed. The new 97 MHz swept-lobe interferometer was discussed in the one page article, written in a succinct but clear style. Bracewell (1950) wrote:

> A new instrument has been developed and put into use by Ruby Payne-Scott and A.G. Little . . . which can determine the place of origin and polarisation of bursts of radio energy from the Sun, virtually instantaneously. It is thus capable of locating the source of the radio energy emitted at the time of solar flares, and following its movement through the solar atmosphere. That this instrument is a triumph for radio astronomy will be readily agreed on contrasting the resolving powers obtainable at optical and radio wave-lengths.

The figure in the publication by Bracewell shows the dramatic motions of a Type IVM event on 17 February 1950. The figure for this event of 17 February 1950, as presented in the publication of Payne-Scott and Little (1952), is shown in Fig. 9.12. At the start of the event, a Type II radio outburst may have accompanied an optical flare. The figure shows that the radio emission began near the flare at a displacement of about 0.5 solar radii; the event was followed for the next 30 min as the radio emission moved to the West, ending up high in the corona after crossing the solar limb. The final position (when the radio emission was no longer detected) occurred at a position more than half a solar radius beyond the solar limb. The polarisation state began as random polarisation (possibly the Type II event) and evolved into left

[50] His advisor was Ratcliffe working on a Ph.D. ("The Propagation of Very Long Radio Waves in the Ionosphere," submitted 1 September 1949). Ronald N. Bracewell (1921–2007) was on the scientific staff of RPL from 1944 to 1955. From late 1955 to 1991 he was a member of the faculty of the Electrical Engineering Department at Stanford University in California; after his retirement in 1991 he was professor emeritus. At Stanford Bracewell developed a 10 cm spectroheliograph, a radio telescope for rapid imaging of the sun. In 1998, Bracewell became an Officer of the Order of Australia. Portions of Bracewell's archive are in the NRAO Archives in Charlottesville, Virginia.

[51] Subsequent articles described the controversy about the nature of the new compact radio sources ("Radio Stars or Radio Nebula") after the large angular extent of some of the new sources (Sgr A and Vela) had been determined at RPL by Mills. The new discovery of Cygnus X (the HII region complex in the Milky Way only a few degrees from Cygnus A) was also reported (Piddington and Minnett). The third papers in the series described the new proto-Mills cross at Potts Hill. The final papers in the series concerned a possible association of meteor showers and rain (the work of Bowen) and the use of low frequency galactic radiation at 18 MHz by Shain to detect sudden ionospheric disturbances (SID) caused by solar flares.

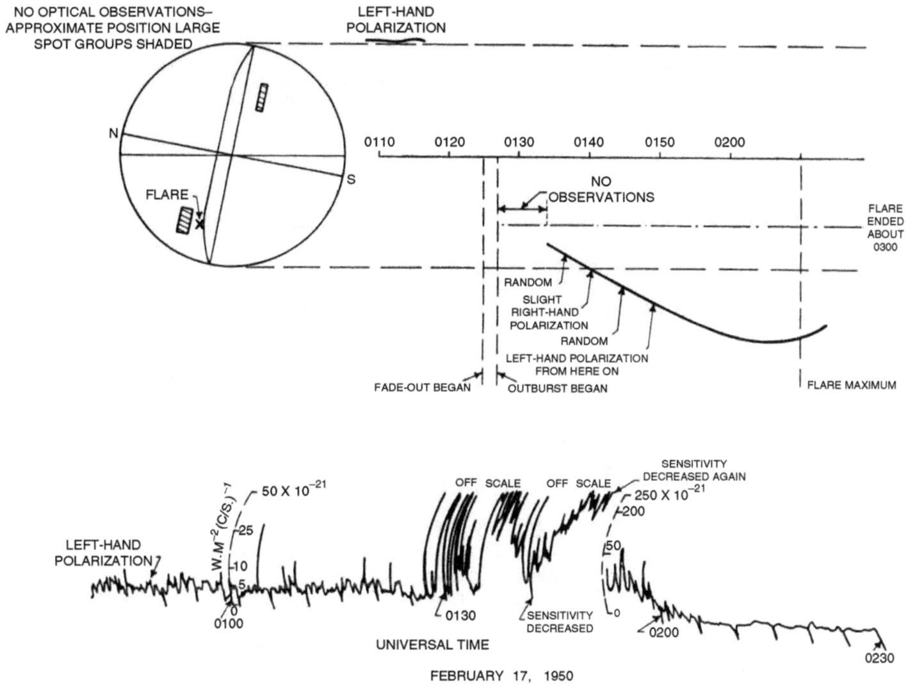

Fig. 9.12 The dramatic outburst (likely Type IVM) of 17 February 1950; Payne-Scott and Little (1952) showed that the solar emission position moved a projected distance of a solar radius in a time span of only 30 min. A prominent flare began a few minutes before the radio observations began. The total power data at the bottom of the figure was observed with a single 98 MHz Yagi (similar to the 62 MHz Yagi in Fig. 9.6a). Their Fig. 1b. *Australian Journal of Scientific Research (A)*, vol. 5, p. 32. CSIRO Publishing, Copyright © CSIRO. See http://www.publish.csiro.au/nid/51/paper/CH9520032.htm

hand circular polarisation (the Type IV event) at a time of 0150 UT. The optical flare reached a maximum at about 0210 UT and ended 50 min later. This data was discussed in detail by Payne-Scott and Little in Paper III; the Bracewell publication was probably effective in providing publicity for a wider audience.[52]

[52] This publication by Bracewell elicited a response from Charles Seeger III (1913–2002), the son of the famous musicologist, Charles Seeger Jr. (1886–1979) and brother of the well known folk singer, Pete Seeger (1919–). Seeger had recently left Cornell and was for a short period in late 1950 at Chalmers University of Technology in Gothenburg, Sweden. Seeger wrote to Payne-Scott on 16 December 1950 with extensive questions about the new swept-lobe interferometer (NAA: C3830, A/1/1, Parts 5 and 6); apparently Seeger was waiting for the winter weather in Sweden to pass before undertaking the design of a similar instrument. Payne-Scott wrote on 1 February 1951 with extensive details (and block diagrams), with emphasis on the new 25 Hz phase changer at the local oscillator frequency. She urged the group in Sweden to build an instrument at frequencies other than 100 MHz (the swept-lobe Potts Hill instrument) or 1,200 MHz (the planned Christiansen cross which was to begin construction in 1951). Apparently, no solar interferometer was constructed in Sweden; shortly, Seeger was to move to Leiden where he remained until the early 1960s.

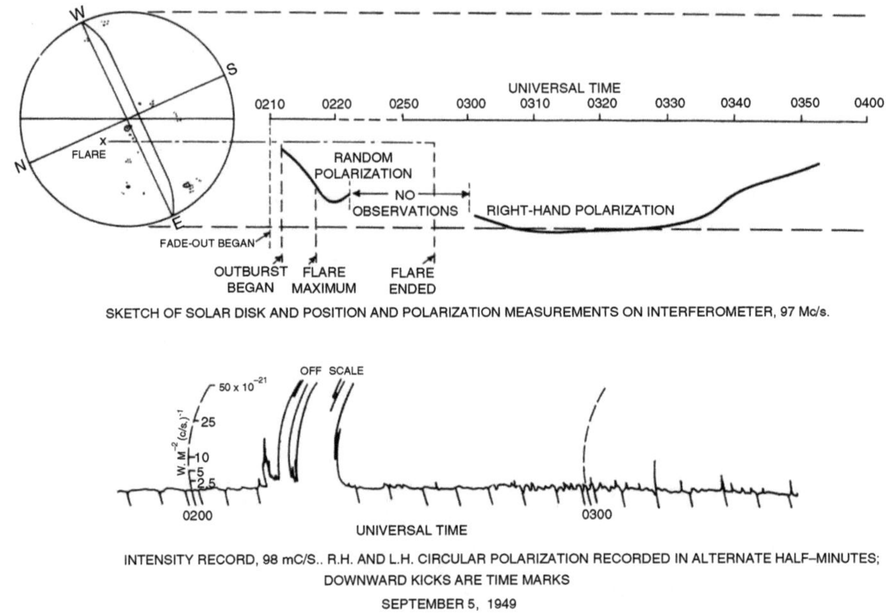

Fig. 9.13 The Type IVM event of 5 September 1949; the motion of the solar emission was from about 0.6 solar radii to a height of over four solar radii during a time period of about 1 hour. From Payne-Scott and Little (1952). The total power over this time frame is shown in the bottom panel. Their Fig. 1a. *Australian Journal of Scientific Research (A)*, vol. 5, p. 32. CSIRO Publishing, Copyright © CSIRO. See http://www.publish.csiro.au/nid/51/paper/CH9520032.htm

Additional observations were reported for five new outbursts. As Pawsey and Smerd pointed out in their review article (in *The Sun*, 1953): "This expectation [the movement of the outbursts towards the edge of the sun for non central flare events] has been beautifully verified in the 'swept-lobe' interferometer observations of Payne-Scott and Little (1952)." One additional example of this data is shown in Fig. 9.13, reproduced from the original publication (Paper III).[53] The entire event from 5 September 1949 lasted for almost 2 h. The right hand portion of the figure shows the east–west displacement of the emission centroid as a function of time. The radio emission began at the position close to the optical flare and moved towards the eastern limb of the sun. The initial event from 0210 to 0220 UT was probably a Type II burst (the polarisation was random as expected) followed by the weaker Type IVM burst from 0300 to 0350 and moved onwards with appreciable circular polarisation. (The data at 98 MHz shown in the bottom of the figure was

[53]This figure was reproduced by both Pawsey and Smerd (in *The Sun*, 1953),Wild (1968) and Kundu (1965). Kundu wrote: "The distinctive features of this observation were an initial outburst followed by a rather smooth and less intense continuum . . . Although part of the initial motion was perhaps due to the associated type II source ... the prolonged movement and late return near the flare position was definitely due to the type IV source."

obtained with the single Yagi shown in Fig. 9.6a.) The inferred height of the event at the start of the outburst was in the range 0–0.6 solar radii above the optical surface. The greatest height (photospheric radii above the photosphere) reached was then 4.2 solar radii. The measured radial velocity was 1,500 km/s, while the radial velocity inferred from the time delay was 1,400–1,700 km/s (the time delay estimate has a large error and was based on the time delay between the visible light flare and the onset of the Type II burst). The expected size of about 10 arc min for Type IV bursts would not have been detected by Payne-Scott and Little due to the limited angular resolution of the interferometer (fringe about 40 arc min). At the end of the outburst, the position usually returned to that of the flare. The authors also suggested the existence of "non-selective fading", the partial disappearance of the burst radiation with correlated behaviour between 62 and 98 MHz. Since the magnitude of the effect was greater at the lower frequency, it seemed likely that this was due to amplitude scintillation in the earth's ionosphere. However, in the publication, Payne-Scott and Little were cautious: ". . . the authors do not wish to make any definite suggestions about the origin of this fading without first considering the results of more detailed observations."

The main conclusion of this important publication was:

> The evidence . . . establishes that outbursts of radio noise are another manifestation of the solar disturbances already known to produce optical flares, radio fade-outs, and terrestrial magnetic storms. The apparent changes in position ... can most readily be interpreted as due to the passage through the solar corona of an exciting agent moving with a speed of from 500 to 3,000 km/s. It seems likely that normally the exciting agent is the corpuscular stream which is assumed to cause terrestrial magnetic storms.

However, Pawsey and Smerd were somewhat less optimistic in 1953:

> The radio evidence therefore favors the hypothesis that outbursts are due to the passage of auroral particles through the solar atmosphere, but it is clearly not yet conclusive. The outstanding deficiency in the evidence is the lack of direct observations, by any other means, of these particles in the vicinity of the sun.

The impact of these Potts Hill observations carried out in this 15 month period in 1949–1950 was evaluated almost two decades later by Paul Wild (1968):

> Following the first spectroscopic results another Pawsy-inspired experiment was put into operation and brilliantly performed by Payne-Scott and Little. The idea was to locate ... the instantaneous position of the dominant source on the sun at any one time.[54]

These 3 years at Potts Hill represented the high point of Payne-Scott's career as a radio astronomer. The work here laid the basis for the interferometer observations made in future years at Dapto and at Culgoora from 1967 to 1984. A typical afternoon tea scene from this period is shown in Fig. 9.14, and includes Payne-Scott, Alec Little, George Fairweather, Alan Carter and Joe Pawsey. The photo is unfortunately damaged with numerous blotches. However, Carter really had a eye-patch, the result of a skiing accident.

[54] 1968 Pawsey Lecture presented by J.P. Wild in Brisbane, Australia, 30 April 1968.

Fig. 9.14 A photo of a number of the RPL staff at Potts Hill circa 1950. From left to right , Ruby Payne-Scott (drinking a cup of tea), Alec Little, George Fairweather (in the doorway, a technician), Alan W.L. Carter and Joe Pawsey. Carter really had an eye-patch as a result of a ski accident. Carter left RPL some years later for a position at the Australian Naval Research Laboratory. The group are in front of one of the trailers used for instrumentation of the radio telescopes at Potts Hill. The photo is slightly damaged with numerous spots, possibly due to ink. Bill Hall family collection, used by permission of Peter Hall.

Collaboration with Chris Christiansen

An additional publication was partly based on Potts Hill data: "Radio Observations of Two Large Solar Disturbances" by W.N. Christiansen, J.V. Hindman, A.G. Little, Ruby Payne-Scott, D.E. Yabsley, and C.W. Allen, 1951.[55] In contrast to other publications by Payne-Scott, this one presented only a series of observations with little interpretation. The observations described two intense solar outbursts of 17 February (see text above as well as Fig. 9.12) and 21–22 February 1950. The impressive frequency range used for the data in this project ranged from 62 to 9,400 MHz, based on observations at seven discrete frequencies. Payne-Scott and Little provided the data at 62 and 98 MHz using the Potts Hill single Yagis to provide the total power data; in addition, the motions were determined with the swept-lobe interferometer at 97 MHz.[56] The 200 MHz data were provided by Allen at Mount Stromlo. For the frequencies above 600 MHz, the intensities of the bursts were higher than ever detected up to this date. For 1,200 MHz, the flux densities

[55]RPP 125. The first draft was dated 30 June 1950 and the paper was submitted to the *Australian Journal of Scientific Research* on 28 September 1950.

[56]See Fig. 9-12 for the motion of the radio event in an eastward direction to a position 10 arc min off the solar limb; the time scale for the motion was about 30 minutes. The origin of the 97 MHz emission began near the position of the optical solar flare.

were in excess of 10^8 Jy. A modern interpretation of these events[57] suggests that at frequencies below 1,200 MHz, the majority of the emission was Type II, III and even IV radiation. (On both dates Payne-Scott and Little observed the motions of the moving Type IV events with displacements in excess of a solar radius in a time period of 30 minutes) At higher frequencies the radiation could probably be associated with impulsive incoherent gyro-synchrotron emission. There were no dynamic spectra and a detailed classification remains uncertain. These large events were eruptive in nature since radio fade-outs were observed using 9.6 MHz short wave broadcasts from Brisbane; in addition, a magnetic storm was observed at the Watheroo Magnetic Observatory in Western Australia (35 h later for the 21–22 February events). The timing of the events on 17 February suggested a Type II event (see above), while the 21–22 February events (not covered with the swept-lobe interferometer until about 30 min after the start of the outburst) may well have been Type III bursts. The uncertain state of the possible interpretations illustrates why the swept-frequency observations of Wild and McCready (1950) were more favourable for the classification of major solar events such as these of February 1950.

In 1951, Payne-Scott's career as a solar radio astronomer came to an end. She had had remarkable success ever since starting it in March 1944. Her contributions to the discovery of Type I, II and IV bursts and the key role she played in elucidating the properties of Type III bursts, especially the second of time delays, remain an impressive achievement. She provided the first evidence for the direct determination of the motions of solar outbursts based on the swept-lobe observations of 1949–1951. Her contribution to the recognition of the principles of radio aperture synthesis was also significant.

[57]Provided by Tim Bastian (NRAO), August 2006.

Chapter 10
Payne-Scott and URSI, 1952: Her Last Experience as a Radio Astronomer

Payne-Scott experienced a postscript to her career as a radio astronomer in August 1952, 13 months after her resignation which had occurred a few months before the birth of her son, Peter Hall, in November 1951 (Chap. 4). The first international congress to be held in Australia had been the Second Pan-Pacific Science Congress almost 30 years earlier, in August 1920, in Melbourne and Sydney.[1] One of the first international congresses, however, that had ever been held outside the US and Europe, the tenth URSI (International Scientific Radio Union) General Assembly, took place at the University of Sydney from 11 to 21 August 1952.[2] On 22 and 23 August there was an official visit to Canberra where Sir Edward Appleton gave another lecture.

[1] A number of authors pointed out that the URSI Assembly of 1952 was the first international scientific union outside the US or Europe (e.g., Haynes et al. 1996; Robertson 1992). Already in 1950, Professor F.J.M. Stratton (prominent UK astronomer and the Secretary of the International Council of Scientific Unions) asserted that "no Union so far has ever had a meeting outside Europe and the U.S.A." (NAA: C3830, C6/2/4A, Part 1). However, there had been earlier meetings with an international character in Australia. A meeting of the British Association for the Advancement of Science was held in Australia in 1914 at the beginning of World War I (BAAS publication in 1915). This event was a series of lecture tours throughout Australia, with Eddington, Lodge and Rutherford as eminent speakers; the event was essentially British and Australian in character. A more likely candidate for the first international conference in Australia was the second Pan-Pacific Science Congress of 1923, which had more of an international nature with participants from the US, Canada, Britain, Japan, the Netherlands, and New Zealand see Hobbs (*Science*, 1923); the geology of the Pacific was a major topic. A new US warship (the cruiser *Milwaukee*) was a big attraction at the Sydney meeting, because it had a new sonic depth finder which provided accurate soundings from Seattle to Sydney. The following Pan-Pacific Science Congress was held in Japan in 1926, and the Fourth was held in Batavia (Jakarta), Indonesia in 1929 (see A.M. Goss 2009). Professor Gary Tee of the University of Auckland and Professor W.T. Sullivan, III of the University of Washington have provided numerous comments on this topic.

[2] Haynes et al. (1996) have described the Assembly, including the trepidation of the Australian hosts in preparing suitable coffee for the overseas visitors. The circumstances of the Assembly have been reconstructed, based on programs and documentation provided by Bracewell in 2006–2007 and Madam Inge Heleu, Executive Secretary of URSI, in February 2007. McGee was an attendee.

W.M. Goss and R. McGee, *Under the Radar*, Astrophysics and Space Science Library 363, 185
DOI 10.1007/978-3-642-03141-0_10, © Springer-Verlag Berlin Heidelberg 2010

Fig. 10.1 URSI 1952 begins with the arrival of the VIP guests abort the ship RMS Srathmore. The welcome of URSI guests aboard the ship in Sydney. Left to right are D.F. Martyn, Col. E. Herbays (Secretary of URSI, Brussels), Sir Edward Appleton, E.G. "Taffy" Bowen and J.L. (Joe) Pawsey. Based on a letter from Appleton to Bracewell (the Organizing Secretary of the Conference) dated 18 June 1952 (Bracewell archive NRAO), the Strathmore was predicted to arrive on Friday 8 August 1952 in the morning. Newspaper accounts from the Sydney newspapers in August 1952 agree with this date. That evening at 8 p.m. Sir Edward gave an address sponsored by the Sydney University Engineering Club (Twenty-Third Annual War Memorial Lecture) in the Great Hall of the University of Sydney, "Science and the Public." ATNF historical photographic archive: B2842-1

Figure 10.1 shows the arrival in Sydney by ship (the P&O RMS *Strathmore*) of Appleton, the URSI President and Vice-Chancellor of the University of Edinburgh (third from left) being welcomed by the Australian hosts.

From the left the others were D.F. Martyn from the Radio Research Board, Canberra (URSI Vice President and Chair of Commission V, Radio Astronomy), Colonel E. Herbays (Brussels, Secretary of URSI), E.G. Bowen (RP Chief), and J.L. Pawsey (RP Assistant Chief). Many of the European guests had spent more than a month on board ship across the Indian Ocean. Brian Robinson (2002) has suggested that many of the delegates to the conference must have spent many days in discussions during the trip to Sydney.[3] Many well known overseas guests attended the Assembly, including Sydney Chapman from Oxford, J. Ratcliffe

[3] In fact, Appleton had been persuaded to give a lecture on board, entitled "The Challenge of Scientific Progress". His younger daughter, Rosalind, accompanied Appleton and his wife to Australia; Miss Appleton met the purser of the *Strathmore* on the voyage and they were married later that year in Edinburgh (Clark 1971). During the URSI Assembly, a number of the Sydney newspapers carried articles about the Appleton family (including several on the glamorous Rosalind). On his arrival in Sydney, Appleton was asked by a reporter from *The Daily Telegraph* about the honorary degree that the University of Sydney was to present to him during his visit. When asked about the number of honorary degrees he had received, he responded, "I don't know –

from Cambridge, B. van der Pol from Geneva (CCIR, the International Radio Consultative Committee), A.H. de Voogt from The Hague, M.L. Oliphant from the Australian National University in Canberra, S. Silver from the University of California, Berkeley, C.R. Burrows of Cornell, John Dellinger of RCA in the US, and Sir John Madsen of Sydney University (also Chair of the Australian Organising Committee). About sixty overseas delegates from thirteen countries were in attendance as well as more than 250 Australian attendees; the conference included toursof Sydney, the local beaches, a harbour cruise and an excursion to Wollongong on the South Coast, which included a tour of the Dapto field station. The 2 weekends were spent at Jenolan Caves and the Federal Capital, Canberra (with visits to Mount Stromlo Observatory).

The Assembly began on Monday, 11 August 1952 at the University of Sydney. Bracewell, who had only returned to Sydney in late 1949, has written (1984, in Sullivan, 1984) about his experiences as the Secretary of the Australian Organising Committee. He has reported that his enjoyment of the conference was compromised by a serious illness which forced him to miss many of the scientific sessions. There was, however, a silver lining; he recuperated in the Sydney home of his future wife, Helen, and was duly impressed by the impressive collection of books of his future father-in-law. Bracewell designed the URSI Australian pin shown in Fig. 10.2. This

Fig. 10.2 The URSI pin from 1952 designed by Ron Bracewell. Bracewell gave his pin to Goss in January 2007. Goss photo 2007

---

but I've got more degrees than a thermometer" (Bracewell archive, National Radio Astronomy Observatory).

pin can be seen in many of the photographs.[4] There was also an URSI flag designed by Bracewell and made by Frank Kerr (which has since disappeared).

In the Administrative Opening Session on 11 August, Appleton gave a lengthy reply to the welcome by Sir John Madsen, the Chairman of the Australian National Committee of URSI. He first paid "tribute to the distinction which Australian workers have won for themselves in the field of scientific radio". This was followed by a detailed history of the Radio Research Board which had started in 1927 under Madsen's leadership. Then he gave a short summary of the War time successes of the Division of Radiophysics Laboratory. He described the recent discovery by Bolton and Stanley of the small angular size of Cygnus A as determined at Dover Heights.[5] Next he described David Martyn's work on tides in the ionosphere. The recent decision to fund the 250 ft telescope at Jodrell Bank in the UK led him to make a prophetic statement: "... [I] would like to see, in due course, a similar instrument at the disposal of your radio astronomers here in the southern hemisphere." He ended with a discussion of the recently discovered hydrogen 21 cm line (25 March 1951 by Ewen and Purcell, 1951) as predicted by Henk van de Hulst in 1944, and he stressed the importance of frequency protection for this frequency near 1,420.4 MHz, "... so that this new and important discovery may be developed without trouble due to radio traffic interference".[6]

The grand opening of the Assembly was on the following day, with welcoming addresses by the Rt. Hon. R.G. Casey (1890–1976),[7] then Minister for External Affairs and the Minister of the CSIRO. Appleton, as President, replied with words of praise for the "young" Australia: "For in one thing, we can all claim to be Australians ... so we can ourselves believe that our own subject of radio has vast and undreamed-of possibilities of expansion and development in the years which open out before us ... We must be fortified with the pioneering qualities – the Australia qualities – of courage, enthusiasm, and resource."

The main event of the day was a performance by the Sydney Symphony Orchestra; the choice of Wagner's "The Ride of the Valkyries" so close to the end of World War II was controversial. The program did, however, impress the overseas guests with the very Australian "Corroboree" by John Antill. As Bracewell wrote, "... with bullroarers in the background that solved the problem of how to introduce something Australian without inviting comparison with Beethoven."[8]

---

[4]Interview with Goss, Palo Alto, California, February 2007. Bracewell gave his only remaining URSI pin to Goss after the interview.

[5]In his closing remarks, Appleton confessed that he had only been able to visit the Dapto and Potts Hill field stations. "I am sorry I have not yet been able to visit Dover Heights – which is now an historic site – but I hope to do so before I leave Sydney." We do not know if he did in fact visit this site.

[6]URSI, *Proceedings of the General Assembly*, Vol. IX, Fascicule 1, Administrative Proceedings 1952, Brussels, p. 9.

[7]Former Ambassador to the US during World War II and later Governor General of Australia, 1965 to 1969.

[8]Bracewell (1984) (in Sullivan 1984).

Fig. 10.3 Inaugural ceremony of URSI 1952 at Sydney University, Tuesday 12 August 1952. Concert by the Sydney Symphony, followed by addresses by Prof. H.K. Ward (Chairman of the National Research Council), the Right Hon. R.G. Casey (Minister of State for External Affairs and Minister-in-Charge CSIRO) and Sir Edward Appleton, President of URSI. The Great Hall of the University of Sydney. ATNF historical photographic archive: B2842-35

The audience in the Great Hall is shown in Fig. 10.3. Just before the conference opening the attendees gathered for a group photo (Fig. 10.4). Brian Robinson has published an impressive key with identifications of 174 of the attendees.[9]

This international conference was a great opportunity for Australian scientists to meet their colleagues from overseas;[10] in addition, the Australians could show off their successes in radio science, and radio astronomy in particular, to the rest of the world. Commission V (Radio Astronomy), with D.F. Martyn as President,[11] had six sessions of 2–3 h' duration. Nineteen scientific talks were presented (nine of them by Australians) at the four scientific sessions, the special interest on Monday,

[9]"Recollections of the URSI Tenth General Assembly, Sydney, Australia, 1952", *The Radio Science Bulletin.*

[10]Due to continued ill will as a result of WWII, the Germans and Japanese were not invited. Martyn did go to Japan after the Assembly, however, to transmit news of the proceedings (Gillmor 1991). Japan submitted a national report in the published URSI proceedings; Germany did not. Also, as the Cold War had begun, there were no participants from the USSR or other East European countries.

[11]Martyn had been responsible for the invitation to URSI to be held in Sydney in 1952.

Fig. 10.4 Group photograph of URSI, Sydney 1952, on Monday 11 August 1952 outside Wallace Theatre at The University of Sydney. Brian Robinson (2002) has published a finding chart for 174 of the participants. This is the only photo during the URSI assembly in which the authors have found an image of J.L. Pawsey. ATNF historical photographic archive: B2842-15

18 August 1952, being a session on "radio emission from interstellar gas". Professor Schilt from Columbia and Yale (visiting Mount Stromlo) was the Chair of this session; the newly discovered HI line at 21 cm was the topic with the opening talk by Harold I. ("Doc") Ewen of Harvard, one of the discoverers, together with Purcell, of the line. The solar radio noise session was held as the second one on 13 August. Presentations were given by Christiansen (RPL), F.G. Smith (Cambridge), Steinberg (CNRS, Paris), Hagen (Naval Research Laboratory, Washington, D.C.), Laffineur (Institut d'Astrophysique, Paris), Piddington (RPL) and Smerd (RPL). Payne-Scott was probably present at this session, but did not give a paper.

The fact that she gave no presentation is surprising, since Pawsey had asked her to give a paper. On 11 March 1952, Pawsey sent a memo to all members of the radio astronomy group with a description of the process for making presentations to the URSI Assembly. A multistage selection process was involved having first RPL and then Australian national committee approval. Proposals for papers (a brief half page outline) were to be given to Bracewell by 18 March. The papers were to represent "an outstanding contribution" and also had to be "reasonably novel". Pawsey sent a copy of the memo and a personal letter to Payne-Scott's home (to Mrs. W. Hall, 120 Woronora Parade, Oatley). He wrote: "In my provisional list, unofficial, I include your work with Alec [Little] as one of our star efforts. Further, if such a paper is to be presented verbally at the Conference, I think you would do it excellently." After a request that she discuss this plan with Alec, he asked that she let him know what she would prefer.[12]

[12]The correspondence for URSI 1952 is from NAA: C3830, C6/2/4B.

On 20 March 1952 she wrote a letter to Pawsey, including an outline of her planned presentation, "Relation Between Solar Radiation at Metre Wave-Lengths and Other Solar Phenomena", with co-authors Alec Little and Paul Wild. She saw no chance to have a draft ready by the deadline of 7 April, which is understandable as she had a 6-month-old son and was no longer at RPL. She had tried to contact Little, but he was on holiday. She ended the letter by inviting Pawsey and his family to visit her family in Oatley; this would have been much easier for her than to go to see Pawsy, since the Pawsey family had a car while the Hall family did not. She signed the letter "Ruby Hall".

The outline written by Payne-Scott discussed "noise storms" (Type Is) and "outbursts" (likely Type II and IV). (She did not use the new nomenclature, although Wild was probably already using it.) The noise storms could be associated with large sunspots and had appreciable amounts of circular polarisation, with a polarity determined by the associated sunspot. An origin high in the corona was suggested since the radiation was displaced to the limb of the sun. Then the "outbursts" were usually accompanied by large flares, often associated with radio fadeouts. The events were randomly polarised, with circular polarisation occurring towards the end of the event (now known to be the associated Type IVM events following a Type II event). The movement through the solar atmosphere had been inferred by both the frequency drift (the swept-frequency spectrograph by Wild and McCrady) and the direct determination of the motions with the swept-lobe interferometer at Potts Hill. An exciting agency moving with the velocity of the "magnetic storm particles" was suggested. There were still questions as to how the two types of evidence (angular motion and frequency drift from high to low frequency) fitted together. The conclusion was that these Type II and IV events were probably caused by particle streams which also caused the magnetic storms on the earth about a day later.

Payne-Scott showed that she had kept up with developments in the field, even though she must have been extremely busy as a new mother. Pawsey replied positively ("Dear Ruby") on 26 March 1952. He had discussed the proposal of a paper with Paul Wild, who "favours the idea". Pawsey said that he had given a similar paper at the 1950 URSI in Zurich but

> [t]his is a proper follow-up and I should like you or Paul [Wild] to attempt it. As I see it, you have the more recent work and hence the right to do it. If you did not feel able to, I should suggest Paul do it. The broad objective is an integrating paper, collecting the known facts and rejecting fallacies, concerning the relations between optical and radio phenomena. It should set this out as an objective.

Pawsey suggested that the three of them meet at Ruby's house, when Paul returned from Dapto; we have no evidence as to whether the visit occurred.

In the end, neither Wild nor Payne-Scott gave a presentation at the URSI Assembly. In the solar session on 13 August 1952, RPL presentations were given only by Christiansen, Piddington and Smerd. The omission of a presentation on metre wave solar bursts seems surprising. Perhaps the organisers wanted to avoid the impression of domination of the solar session by the Australian group from RPL.

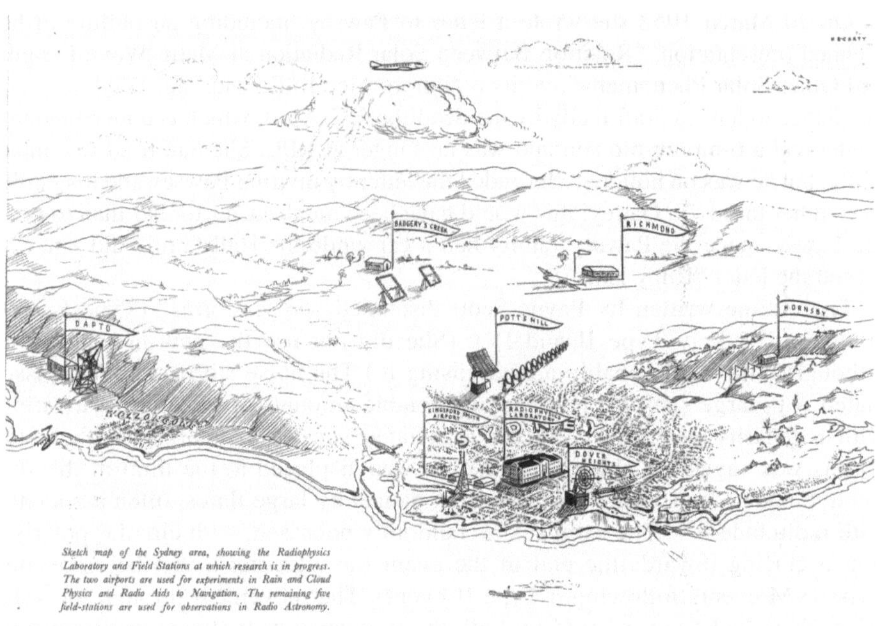

Fig. 10.5 Sketch map of the Sydney area showing RPL and the Field Stations (see Fig. 1.4). A booklet was published by RPL for URSI (1952): "Research Activities of the Radiophysics Laboratory." (RPL Archive, from Christine Vanderleeuw, CSIRO Radiophysics Library, 2007)

During the conference there were at least two excursions[13] to the RPL field stations at Potts Hill (Chap. 9) and the new solar station at Dapto, where the swept-frequency analyser had recently been installed. These stations and other activities of the laboratory were summarised in an idealised drawing for the booklet, *Research Activities of the Radiophysics Laboratory*, prepared for distribution to the USRI participants. The map in the booklet is shown in Fig. 10.5. Seventeen of the 37 pages were devoted to descriptions of the radio astronomy and ionospheric research activities of the laboratory; cloud and rain physics, automatic computation (a computer project) and radio navigation were also included. The swept-lobe interferometer was described under the rubric "Locating Sources of Transient Solar Disturbances". (The instrument was no longer being used in a systematic fashion in August 1952.)

On Thursday 14 August 1952, there was an official afternoon tour of the RPL in the University of Sydney grounds, followed by a trip to Potts Hill.[14] Many photographs of the inspection of the various instruments were taken. One example is Fig. 10.6, with Christiansen showing Appleton (dark hat) and van der Pol (lighter

[13] Apparently these visits to the field stations were organised around detailed briefings by RPL staff. Appleton was present at both of these excursions.

[14] Bolton (1953) described the official field trips to Dapto and Potts Hill. The trip to Hornsby is not mentioned, but Bolton did say that "private visits were made to other field stations engaged on galactic work". Surprisingly there is no record of any trips to Dover Heights; we do not know if Appleton had requested a visit to this "historic site".

Fig. 10.6 URSI tour of Potts Hill Reservoir Field Station, August 1952. Here Chris Christiansen (left) describes the new 32 element E–W grating array, opened November 1951. Sir Edward Appleton, President of URSI, points to one of the 1.7 m antennas. To the extreme right is Prof. Balthasar van der Pol, Director of CCIR in Geneva and prominent Dutch radiophysicst. Note the URSI pin on the jackets of Appleton and Christiansen (Fig. 10.2). ATNF historical photographic archive: B2842-R6

coloured hat) the E-W 20 cm solar grating array. The former Chief of RPL, Fred White, who was the CEO of CSIRO by this time, was also at URSI in 1952. He was in the group with Appleton on this tour (Fig. 10.7). A fascinating photo is Fig. 10.8 with Alec Little explaining the workings of the swept-lobe interferometer, the instrument used so effectively by him and Payne-Scott to determine the motions of solar events (Chap. 9). Clearly RPL was well prepared for this excursion. It is unlikely that Payne-Scott was present.

There was also an unscheduled excursion to the Hornsby site (Chap. 8), possibly on 14 August. Many of the well known participants (Graham-Smith, Hanbury-Brown, Warbarton, Higgins, Hindman, McGee, Mills, Slee, Shain, as well as some others) are shown in Fig. 10.9.

Both before and after the conference there were at least three reports written in Australian and international publications summarising the URSI Assembly. Two months before the conference, Bracewell (in his role as Secretary of the URSI Assembly) wrote an article in the *Australian Journal of Science*.[15] This announcement of the upcoming meeting was full of optimism:

[15] 21 June 1952.

Fig. 10.7 Continuation of the tour at Potts Hill during URSI. Again the group is inspecting the 32 element E–W grating array. This 21 cm continuum instrument was used to provide high resolution radio 1-D image of the sun. From right to left: Bowen, Appleton, van der Pol, Fred White. At extreme left is Chris Christiansen. ATNF historical photographic archive: B2842-R65

> That URSI has honoured us by its present choice is tangible international recognition of the status of scientific radio in Australia ... it tends to raise Australia's prestige abroad to the level merited ... The Australian National Committee on Radio Science is under an obligation to ensure that foreign delegates to the forthcoming Assembly will find it rewarding, and it is hoped that a successful outcome will lead to fuller Australian participation in these important scientific activities, in radio and in other fields as well.

Bracewell voiced the hope that the Prime Minister (Sir Roberts Menzies) would address the opening session; unfortunately the Prime Minister could not attend.

Two additional reviews of radio astronomy at URSI appeared in early 1953, by Bolton in the British journal, *Observatory* (February 1953), and by Kerr in the popular US astronomy journal, *Sky and Telescope* (January 1953). The latter contained six figures, including Fig. 10.10. Both articles reflected the interests of the author; thus the description by Bolton had an emphasis on radio stars, while the Kerr article had a galactic HI structure slant. This latter article described in detail the HI investigations carried out in the Netherlands by J.H. Oort and his colleagues, as well as the new determinations by both Australian and British groups on the sizes of radio sources. Since none of the compact sources had stellar characteristics in terms of size, he proposed using the new terminology: "... the former term *radio stars* is now falling into disuse, in favor of the more general phrase, *discrete*

Fig. 10.8 URSI tour of Potts Hill. Alec Little describes the swept-lobe interferometer that he and Ruby Payne-Scott used from 1949 to 1950. Note the trailer in the background (Fig. 9.15). In the background near the car J.P. Hagen (hat) and H.S.W. Massey (UK) are shown. At the extreme left is C.A. (Lex) Muller from the Netherlands. We think that it is unlikely that Payne-Scott was on this tour. ATNF historical photographic archive: B2842-R58

Fig. 10.9 An unofficial tour of the Hornsby instrument (Chap. 8) was also organized during URSI. Note the primitive outhouse to the left! From the right the attendees are Dick McGee, two unknown, L.W. Davies, Joe Warburton, F. Graham-Smith, Alex Shain, Jim Hindman and unknown. From CSIRO Archive (Rob Birtles) Canberra: 163.140

Fig. 10.10 Most of the radio astronomers at URSI 1952 are shown here. Front row left to right: Chris Christiansen, F. Graham-Smith (UK), B.Y. Mills, S.F. Smerd, C.A. Shain, R. Hanbury Brown (UK), Ruby Payne-Scott, A.G. Little, M. Laffineur (France) and J.G. Bolton. Second row: J.P. Wild, J.L. Steinberg, J.V. Hindman, F.J. Kerr, C.A. Muller (Netherlands) and O.B. Slee. Third row: C.S. Higgins, J.P. Hagen (USA) and H.I. Ewen (USA). Back row: J.H. Piddington, E.R. Hill and L.W. Davies. Individuals with no country designation are Australian. Surprisingly Pawsey, the leader of the radio astronomy group at RPL was not in the photo. Likely day for the photo is 11 August 1952. ATNF historical photographic archive: B2842-43

sources." The article gave a short summary of the solar work, concentrating on the new 21 cm continuum imaging of the sun by Christiansen at RPL with a resolution of 3 arc min. Bolton's review mentioned the recent Fourier imaging of the quiet sun at Cambridge and well as the results of two eclipse expeditions to the Sudan in 1952 over a wide wavelength range (shortest 8 mm). Bolton gave mixed reviews of the session on "Dynamics of Ionized Media" organised by Professor Harrie Massey: "... [it] gave a very able summary of the present theoretical position ... The writer was, however, left with the impression that a theoretical interpretation of some cosmical phenomena, in particular the radio noise from the disturbed Sun and the galaxy, was far from being achieved, except in the most general terms."[16] Bolton

[16]Kerr also had doubts about the status of theory in explaining emission processes in radio sources: "The discussion showed that this is a very difficult subject, and a thorough understanding of the nonthermal processes that are important in radio astronomy is still a long way off." Indeed the recognition of the importance of synchrotron emission was not to occur until the mid-1950s. The importance of the paper, "Cosmic Radiation and Radio Stars", by Alfvén and Herlofson, who proposed synchrotron emission, was not recognised in 1952 (see also Chap. 7). Nicolai Herlofson from the Royal Institute of Technology in Sweden was at the URSI conference in 1952.

also criticised the use of the term "radio star"; "it seems that the term'"radio star' may be a misnomer". Indeed in a few years the use of this term did disappear.

A group photo of the prominent radio astronomers at URSI was taken, perhaps on one of the days of the sessions for Commission V, Radio Astronomy. Since

Fig. 10.11 The registration process at URSI, possibly on Monday 11 August 1952. Sally Atkinson is seen assisting an overseas delegate. At this time she was the RPL Chief's Secretary. William E. Gordon of Cornell University (later Director of the Arecibo Observatory) is the second person from our left, wearing a bow tie. From CSIRO archive (Rob Birtles) Canberra: 163.129

Fig. 10.12 During the URSI Tenth General Assembly a number of photographs were taken of the audience. Ruby Payne-Scott is seen on our extreme right, possibly asleep or just stretching. Her son, Peter, was only about 9 months old. McGee, with a big smile, is seen just to the left of the standing man's nose. The late Brian J. Robinson is just to the left of Ruby's right elbow, in the row behind. From CSIRO archive (Rob Birtles) Canberra: 163.108

Payne-Scott was present, a possible date was 13 August 1952.[17] In the historic photo shown in Fig. 10.10, Payne-Scott is surrounded by 21 prominent colleagues from Australia, Europe and the US. She is the only woman, and is not wearing the URSI pin.[18] We have been told by Ms. Sally Atkinson that she and a colleague at RPL in the administration, Sylvia Mossom, were the baby sitters this day, looking after the 10-month-old Peter Hall while his mother attended the URSI sessions.[19] Sadly, Pawsey was not present for this photograph; Kerr, in the *Sky and Telescope* article, also mentioned Pawsey's absence from the group photo. Sally Atkinson was most likely efficiently running the local administration of URSI in 1952. In Fig. 10.11, she is assisting one of the foreign delegates at the time of registration. Nearby we see W. E. (Bill) Gordon of Cornell, with a bow tie.

There is another photo (Fig. 10.12), taken during the URSI Assembly while Appleton was giving a lecture, in which a figure, who appears to be sleeping, may be Payne-Scott. The new mother might have been catching up on sleep during one of the URSI sessions; it is possible she could have only been stretching! A very cheerful McGee is also in this picture. In the Australia Telescope National Facility archive there is another photograph (B2842-41) taken during this lecture; Payne-Scott is wide awake in this image! This international Congress represented the last formal involvement that Payne-Scott had with the Division of Radiophysics and the CSIRO.[20]

Certainly URSI 1952 in Sydney was an encouraging opportunity for the Australian scientific community to show that they were among the world leaders in radio astronomy. A few months later (27 November 1952), a British reporter for the *Manchester Guardian*, Douglas Wilkie, wrote with inflated praise:[21]

> A visiting American scientist remarked that Australia's beer, beaches and radio astronomy were the best in the world ... Why do Australians have a gift of scientific excellence? ... It is certainly at variance with their frequent reputation for cultivating the good-enough as a way of life. The best explanation, it seems, lies in that pioneering heritage . . . Now seeking new worlds, they find a challenge in scientific exploration.

---

[17] The RPL photo is dated 8 August 1952, but this date is before the beginning of the Assembly on 11 August. The first session of Commission V (Radio Astronomy) was 13 August 1952.

[18] Although Payne-Scott has no pin, she was a registered participant; she appeared in the list of participants with profession "physicist", address 120 Woronora Parade, Oatley. Her name was listed as "Ruby Payne-Scott (Mrs. W. Hall)". Only three women were registered for the URSI Assembly.

[19] Based on an interview with Sally Atkinson, February 1999, Epping, NSW. For information about Sally Atkinson, see Chap. 5, footnote 7, as well as the Preface. Sylvia Mossom Blackwood (1914–987) worked in the RPL administration during World War II, starting in 1941; she was also an infrequent baby sitter for the Hall family in the 1950s (letter from her husband, Fred Blackwood, July 1999). In the late 1950s, Mossom was a literary assistant to D.P. Mellor as he prepared the influential volume, *The Role of Science and Industry*, in the series *Australia in the War of 1939–945. Series 4, Civil* (1958), sponsored by the Australian War Memorial.

[20] Payne-Scott did, however, maintain contact for some years with Christiansen and Mills. We do not know if she stayed in touch with Pawsey.

[21] From the Sullivan archive.

Chapter 11
Reminiscences and Anecdotes of Ruby Payne-Scott as Told by Friends and Colleagues

Ruby Payne-Scott had numerous friends and colleagues who outlived her. Over the years, they have related stories, anecdotes and events about Payne-Scott, which the authors have collected below.

Joan Jelly, née Freeman (1918–1998)

Joan Freeman, a well known physicist in the UK, interacted with Payne-Scott on many occasions during World War II at RPL; the nature of these interactions has been outlined in a fascinating manner by Freeman in her autobiography, *A Passion for Physics* (1991).[1] Freeman joined the RPL group in Sydney working on radar in June 1941, barely 2 months before Payne-Scott began her employment in August 1941. Freeman left Australia on a CSIR Fellowship in August 1946; after gaining a Ph.D. in Nuclear Physics at the University of Cambridge, she had an illustrious career at the Atomic Energy Research Establishment, in Harwell, UK. In 1976, she shared the Rutherford Medal of the Institute of Physics with R.J. Blin-Stoyle. Freeman married the physicist John Jelly (of Cerenkov radiation fame) in 1958; the two had met in Cambridge in 1948. John Jelly died in 1997, preceding his wife's passing by 8 months; Joan Freeman died on 18 March 1998 in Oxford. Nessy Allen (1990) has written a fascinating account of the lives of Freeman and Makinson.[2]

Two anecdotes involving Payne-Scott are recounted in Freeman's book. The first, a humorous conflict, involved both scientists in a battle with the Librarian of

[1] Freeman's assessment of Payne-Scott's personality and political views are summarized in Chaps. 13 and 15, her bush walking activities in Chap. 11, and the marriage controversy in Chap. 4. On 24 October 1952, Joan Freeman wrote Ron Bracewell congratulating him on his engagement to Helen Eilliot. Joan may have regretted that she left the field of radiophysics as she had begun a new career in nuclear physics: " I often regret very much that I somehow worked myself out of the radio line, it seems to be very profitable and quite exciting these times." (Bracewell archive, July 2009)

[2] "Australian Women in Science – A Comparative Study of Two Physicists" (1990).

W.M. Goss and R. McGee, *Under the Radar*, Astrophysics and Space Science Library 363, DOI 10.1007/978-3-642-03141-0_11,

the National Standards Laboratory, a facility shared with RPL. Freeman said that this occurred in late 1941 or early 1942. The Head Librarian, Mrs. Enid Eastman, was previously the Secretary of Professor Sir John Madsen, Professor of Electrical Engineering at the University of Sydney and the Head of the Radio Research Board. Mrs. Eastman apparently saw herself as the arbiter of female behaviour at the National Standards Laboratory and RPL; Rachel Makinson reported to us that Mrs. Eastman was known as "a bit of a dragon" at RPL. The Head Librarian was very upset by the fact that Payne-Scott wore shorts and that Freeman was a smoker. Payne-Scott found the shorts quite appropriate attire since she was climbing up ladders to work on radar antennas; naturally many of the men also wore shorts, especially during the hot summers. Freeman had taken up smoking while she was at university. Mrs. Eastman wrote a peremptory note to the two telling them to shape up! Payne-Scott reacted in an angry fashion and continued to wear shorts and urged Freeman to continue with her smoking since most of the men did so. Both were finally summoned to attend a meeting in the library with Mrs. Eastman at which time the proper behaviour of the females in the laboratory would be discussed. On that day Payne-Scott deliberately changed into shorts for the inquisition and Freeman refused to attend.[3] Payne-Scott must have got into a shouting match with the Librarian since she soon returned to her office; she had been dismissed by the disciplinarian. A stand-off continued and was only resolved when a new librarian arrived, the famous Australian novelist, Marjorie Barnard.[4] As Freeman put it, Barnard, an enlightened colleague, was not at all "concerned about Ruby's shorts or my cigarettes".[5]

The second story also involved Freeman and Payne-Scott; in this case it was Freeman who was more upset. Every few weeks during the early period of World War II, secret radar reports would arrive by air from Britain. The medium was microfilm; since there were no copying facilities at RPL, these were taken to the

[3]In a few publications in Australia during the last 10 years, the claim has been made that Payne-Scott was also a smoker; we think this is quite unlikely.

[4]Marjorie Barnard (1897–1987), a famous Australian author, was the Librarian at RPL and National Standards Laboratory from 1942 to 1950. With Flora Eldershaw, Barnard wrote five major works of fiction using the single pseudonym, M. Barnard Eldershaw. Their collaboration included *A House is Built* (1929) and the futuristic novel, *Tomorrow and Tomorrow and Tomorrow*. (This novel was censored on publication in 1947.) Barnard wrote an unpublished manuscript in 1946 (the date is uncertain) entitled *One Single Weapon*, a history of the RPL radar effort. This manuscript of over 300 pages is in the Basser Library of the Australian Academy of Science in Canberra as well as in the Mitchell Library in Sydney. An accompanying note with the manuscript at the Basser Library states: "This manuscript of the history of radar was commissioned by [RPL] after the Second World War, and owing to some disagreements on factual matters between the author and some members of the Division was never published." In the same note, Bowen also was reported as being uncertain about some of Barnard's conclusions. Like Payne-Scott, Barnard had also attended Sydney Girls' High School and the University of Sydney (first class honours in 1918).

[5]From Freeman's autobiography. Rachel Makinson has told us (2007) that the women at RPL during the War were seldom allowed to work at the field stations due to the lack of toilets for women! In particular, they were not permitted to visit RAAF bases in the Sydney area.

Sydney Public Library for photocopying. Since the reports were top secret, someone with security clearance had to take them to the Library and be present during the entire copying procedure as well as ensure that all work, including spoiled sheets, was brought back to the laboratory. Payne-Scott and Freeman were told to alternate in accompanying the films. Each session lasted several hours while the attendant had to watch the copy process in an almost darkened room. Freeman despised the tedium and the lost time for her own projects; she felt that some of the junior men should also have been given this task. The worst part of the chore for Freeman was the endless chatter of the young woman doing the copying. Payne-Scott would not join Freeman in making a "fuss;" she did not mind the respite as she could knit in the dark, a skill that Freeman had no intention of acquiring.

Freeman began an exchange of letters with the authors in 1997 about Payne-Scott. She had lost touch with Payne-Scott (See Chap. 15 for a discussion of a letter from Freeman to Payne-Scott in February 1976) after she left for Cambridge. Freeman wrote Goss in 1997:

> Yes, I am sure Ruby was the first woman radio astronomer in the world. It all started with Joe Pawsey who was a brilliant scientist and did not himself receive the recognition he deserved. It was quite shocking that the famous paper [*Proceedings of the Royal Society*, 1947, see Chap. 7] was "sat on" for so long by the referees. Yes, I feel sure that Ruby made her own original contributions to that paper too ... I suppose her family life was a strong competitor to her scientific interests.

In early 1998, when a second exchange of letters had begun, colleagues in the UK reported Freeman's death. She was made a posthumous Officer of the Order of Australia on 26 January 1999.

K. Rachel Makinson, née White (1917–)[6]

Kathleen Rachel Makinson was born 15 February 1917 in the UK and moved to Australia in 1939 after completing a Physics undergraduate degree at Newnham College at Cambridge. She married Richard E.B. Makinson (1913–1979) in 1939[7] and worked at the School of Physics, University of Sydney, during World War II with some time off for the birth of her elder son, David, in 1941.[8] During the War Makinson held positions on an annual, temporary basis as both Research Assistant

[6]Based on extensive interviews, 2 July 2003 and 21 February 2007, together with correspondence from 2003 to 2007 with Dr. Rachel Makinson and her son, Dr. Robert Makinson.

[7]Freeman has written about a party that she attended on the night of 16 August 1939 for the wedding of Rachel and Dick Makinson. Rachel had just arrived by ship from England; the wedding occurred some hours later in a Registry Office in North Sydney. Dick was Freeman's quantum mechanics lecturer; her claim that Dick, an Australian, was an especially poor lecturer is not accepted by Rachel Makinson. Rachel and Dick had met while he was doing his Ph.D. at the Cavendish Laboratory of the University of Cambridge in the late 1930s.

[8]David was born in August 1941, the younger son, Robert, in January 1956. See also footnote 11 this chapter.

and Research Scholar. In the early part of the War she worked at the University of Sydney on a rapid on-off radar switch (for transmit-receive) under the direction of V.A. Bailey (grant from RPL). (This project was described in RP 222/1, a secret report from October 1944: 'Magnetically Controlled Gas Discharges - Possible Application to Modulators' by R.E.B. Makinson, J.M. Somerville, and K.R. Makinson. Her affiliation was listed as 'late Commonwealth Research Fellow, now CSIR Officer.)' She was involved in the "Bailey Boys" (after Professor V.A. Bailey of the School of Physics) courses in electronics and radar techniques for military personnel (mainly the RAAF), in which she gave some of the lectures on vacuum tubes, cavity resonators, wave guides and electromagnetic horns. In Fig. 11.1, a group of RAAF airmen is shown with some of the academic staff, including Rachel Makinson, in the back row. Even though she was working at the University of Sydney with the "Bailey Boys", she was being paid by the CSIR Division of Radiophysics from 1944 to 1945. After the War, she began a career as a physicist of wool fibres, and from 1953 to 1982 she was employed at the newly created CSIRO Division of Textile Physics (originally Wool Textile Physics). Her remarkable career has been summarised in two publications by Nessy Allen (1990 and 1993).

Makinson told the authors about a number of interactions she had with Payne-Scott during the years 1941–1950 which included letting the Hall family live in their house for a year while Bill and Ruby Hall were building their own house in Oatley.[9] She also said that Payne-Scott was in the RPL branch of the Communist

Fig. 11.1 "Bailey Boys" at the Number 3 Radiophysics Training School for RAAF personnel August–March [sic, the designation on the photograph. Perhaps March to August ?] 1943. Back row from right Flying Officer Watson, Dr. F. Fraser, Rachel Makinson, Mrs. J. Bailey, Prof. V.A. Bailey, J. Somerville, P.G. Guest, A. Pollard and unknown Flying Officer. (Most of the names of the trainees are not legible on the University of Sydney Archive photo, G3/224/1799.) Used by permission of Julia Mant, Reference Archivist, the University of Sydney

[9] In late August 1951, only a few months before Peter was born, Bill and Ruby Hall moved into their partially completed Oatley house which had as yet no water or electricity. NAA: A1/1/1, Part 6.

Party. Because of her passionate left-wing views, Payne-Scott was known as "Red Ruby" around the RPL and NSL; no one was brave enough, however, to use this term in Payne-Scott's presence!

Makinson continued to work at the University of Sydney in 1942 when David was an infant. On a Saturday – the work day ended at 12:30 – when David was ill, she was quite late arriving for work; Payne-Scott chastised her for being late: "The men here must not think that we women cannot cope ..." Makinson bit her tongue but was tempted years later, when Payne-Scott began a family to say, "See – I told you so". Not surprisingly she did not. Allen (1990, 1993) has described how "devastatingly difficult" Makinson found her life during this period;[10] for help in the home she was paying out more than her own salary; to take the baby to child care and then go to the University of Sydney campus, she had to use five different public buses.

Makinson and Payne-Scott were both quite active in left-wing politics[11] and in the CSIR(O) OA. In this trade union they were instigators of a committee relating to equal pay for women (see Chap. 4 and Appendix H). Makinson recounted a tragedy that happened at a meeting of the OA sometime in 1945–1947 when Payne-Scott was suddenly taken ill and was taken to hospital as a miscarriage began (see also Chap. 7). Makinson did not witness this event but heard about it within a day. Joan Murray, who worked in the CSIRO administration during these years, also remembers the miscarriage which began during a meeting in the NSL Library.[12] Fortunately, in 1951, Payne-Scott had a successful pregnancy; Peter was born in November of that year.

In Fig. 11.2, a photograph of Rachel Makinson, taken by Goss in February 2007, in Turramurra (a North Sydney suburb), is shown; at this interview Rachel Makinson told the authors that she was "very proud" of the achievements of Payne-Scott. In spite of pronounced discrimination because of their gender, both had successful careers in CSIRO.

---

[10]She lost 3 stone (42 lbs or 19 kg) during this period.

[11]Richard E.B. Makinson (1913–1979) was a well known Australian physicist (condensed matter) who was quite controversial in the 1940s–1950s. Phillip Deery (2000) has written a gripping article, "Scientific Freedom and Post-War Politics: Australia, 1945–1955", describing the price that Dick Makinson paid for his left-wing and publicly admitted Communist views. He was never considered for promotion to professor at the University of Sydney School of Physics, possibly due to the influence of ASIO. Later, he was promoted to Associate Professor at Macquarie University, 1968–1978. Dick Makinson was also associated indirectly with the Tom Kaiser affair (starting with the Kaiser demonstration in London on 27 July 1949) which involved the CSIRO, MI5, the US Department of State and ASIO (see Deery, "Science, Security and the Cold War: an Australian Dimension", 1999; also see footnote 17 Chap. 7 and footnote 78 Appendix H). Makinson was the founder of the "Kaiser Protest Committee" in Sydney, in September 1949. His most vocal opponent was the famous Liberal Party member of the Australian parliament, W.C. Wentworth (1907–2003), who said in a debate on 6 June 1952: "This man [Makinson], who organised what was in effect treasonable conspiracy, remains a lecturer in physics at Sydney University . . . and I believe that Dr. Makinson is an embryo Dr. Fuchs [the confessed atomic spy who had been a British participant in the Manhattan project when he passed information to the Soviets]."

[12]Letter from John D. Murray and Joan Murray, 26 January 2004. In Chap. 7 we suggest that the miscarriage may have occurred in late 1946.

Fig. 11.2 K.R. (Rachel) Makinson, February 2007 with Libby Goss at Makinson's home in Turramurra NSW, Sydney. Photo by Miller Goss

Harry C. Minnett (1917–2003) (Chief of RPL, 1978–1981)

Harry Minnett[13] has remained an invaluable source on Payne-Scott, as he knew her at the University of Sydney in the years 1936–1938, while he was studying Physics II and Maths II; he was pursuing both an engineering and science degree (mathematicss and physics). Minnett joined the RPL in April 1940 (Thomas and Robinson 2005). He wrote to Goss in 1998:

> To my surprise, Ruby turned up in mid-1941, just after Joan Freeman had joined ... They were the only two women scientists on the staff and made a considerable contribution to the work, as well as adding a much needed feminine touch to the community. It was a great opportunity for them to work on state-of-the-art techniques, particularly as employment was not easy for women scientists before the war. Both went on afterwards to distinguished research careers. ... I worked in association with Joan from time to time on microwave projects [they both developed the TR-transmit-receive switch for the LW/AHW Mk II 25 cm system, Minnett having developed the TR switch for the famous 200 MHz radars in 1941–1942] ... but never with Ruby. Our discussions were more by chance at coffee breaks or the lunch room or in colloquia. ... I would like to contribute to your queries about Ruby. She was a very pleasant but determined girl, especially when debating her ideas. I remember well how in a discussion, if a contrary point of view started an argument, her voice could rise in pitch and intensity to over-ride and then silence the opposition. Although such discussions were very wearing, she was always responsive to the logic of an opposing point of view and, if convinced, would readily concede.[14] Her ideas were always worth listening to and I learned much from her.

[13]Minnett is the fourth person from the right in the front row in the December 1945 radar conference photo (Fig. 6.1 in Chap. 6).

[14]Other colleagues have suggested that Payne-Scott was not quite so open-minded.

Minnett remarked on the relationship between Pawsey and Payne-Scott:

> ... Pawsey appreciated Ruby's abilities as a physicist, evident in her research and in colloquia. As well, her wartime work on small-signal visibility on radar displays and the accurate measurement of receiver noise factor was attractive to Pawsey, who had become keenly interested in the thermodynamics of external noise fields, as expressed by antenna temperature, and its relationship to the internal noise of receivers, as expressed by the noise factor. It was therefore natural for Ruby to assist him in early wartime searches for solar radiation. [See Chap. 6 for a description of the 1944 solar observations from RPL by Payne-Scott and Pawsey.]

After the War, therefore, Payne-Scott was a valuable part of the team, which had been joined by Lindsay McCready, and which made the first important discoveries on "sunspot radiation". Minnett also confirmed that the aperture synthesis formulation probably resulted from intensive discussions between Pawsey and Payne-Scott (Chap. 7).

Professor B.Y. Mills (1920–)

The collaboration during the War between Mills and Payne-Scott has been described in Chap. 5. Mills has given his impressions of Payne-Scott:

> Her contributions to solar radio astronomy were crucial, starting with the classic paper with Joe [Pawsey] and Lindsay [McCready], proceeding with the interpretation of the frequency–time characteristic of solar bursts as an outflowing disturbance and finally demonstrating the actual position shifts of these bursts.[15]
>
> ... so I will end with my assessment of her as a person. Her most obvious characteristic was that she was very forthright and outspoken. She held opinions firmly, both scientific and social, and was always ready to defend them and to expect the same from people who disagreed. This did not endear her to some who regarded her as very aggressive! However those who knew her well respected her integrity and honesty, particularly in scientific matters. In this she was similar to Joe Pawsey although they were so different in other respects. As a scientist I think of her as extremely competent and knowledgeable, always ready to embrace new ideas if they appeared valid but not, I think, outstanding as an originator.[16]

In late 1948, Mills joined the radio astronomy group, transferring from a group working on producing pulsed X-rays using a magnetron at 1200 MHz.[17] This apparatus could be used for the radiography of moving machinery. (The project is described in RPR 77, January 1948 by B.Y. Mills, 'A Million-Volt Resonant Cavity X-ray Tube.' Mills's first project was to assist in the solar eclipse observations of 1 November 1948 (Christiansen et al. 1949); Mills looked after the observations in

[15] Partly based on an interview with Bernard and Crys Mills in Roseville (Sydney) on 1 April 2007.

[16] Letter to Goss, 14 September 1997.

[17] This transfer was discussed by Pawsey and McCready by letter during Pawsey's overseas trip in 1947 and 1948. NAA: C4659, 8.

Fig. 11.3 B.Y. Mills, 1 April 2007, Roseville, NSW, Sydney. Photo by Goss

the Sydney area at Potts Hill (see Orchiston et al. 2006, for a description of this campaign). At the time of joining the radio astronomy group, he was aware of conflicts between Payne-Scott and Bolton: "... Ruby was quite forthright with some of her comments to Bolton" at the bi-weekly meetings of the Propogation Committee (Radio Astronomy Committee after 1949).

Figure 11.3 is a photograph of Bernard Mills on 1 April 2007 in his home in Roseville, a northern suburb of Sydney.

Wilbur Norman "Chris" Christiansen (1913–2007)

Chris Christiansen was a colleague of Ruby Payne-Scott at AWA before the War. He was an antenna expert and after the War he moved to RPL and became a leader in solar astronomy. He and Ruby worked together at Potts Hill during 1948–1951. In 1960 he moved to the University of Sydney where he became a professor of electrical engineering. The famous image at Potts Hill field station, with Payne-Scott, Christiansen and Little, is shown in Fig. 1.2.

In a letter to Goss, Christiansen related a memorably ornery exchange at the end of a meeting of the Propagation Committee on 28 July 1949. At the meeting, Bolton and Kevin Westfold presented a progress report on their observations of the entire southern sky at 100 MHz with a resolution of 17°.[18] Bolton summarised the observations while Westfold followed with a lengthy discussion of the observed radio background emission of the Milky Way galaxy. At that time, before the emission mechanism was understood to be synchrotron emission, the intensity of the background radiation was unexplained. Westfold demonstrated that free–free emission could not provide an interpretation. He then tried to provide an explanation using radiation from *visible* stars, which was rejected because of the implausibly high effective temperatures that would be required due to the small filling factors. The final model, which had been favoured by Westfold and Bolton, explained the emission as originating from galactic radio stars with large angular sizes.[19] This model could apparently fit the data but strong objections were raised at meeting. The secretary of the committee, Chris Christiansen, wrote in the minutes:

> The detectability of [radio] stars was to be treated but the validity of the initial assumptions of the analysis were challenged by the members of the Committee and the section was withdrawn for modification.

Mills and Christiansen have both provided us with parallel accounts of what happened. According to Christiansen's letter,

> Ruby said "Kevin, where did you get that." Kevin said, "from John Bolton." Ruby said, "well, it is utter nonsense" upon which Kevin said, "then so is my paper" and sat down. The meeting ended.

Mills has confirmed this exchange in detail in his own letter to Goss.[20]

---

[18]Described in three papers in the *Australian Journal of Scientific Research* (Bolton and Westfold 1950a, b, 1951).

[19]From the PC minutes (prepared by Christiansen); the dilution factor for radio stars was assumed to be 10^{-3} compared to 10^{-13} for visible stars.

[20]Mills told Goss on 1 April 2007 that the exchange between Ruby and Kevin went as follows: Kevin responded "Well, that's what John Bolton told me." Then Ruby was quoted by Mills as saying: "Well, that doesn't matter if it's wrong." Remarkably, there was extensive correspondence in September 1950 between Bowen and Pawsey about this PC meeting of July 1949 (NAA: C3830, F1/4/PAW/2). Pawsey was in Europe and met Bolton at the URSI Conference in Zurich. There was a conflict brewing between Bolton and Piddington about the interpretation of all sky radio continuum images based on distributions of radio stars, combined with thermal emission throughout the Milky Way. During this discussion, the details of the July 1949 presentations by Bolton and Westfold were reconstructed. Pawsey wrote to Bowen (9 September 1950) about Westfold: "The colloquium took place just before Westfold left [for Oxford] and may be remembered from a criticism by many of us of some other allied data which Westfold put forward and finally withdrew." Bowen replied on 22 September 1950 to Pawsey summarising the colloquium given by Bolton and Westfold: "... Bolton confined himself more or less to a statement of results. Westfold, however, embarked on some ideas about the mechanisms ... and ran into some criticisms ... These criticisms prevented him finishing what he was going to say, and this appears to have been the crucial point." The Westfold and Bolton papers are summarised in footnote 18 this chapter; Piddington's rival paper, "The Origin of Galactic Radio-Frequency Radiation" appeared in 1951.

Elizabeth ("Betty") Kate Hall, née Hurley (1919–)[21]

Elizabeth Hall has provided the most reliable long-term account of the life of Payne-Scott; Betty Hall knew her from 1946 to the time of her death in 1981. Ruby and Bill Hall befriended Betty when she was a new arrival in Australia in the post World War II years.

Elizabeth Kate Hall, *née* Hurley (no relation to Bill Hall), was born in London on 5 June 1919 and grew up over a florist shop run by her parents, who had inherited it from Betty's grandparents. As an only child, she became an orphan after her father's death when she was seven and her mother's when she was seventeen; she ran the florist shop for several years with the help of the manager. Betty at the age of 17 is shown in Fig. 11.4. She was conscripted into the Royal Navy in early 1944 when she was 25 and the shop was then closed. In the Women's Royal Naval

Fig. 11.4 Elizabeth Hurley (Betty Hall) at age 17 in London, circa 1936. Photo from and permission of Betty Hall (Dr. Elizabeth Hall)

[21] Based on letters dated 16 February and 19 March 1999, plus an interview on 12 February 2007 with Betty Hall and two of her daughters, Sue Brian and Jan Christensen.

Fig. 11.5 Elizabeth Hurley. Immediately after World War II in Sydney with friends. Photo from and permission of Betty Hall

Fig. 11.6 Elizabeth Hurley. Postwar in Sydney. Betty changes a tyre on a military truck. Photo from and permission of Betty Hall

Service (WRNS), she became a transport driver. She volunteered for overseas service and was sent to Australia just as the War was ending in 1945. Two images of Betty in the Navy are shown in Figs. 11.5 and 11.6; the latter is an impressive scene as she changes a tyre on a military truck. In Sydney she was attached to the civil engineers as a driver and saw a lot of the city. She was discharged on 5 July 1946, and having decided to stay in Australia, found a job while she lived in a single room in Bondi, an eastern beach suburb in Sydney. With no ties back in London, Betty remained in Australia.

Betty began bush walking with another young woman. On one of their first trips in the Burragorang Valley they returned quite late, missing the last bus from the top of the track. They were rescued by a man in a truck, who told them about the

Sydney Bush Walkers (SBW); since she knew she needed to develop a network of friends, she became a prospective member in December 1946 and a full member on 30 April 1947. During the 1946 Christmas period, she went on a Mt Kosciuszko trip before she was a member. She wrote:

> On this first trip I took the wrong kind of billycan – a tin one which makes the tea taste terrible – I was teased by Ruby for taking "Clive of India" brand curry powder, thus betraying my Imperialist tendencies, and my sleeping-bag was too thin for Kosciuszko weather. Bill and Ruby forgave me the billycan, used the curry-powder, insisted that I sleep between them to keep warm, and during the trip taught me many of the things a bushwalker in Australia needed to know. Listening to their talk around the campfire gave me my first introduction to Australian left-wing politics. Ruby was a forceful arguer and antagonised some people, but she was not an intellectual snob. Although I was no match for her, she never made me feel like a fool.

After the probationary hikes (see Chap. 12), she came before the full SBW committee to be considered for membership. There was some doubt about her because she had been hitchhiking, an activity frowned upon; but Payne-Scott spoke up for her saying that Betty "was good at washing up". In Fig. 11.7, she is shown on a bushwalking trip to Era Beach in the Royal National Park near Sydney in 1949. Betty has stated that "the friends that I made in the SBW, including Bill and Ruby, influenced my decision to stay in Australia".

> As to Bill and Ruby's adopting me, they did assume a somewhat parental role in helping me to settle into the club and find my level ... So far as Ruby's matchmaking was concerned, this was minimal but effective ... Because he [Phil Hall] was 4 years younger than me I was refusing to take him seriously and it was only when Ruby indicated we would make a good pair that I began to consider him as a possibility.

Remarkably, Betty Hall knew little of Ruby's professional life. As the two Hall families began to raise their children (Ruby with two and Betty with four, having

Fig. 11.7 Elizabeth Hall, circa 1949, Era Beach, Royal National Park, Sydney. Photo by Phil Hall, provided by Betty Hall. Used by permission of Betty Hall

Fig. 11.8 Betty Hall, recent photograph, provided by Betty Hall, 2007. Used by permission of Betty Hall

married Phil Hall in early 1950),[22] they had little contact until Payne-Scott fell ill in the 1970s. Betty Hall did not know that Payne-Scott was in the CPA; in fact Betty had tried to recruit Payne-Scott into the Oatley branch of the party, with no success![23] Betty remembers that Bill Hall (Ruby's husband) insisted that when she delivered the CPA newspaper, *Tribune*, to them in the 1950s, she not leave the paper at or near the postbox by the street, but take it directly to their house. He was afraid that the neighbours would notice the title of the publication!

Betty Hall's impressions of Payne-Scott are of lasting value in assessing the character of her friend:

[22] Betty Hall married another Bill Hall who was not related. Since both "Bills" were in the SBW, each was asked to use his middle name in the organisation. Payne-Scott's husband refused since his middle name "Holman" was the name of a man who had disgraced the Labor Party. So Betty Hall's husband-to-be (William Phillip Hall) adopted his middle name and became "Phil Hall". Betty's children are Marion Bagot (born 10 February 1951), Susan Brian (born 9 June 1952), Janet Christiansen (born 9 April 1954), and Geoffrey Hall (born 12 April 1958).

[23] Betty and Phil Hall left the Party in 1956, after the Soviet invasion of Hungary.

Like the rest of us, Ruby may have had her faults, but I have the warmest recollections of her. Although she was not in general demonstrative it still moves me to tears to remember how she referred to Bill as "my Billy" and how her face softened when she spoke to him.

A recent photograph of Dr. Elizabeth K. Hall is shown in Fig. 11.8.

O. Bruce Slee (1924–)

Bruce Slee made his entrance into the world of radio astronomy while serving in the Royal Australian Air Force. He joined the RAAF in 1942 and towards the end of the War he was transferred first to a radar station on Melville Island (north of Darwin, Northern Territory) and then to Radar Station No. 59 near Darwin; here he was the Sergeant in charge of the station (Orchiston 2004). At this station in late 1945, he discovered severe radio interference at sunset and concluded that this was due to radio emission from the sun. After this 200 MHz detection, Slee submitted a report to RPL and was subsequently offered a job by J.L. Pawsey, which he began in November 1946. His first position was as a technical assistant in the group at Dover Heights, working under the supervision of J.G. Bolton. In succeeding years Slee played a major role in the early "radio star" research with Bolton and Stanley. The remarkable career of Slee has been described by Orchiston (2004).

Slee got to know Payne-Scott while they both worked at the Dover Heights station in 1947. They overlapped for some months until Payne-Scott left for the Hornsby site in late 1947–1948 (Chap. 8). Slee recounted his impressions in an interview for an ABC (Australian Broadcasting Corporation) Television production of the series REWIND. The episode was recorded in May 2004 and broadcast in February 2005, but unfortunately the Slee interview was not shown. Figure 11.9 shows a photograph of Slee and Goss taken after the interview with Slee at Dover Heights (Sydney).

Slee has described the complex interactions at Dover Heights during this period. For daytime observing, there were two solar groups, Payne-Scott's and Bolton's. During late 1946, Bolton began his survey of circular polarisation of solar bursts, but the sun showed no activity at the end of 1946. Bolton continued his solar observations in early 1947, but by early 1948 he had lost interest in working on radio emission from the sun.[24] He and his group stopped their solar projects, while Payne-Scott continued. In October 1947 his team began the "radio star" observations in earnest, which led to the discovery of Taurus A on 6 November 1947.[25] This was probably the source of much of the conflict as Bolton's group wanted to

[24] Much of Bolton's solar noise work on the circular polarisation of bursts and outbursts was in fact later published by Pawsey in his review paper of 1950a (see Appendix E). See footnote 13, Chap. 8.

[25] The Cygnus A "radio star" had been detected on 17 June 1947 and additional observations were made in July, August and September. During this period, Virgo A (although then called Coma Berenices A) and Centaurus A were also discovered.

Fig. 11.9 Bruce Slee and Miller Goss at the time of the ABC Television filming of the "History Detectives" programme at Dover Heights May 2004. Slee had worked with Payne-Scott at Dover Heights in the late 1940s. An abbreviated version of the TV production (without the Slee interview) was broadcast in February 2005 in the "Rewind" series. Photo by Laura Critchley. Used by permission of Laura Critchley

test the equipment in the day time (the Yagi antennas at 100 and 200 MHz), at the same time as Payne-Scott's team wanted to carry out solar observations. Slee has used the terms "competing, tension, and confusion" to describe the relations between Payne-Scott and the others during this period. We assume that the nature of these tense conflicts was one of the reasons for Payne-Scott's transfer of her equipment to Hornsby in late 1947 (see Chap. 8).

As Slee has said, "Ruby always had lots of stamina and her own inner strength." Slee also mentioned that Payne-Scott was a mentor for his development; she was 12 years older than Slee with a thorough background in physics and mathematics. She referred him to W.M. Smart's classic astronomy monograph, *Text Book of Spherical Astronomy*,[26] which was very useful to the electrical engineers and physicists, who needed to learn about astronomical coordinate systems, time, refraction, etc. In addition, Payne-Scott took Slee to plays at the Independent Theatre in North Sydney, then managed by Doris Fitton.[27] Slee never met Bill Hall and knew little of Payne-Scott's private life. He knew about her left-wing views and witnessed her verbal conflicts with more conservative colleagues. In the ABC interview in 2004, he mentioned her "socialist bias" in discussions of world events of the late 1940s. In summary, Slee said of Bolton, Stanley and himself at Dover Heights, "We were timid compared with the forceful Payne-Scott".

[26]This book went through five editions from 1931 to 1962 and was later reprinted in 1965 and 1971.

[27]Doris Fitton ran the Independent Theatre from 1939 to 1977, at 269 Miller Street, North Sydney.

Gordon J. Stanley (1921–2001)

Although Slee got on well with Payne-Scott at Dover Heights, Gordon Stanley did not. Ruby was strong-minded, forthright and made her opinions known, on both scientific topics and daily discussions of current news. Slee reported that she "did not suffer fools lightly" and the rivalries with Stanley, as well as with John Bolton, were intense. She did not give in but held her own. An ugly incident took place in this period which is reported by Kellermann et al. (2005) in their description of the life and career of Gordon Stanley. Stanley painted "Men Only" on the only toilet door. Payne-Scott was not intimidated, however; she just ignored the sign and went into the toilet, laughing.[28]

In later years Payne-Scott's colleagues recognised her brilliance. In a letter to Goss dated 9 October 1997, Stanley wrote:

> She was part of my early education on women's issues, and despite early insensitivities on my part, I grew to have a great respect and liking for her.[29]

Lyn Brown (1918–)

Payne-Scott was closely associated with Lyn Brown, a librarian at RPL on the University of Sydney campus. Lyn Brown is an especially important contact due to the connections she had with Payne-Scott at work, in the Sydney Bush Walkers, and as a neighbour in Oatley.[30]

Brown heard about Payne-Scott while she was a student in the Faculty of Arts at the University of Sydney in 1936. Payne-Scott was already a prominent science graduate and later a Demonstrator in Physics. In 1943, after working for a few years at the CSIR's Division of Animal Husbandry at Badgery's Creek to the west of Sydney (near the location where Mills and others were to carry out some pioneering radio astronomy in the early 1950s), Brown transferred to NSL/RPL as a typist for

[28] Slee has reported (1994) that the chemical toilet "needed our frequent attention"; it was necessary to construct ditches to channel the rain water away from the block-house as well as to plant grass to stabilise the sandy soil. On another occasion, the men at Dover Heights were highly amused by Payne-Scott's inexperience in applying paint to an old desk at the field station. In that era, oil paint had to be thoroughly mixed before application, with the carrier (e.g., linseed oil) being well distributed throughout the paint. Payne-Scott forgot to mix the paint and cheerfully applied it in an unmixed fashion. She was not allowed to forget this oversight; and was subsequently accused of being "impractical". This trait was not, however, evident in her maintenance of the radio equipment.

[29] Practical jokes had also been played on Payne-Scott during the War; her work-issued mat (foot rest) and radiator would be "stolen" on cold winter days, probably in retaliation for her aggressive attitude to women's issues.

[30] Based on a recorded statement of 17 min' duration on 17 February 1999, an interview in August 2003, and letters from Lyn Brown and Fred Brown (27 November 2007 and 4 December 2007).

D.H. Briggs.[31] After a few years, Brown became an editor of scientific papers. During the War years, Brown's work could be characterised as "house keeping for scientists", to use the terminology of Marjorie Barnard.[32] In 1947–1948, Brown (2001) was in charge of the "Technical Records" library, a collection of classified or secret documents from World War II. She was involved in the declassification of these, with the "secret" documents and books being sent to the military, while the others were given to the NSL/RPL library. At the conclusion of this project, Brown was transferred to the main library under the supervision of Marjorie Barnard. During the period 1943–1951, Brown knew Payne-Scott well. Ruby was a "distinguished member of the scientific staff" working on radar and later radio astronomy; since she used the library frequently, she often met Lyn Brown.[33]

At the beginning of the War, Lyn joined Ruby as part of an informal group from the RPL on weekend walks in the Blue Mountains or the Royal National Park south of Sydney. Saturday was a half work day during the War, except for 1 day a month when the entire Saturday was free. On the 2 day weekends, the group would camp overnight, enjoying a 2 day walk. Later in the War years Lyn Brown decided, with a few other RPL friends, to join the well established SBW in order to participate in their well planned series of walks. Here Brown also met Bill Hall, probably before he and Ruby were married in 1944. Brown has especially fond memories of 25 km walk led by Bill Hall from Waterfall to Bundeena (in the Royal National Park south of Sydney); the walk was characterised by views of pink flowers and the dark blue sea in the background. Ruby Payne-Scott was wearing "brief and practical shorts" and looking very athletic.

Lyn Brown married Fred Brown[34] (her maiden name was also Brown!) in early 1951, leaving CSIRO before her son Paul was born in early 1952. By chance, the Brown and Hall families came to live in Oatley; the Browns' two oldest children were almost the same age as Peter and Fiona Hall and went to the same primary school (West Oatley Primary School) as well as the same high schools (Sydney Technical High School at Bexley and Penshurst High School). The two families were connected via the children who played together as toddlers. By the mid to

[31]D.H. Briggs (1893–1987) was Head of Physics at the NSL (1939–1945), and later Chief of the Division of Physics (1945–1958). His wife was Edna Sayce, the first woman to graduate with a B.Sc. in Physics from the University of Sydney in 1917.

[32]In Brown's 1970 bibliography of Marjorie Barnard's written works, she elaborates on Barnard's turn of phrase: "Marjorie Barnard said once that what we Arts graduates were doing at CSIRO was 'a sort of housekeeping for scientists'. I would say that her brand of housekeeping came closer to homemaking." Brown (2001) has published a collection of poems; in several of these poems Marjorie Barnard is mentioned.

[33]On 18 September 1947 a meeting was held at RPL, consisting of Pawsey, Kerr, McCready, and Smerd with Arthur Higgs (Divisional Secretary) as the Chair, to discuss with Barnard and Brown how to organise the vast amount of new material on "solar noise" in the NS/RPL library. A complex system with a card index was instigated: NAA: C3830, A1/1/1, Part 2.

[34]Frederick Charles Brown worked at the CSIRO Division of Electrotechnology and later at NSL after its move from the University of Sydney to Lindfield, NSW. He retired in 1980 after 39 years in the CSIRO; he had known Payne-Scott at AWA in the pre War years.

Fig. 11.10 Lyn Brown in the 1960s. Photo by Fred Brown. Used by permission of Fred Brown

late 1960s, however, the two families had very little contact. Lyn Brown visited Payne-Scott in the nursing home in the year before her death, 1981.[35] An image of Lyn Brown in the 1960s is shown in Fig. 11.10.

Brown's assessment of Payne-Scott is striking:

> I remember Ruby as a clever, forthright, honest, generous and outspoken individual. She could come across as forbidding due to her strong opinions with no fear to express them. But she was a warm-hearted individual who seemed to have little sense of humour. We had quite different philosophies; I am a Christian while she was as an honest agnostic. We never discussed politics, except in so far as it affected our children in schooling . . . Ruby had a continual concern for the status of women in the work force . . . I am very glad that the life and achievements of Ruby are being talked about and hope that she will be remembered as an Australian woman who was very interesting, who made valuable contributions to knowledge . . .

[35]Like Payne-Scott, Brown returned to teaching in the 1960s; she taught French and German at Bankstown Girls' High School for about 2 years and in 1968 transferred to St. George's Technical College (as Payne-Scott had earlier) as a French teacher. She finished her career as a teacher of French in an adult education programme in Oatley.

Chapter 12
A Remarkable Family: Bill and Ruby Hall

All evidence points to a happy marriage between Ruby Payne-Scott and her husband, Bill Hall. Their two children, Peter Gavin Hall and Fiona Margaret Hall, are prominent, successful Australians.[1]

Bill Hall

William "Bill" Holman Hall was born in Inverell, New South Wales, Australia, on 22 August 1911. He died on 21 July 1999, just before his 88th birthday, in Wollongong, NSW. As was the case with Ruby Payne-Scott, Bill Hall came from families that emigrated from the United Kingdom in the late nineteenth century.

Agnes Paterson, Bill's mother, was born in about 1870 in Glasgow, Scotland and died in Oatley, Sydney, in 1963. She emigrated as a young girl, with her family which included five siblings. They departed from Glasgow on 15 March 1884 and arrived in Sydney exactly 3 months later.[2] Agnes, then 14, was the eldest child; the ages of the children ranged from an infant sister to a brother aged 11.

[1]This text is based on interviews with Peter Hall in Socorro, New Mexico in May 1998 and with Peter and Fiona Hall in Canberra in February 1999. Additional telephone interviews took place with Peter Hall in February 2007. Extensive interviews in March 2007 in Adelaide were carried out with Fiona. We have had extensive correspondence with both Peter and Fiona in the period 1998–2008. There have also been numerous letters and interviews with Dr. Elizabeth Hall, 1999–2008.

[2]Two hundred and ten passengers were on board the iron sailing ship, *Bann*. Agnes's parents were James and June Paterson from Lanark, Scotland (about 30 km from Glasgow). The family was Presbyterian. The father's occupation was listed as agricultural labourer (later documents listed him as miner and later still as railroad worker). At the time of Agnes's birth, James worked in the Glasgow railway yards. After working in either Inverness or Aberdeen, he decided to emigrate due to the higher wages paid in Australia. Each adult paid £14 for the voyage. Dr. Elizabeth Hall provided many of these details about the family of Bill Hall.

W.M. Goss and R. McGee, *Under the Radar*, Astrophysics and Space Science Library 363,
DOI 10.1007/978-3-642-03141-0_12,

Bill's father, Sydney Hall, was born in Deal, close to Dover in Kent, in 1872 or 1873; the date of his death is not known but was probably before 1950. Sydney and Agnes were married on 25 August 1902 in Inverell, a town in northern NSW, 230 km west of Grafton, the birthplace of Ruby.

Agnes and Sydney had eight children; Bill was the sixth, born in 1911. Of the six sons, five followed in their father's footsteps in the occupation of butcher. There were two sisters. Ivy (1902–1973) never married and lived near Bill and Ruby in Oatley East. She cared for her mother, Agnes, for the last decades of her life. Mary (married name Pilgrim) was one of the witnesses at Ruby and Bill's wedding in 1944, and remained close to their family while the children were young.

Bill Hall's family probably moved to Sydney during World War I; Bill remembered his father showing him the abattoir (slaughterhouse) at Homebush[3] where he worked as a butcher and warning the children to keep away from the abattoir, fearing that they would be distressed by the slaughter of the lambs.

In 1997, 2 years before his death, Bill related a remarkable experience from the early 1930s to his daughter, Fiona. Bill was his mother's favourite son, probably because he was one of the younger sons and still at home with her after the older sons had left, and because he defended his mother against her abusive and probably drunkard husband, Sydney. At the instigation of his mother, Agnes, Bill went by ship to Ireland to look for relatives with whom the family had lost contact. This suggests that the Paterson family may have moved from Ireland to Scotland during the nineteenth century. In any case, Bill had no success and soon returned to Australia. This trip was the first of two overseas trips that Bill undertook (see Chap. 15).

The details of Bill's schooling are unknown; we can surmise that Bill left school at an early age, probably in his early teens. He was trained as a French Polisher, a wood finishing technique known to require great patience; this training also gave him wide ranging carpentry skills, which he used extensively for the rest of his life. For example, he made furniture, including beds, for the family home. This furniture lives on in the homes of his children.

Although he was 28 at the start of World War II in 1939, he did not go into military service during the War, perhaps because of his colour blindness. In addition, Bill Hall had a "reserved occupation" during the War, working at the Garden Island Naval Base, the principal east coast base of the Royal Australian Navy.

Later on he became a telephone mechanic for the Australian Postmaster General (PMG), the government organisation that ran the posts, telegraphs and telephone service.[4] To qualify as a telephone mechanic, Bill attended a technical college; during this course, his wife helped him with his mathematics.

---

[3]Homebush would become the site of the 2000 Olympic Stadium.

[4]Merle Waltman (letter to E. Hall, 1999) remembered that Bill worked at the Sutherland PMG exchange, while Arthur Gilroy (letter to E. Hall, 1999) suggested that Bill was in charge of the Blakehurst exchange. Peter Hall remembered that in the late 1950s his father occasionally worked on a night shift, leaving home at 6 p.m. The increased wages were a welcome supplement to the family income.

Sydney Bush Walkers

A key aspect of Bill's life, that led to his meeting Ruby Payne-Scott, was his avid participation in bush walking.[5] On 6 November 1936, at age 25, he joined the Sydney Bush Walkers (SBW), an organisation founded in 1927 by Jack Debert.[6] Bill was clearly an accomplished bush walker. In this period just prior to World War II, he was a member of the "Tigers", a group of 10–12 high powered walkers which included Dot (*née* English) Butler, Alex Colley, Hilma Galliott Colley, Max Gentle and Debert (some of this group are in the large scene at Easter 1941 described below). Butler has described this group in some detail;[7] Figure 12.1 is an adventuresome photograph of Dot (Dorothy) and Bill holding a large Brown Snake killed by Bill; the Brown Snake is one of the more deadly Australian snakes with a potent venom. To qualify for this group, the applicant had to prove herself or himself over a 70 mile, 3-day hike in the Blue Mountains with an over 9,000 feet (2,700 m) change in elevation. Butler wrote about herself: "There were those of mighty statute and physique among the Tigers, but also among them was this small, neat girl who, once the going became really tough, could out walk and out climb all of them." The group broke up in World War II, since many members were in the armed forces.

Ruby Payne-Scott joined the SBW on 10 January 1941,[8] probably meeting Bill soon afterwards. It was customary to be subjected to a 3 month probationary period; the initiate was required to satisfactorily take part in three 1-day test walks and a full weekend test walk. She would have been judged on walking ability and on a general capacity to "fit-in". By Easter 1941 (13 April), Payne-Scott was on a bush walk with Bill in a large group (see Appendix J and Fig. J.1 and J.2) on the south coast of NSW near Pigeon House Mountain. Alex Colley, a well known environmentalist who joined the SBW in 1936 and was also a "Tiger," has written about how these dissimilar individuals could only have met on a bush walk:

---

[5] 'Hiking' in the US or "rambling" in the UK.

[6] Based on the delightful reminiscence *The Barefoot Walker: A Remarkable Story of Adventure, Courage and Romance*, written by Dorothy Butler (1991). Butler (1911–2008) joined the SBW in 1931 and made a number of major ascents in the Blue Mountains, west of Sydney. The term "bush walking" had only been invented in the late 1920s. Later in her life she went to New Zealand where she made a number of ascents, including Mount Cook.

[7] Based on the delightful reminiscence *The Barefoot Walker: A Remarkable Story of Adventure, Courage and Romance*, written by Dorothy Butler (1991). Butler (1911–2008) joined the SBW in 1931 and made a number of major ascents in the Blue Mountains, west of Sydney. The term "bush walking" had only been invented in the late 1920s. Later in her life she went to New Zealand and made a number of ascents, including of Mount Cook.

[8] Dates of Bill's and Ruby's joining the SBW provided by Bill Holland (SBW archivist) to E. Hall, 19 March 1999.

Fig. 12.1 Dot English Butler and Bill Hall; a large brown snake had been killed by Bill on a SBW (Sydney Bush Walkers) event late 1930s or early 1940s. Bill Hall family collection, used by permission of Peter Hall

> I knew Bill quite well . . . His background was very different to that of Ruby . . . later [he] became a telephone mechanic. He stuttered and looked a bit of a rough diamond – we called him Ben Hall – the bushranger.[9] He was a very hardy bush walker and a very popular member of the club. In normal circumstances Bill and Ruby would never have met, or contemplated marriage. But on a bush walk you share a range of experiences, some very challenging . . . and form close, often life long friendships, irrespective of educational, occupational or social background. Bill and Ruby were very devoted to each other.[10]

Members of the SBW remember that, despite the differences in their circumstances, Ruby and Bill were fortunate to find one another.[11] Ruby apparently made some walks with the "Tigers" but was never an official member. Figure 12.2 is a photograph of

[9]Ben Hall was no relative of Bill. Ben Hall lived from 1837 to 1865; he was an infamous Australian bushranger (outlaw) who was killed by police in NSW in 1865. Numerous films and television programmes have appeared about Ben Hall.

[10]Letter to Goss, 18 March 1999, from Alex Colley. Colley also wrote: "Bush walkers come from all walks of life, but we are all equal on walks."

[11]Elizabeth Hall has written (19 March 1999) that the SBW "was also a very successful marriage bureau. There is nothing like a hard walk to bring out the best and the worst in people; after you have walked with them a few times you have no illusions about your chosen partner!"

Fig. 12.2 Bill Hall and Ruby Payne-Scott. Pretty Plains, Kosciuszko National Park, Christmas 1947. Bill has his hair washed by Ruby. Bill Hall family collection, used by permission of Peter Hall

Fig. 12.3 Overland Track in Tasmania 1949. Bill Hall, Ruby Hall and Val Gilroy. Between Windermere and Pelion West, Cradle Mountain National Park. From Bill Hall family collection. Photo by Arthur Gilroy, used by permission of Lindsay Baudinet, his daughter

Bill and Ruby in 1947 on a Christmas trip to Pretty Plains near Mt Kosciuszko; Ruby is washing Bill's hair. In Fig. 12.3, Bill, Ruby and Val Gilroy are studying a map in Tasmania on the Overland Track in Cradle Mountain National Park between Windermere and Pelion West in 1949; the weather was clearly wet and cold.

R.W. Younger, a teenage member of the SBW during World War II, remembers that Ruby and Bill went out of their way to help new members make friends and adjust to the "onerous activities associated with bush walking. Ruby became a member of the [SBW] committee and took part in discussions at general meeting".[12]

Betty Hall (see Chap. 11) reports that, prior to their marriage in September 1944 (see Chap. 4), Ruby and Bill were living "in sin" in a house in Ashfield, a less common occurrence in 1944 than 60 years later. The marriage certificate listed her address as 5 Fairleigh Street Ashfield, Sydney, while his address was listed as 6 Stanley Street, Arncliffe, which was the existing home of his family. In late August or perhaps early September 1951, Bill and Ruby moved to Oatley,[13] where they designed and built their house, camping out while the construction was going on. They lived elsewhere during the week and camped on the building site during the weekend while working on the house (Figure 12.4a,b shows the house at 120 Woronora Parade, Oatley during this period; Bill can be seen faintly in the background as Ruby takes the picture). Peter and Fiona grew up in this house, from the time they were born, in late 1951 and late 1953, respectively.

Bill Hall, the Father

Friends and neighbours remembered Bill as a "good solid man,"[14] who spent lots of time with his children while they were growing up. With no car and no television, the parents created many opportunities for free and creative learning for them as they grew up in the 1950s and 1960s.

Fiona remembers her father with admiration. Bill was always proud of his wife's illustrious career as a radio physicist, from 1941 to 1951 at the RPL, CSIR and later at CSIRO. He told Fiona in 1987: "Doing well is tough, even more so for women. I saw what your mother went through as she battled discrimination."[15] Not surprisingly, Bill was also a proponent of sexual equality; Fiona has stated that her father did not have "a sexist nerve in his body." Bill told Fiona that her mother's troubles with CSIR and CSIRO were due to the fact that she had the "wrong gender". The last few years of his life were spent at the Fig Tree retirement home in Wollongong.

[12]Letter to Goss, 15 March 1999. See Chap. 13, footnote 12.

[13]NAA; C3830, A1/1/1 Part 6. Payne-Scott wrote to Pawsey on 18 October 1951 (a month before Peter was born): "We moved in here about 6 weeks ago without either light or water, but things are gradually straightening out, although I still seem to entertain sundry tradesmen most days of the week."

[14]Betty Hall interview, February 2007.

[15]This conversation occurred when Fiona's career was blossoming. At a time when she had been invited to visit Japan to exhibit her work in an art gallery, her brother had just received a prestigious prize from a French mathematical society.

Fig. 12.4 (**a**) and (**b**) The new house of Ruby and Bill Hall, likely in mid-1951. 120 Woronora Parade, Oatlley, NSW, Sydney. Bill and Ruby camped out during the construction of the new house. In (**b**) Bill can be seen. Photo by Ruby Hall. Bill Hall family collection, used by permission of Peter Hall

Peter Hall: Mathematician

Peter was born on 20 November 1951, 4 months after his mother left CSIRO (Chap. 4). An early photograph of a young Peter, probably taken by his mother while he played in the garden at their home in Oatley, is shown in Fig. 12.5. In Fig. 12.6, Peter is shown as a young boy on a tricycle with his younger sister, Fiona, born 14 November 1953. The two are shown in a formal portrait in Fig. 12.7, while a striking photo (Fig. 12.8) shows the two with their father on holiday at Jamberoo, NSW, about 100 km south of Oatley. The only photograph we have of the entire Hall family in one image, taken in about 1962 (Fig. 12.9), shows them on holiday at Era in the Royal National Park (at a distance of less than 20 km from their home). (Unfortunately the quality of the photo is not very good.)

Fig. 12.5 A young Peter Hall in Oatley; photo by Ruby Hall. Bill Hall family collection, used by permission of Peter Hall

His mother encouraged Peter to be a scientist, by suggesting a variety of educational and career choices. However, it was probably her passion for the sciences that influenced him most. Peter remembered his mother telling him about her high excitement in early 1946 when she observed the radio emission of the sun at sun rise, as shown by interference fringes from the sea-cliff interferometer in 1945 and 1946 (Chaps. 6 and 7). He recounted:

> [She was exited] by the realisation that the radio emission was associated with sunspots; quite late in her life, the excitement was still with her.[16]

[16]Peter Hall interview, 12 February 2007. Could Ruby Payne-Scott have been recalling her memories of the first sun rise when interference fringes at 200 MHz were observed for the first time on 26 January 1946?

Fig. 12.6 The Hall children, Peter and Fiona, in the 1950s. Photo by Ruby Hall. Bill Hall family collection, used by permission of Peter Hall

As a child, Peter found in the Oatley home tables of solar azimuth and elevation for different times of the year as a function of local time. These were probably left over from the period Payne-Scott was observing the sun at Dover Heights in 1945–1947. He was a keen photographer and used these tables to determine the earliest times in the afternoon when the sun would be appropriate for photography near their home, thus utilizing a complex set of scientific data to guide his artistic inclinations.

In addition, Payne-Scott's experience as a mathematics teacher made a lasting impression on Peter. In the 1960s, while she was the Science Mistress at Danebank School (Chap. 14), she was also an adult education instructor at St. George's Technical College in Sydney. She told Peter that success in mathematics was as much a matter of intuition as it was knowledge of the rules. As an example of this assertion she told the story of a butcher who was a student in her adult education class. This man could neither add nor subtract large numbers, nor even work out

Fig. 12.7 Formal portrait of the Hall children, 1950s. Bill Hall family collection, used by permission of Peter Hall

simple algebra. He was, however, adept in the accurate estimation of fractions. Payne-Scott asked him the source of this proficiency. His answer was simple:

> Well, we have a carcass of meat. One third is gristle and bone and will be thrown away. Two-thirds is the saleable portion, on which I make a profit. I've done this so often that I always work out my profit margin on two-thirds of what I buy from the wholesaler.[17]

Peter also remembered how absent minded his mother was when she was concentrating on mathematics problems or lesson planning for the next day. This inattention could lead to over-cooked, if not burned, items in the kitchen. Because lamb was cheap in the 1960s, a frequent evening meal at the Hall house consisted

[17]Peter Hall interview, 12 February 2007, in which he recount his mother's story of the butcher.

Fig. 12.8 Bill, Fiona and Peter Hall on a family holiday at Jamberoo, New South Wales. Photo by Ruby Hall. Bill Hall family collection, used by permission of Peter and Fiona Hall

of lamb cutlets or chops. When her mind was too preoccupied with other considerations, Payne-Scott would forget to check the oven, with disastrous results. As smoke appeared, she would utter her self-invented swear word, "gordy buggers". In addition, her cuisine was often not of top quality; Bill was the better chef. Once when Peter was a child, Payne-Scott was in hospital for some days, with an attendant improvement in the quality of the fare at home as Bill looked after the cooking. Also in this period, Bill Hall would bring a store-bought pie or cake on his way home from work, something Payne-Scott would never do. Thus, when Peter's mother returned from hospital, Peter noticed that the quality of the cuisine deteriorated again in the Hall household.

Peter had initially planned to study physics at university:

Fig. 12.9 The only photo that we have seen of the entire Hall family in the same image. Taken by a friend at Era Beach in the Royal National Park about 1962. Left to right: Fiona, Ruby, Bill, a friend, Peter and second friend with hat. From Fiona Hall, 2009. Bill Hall family collection, used by permission of Fiona Hall

> ... when I finished [Sydney Technical High School], I was determined to become a physicist and I couldn't have given any better reason than that my mother ... showed me that physics was a wonderful and honourable profession.

When he began at Sydney University in 1970, the first physics lecture was given by Professor Robert May.[18]

> ... but he only gave the first lecture to the advanced class in physics and then he disappeared to do something more uplifting. And from that point on, things went downhill. If the other physics lecturers had been as good as [May], I probably would have been a physicist.[19]

Thus, Peter Hall became a mathematician with an Honours Degree in Mathematical Statistics from the University of Sydney in 1974. Peter spent the years 1974–1976 in the UK at Oxford. He received an M.Sc. from the Australian National University, Canberra (1976) and a D.Phil. from Oxford in 1976. While in the UK in mid-1976, Peter met his parents (Chap. 15) during their trip to Europe from Australia.

Peter Hall has had a remarkable career with numerous honours, establishing himself as a leading mathematical statistician in Australia. After a short period as a Lecturer in Statistics at the University of Melbourne (1976–1978), he was at the ANU from 1978 to 2006 (Professor since 1988). In 2006, he became a Federation

[18]Robert May (now Lord May) was Professor of Theoretical Physics at Sydney University, and later President of the Royal Society from 2000 to 2005.

[19]Peter Hall interview, 12 February 2007.

Fig. 12.10 Professor Peter Hall on a visit to the National Radio Astronomy Observatory P.V. Dominici Science Operations Center, New Mexico, USA, on 12 May 1998. Photo by Miller Goss

Fellow of the Australian Research Council at the University of Melbourne. He was elected a Fellow of the Australian Academy of Science in 1987 and a Fellow of the Royal Society of London in 2000. His publication record is prolific, including four monographs. He has received numerous honours and awards and is the President of the Australian Mathematical Society (2006–2008).[20]

Professor Peter G. Hall is shown on a visit to the NRAO, P.V. Dominici Science Operations Center, Socorro, New Mexico, on 12 May 1998, in Fig. 12.10.

Fiona Hall: Artist

Fiona Hall, the well known and acclaimed Australian artist (Fig. 12.11 in the Greg Weight photograph of 1995), was born on 16 November 1953, almost exactly 2 years after her brother Peter. In 2008, more than 100,000 people visited the exhibit "Fiona Hall: Force Field" at the Museum of Contemporary Art in Sydney. To date this has been the most popular Australian exhibition, based on the number of visitors.[21] The exhibition was a retrospective on the work of Fiona Hall since the 1970s and included photography, sculpture and installation art. In mid-2008 the

[20] Peter Hall married Jeannie Jean Chien on 15 April 1977.

[21] *Wentworth Courier*, Sydney, 18 June 2008.

Fig. 12.11 Fiona Hall. A formal portrait taken by Greg Weight in 1995. The fabric is the same as that shown in Fig. 12.16. (From Ewington 2005, the dust jacket) Photo from and permission from Fiona Hall

exhibition moved to New Zealand, first to the Wellington City Gallery and then to Christchurch from late 2008 to the first months of 2009.

At the age of two or three, Fiona Hall was given materials for simple art work by her mother. She has described the educational philosophy of her parents:[22]

> I think [my parents'] philosophy was you educate your children by example, not telling them what's what. I think I knew as a child subconsciously that I could do whatever I wanted to. There were no expectations about what a girl's later role in life should be.

Fiona Hall has described the semi-austere conditions while she was growing up. The family ate dried fruit with a few store-bought biscuits. Payne-Scott sewed for her daughter, making frilly party dresses for her birthdays. The Hall household was

[22]Karen Pakula in *The Sydney Morning Herald*, 1 March 2008.

full of books and discussions about theatre, the visual arts, literature and especially public affairs and politics. There was little music, however, in their home.

A major event in Fiona's life occurred in 1967 when she was only 14; her mother took her to the Art Gallery of New South Wales to see the landmark exhibition "Two Decades of American Painting", an "outing typical of the expeditions organized by Ruby to develop the children's interests that leaves a lasting impression".[23]

There were sometimes conflicts with her mother while Fiona was growing up; both were remarkably single minded. Fiona was very close to her father; on weekends the two would walk through the bush to visit her paternal grandmother and her Aunt Ivy who, as mentioned earlier, lived with Agnes. Fiona would accompany her father, who called her "my little shadow". A running joke, originating with Bill, was: "Who shall go first through the jungle so the tigers will get you [Fiona] or me [Bill] first?" Fiona had little affinity with her aged grandmother, who was 86 years older than Fiona and showed more affection for her grandsons. Agnes Hall was, however, the only grandparent the Hall children knew; she died in 1963 when Fiona was 10. By contrast, Fiona was close to her Aunt Ivy, who died in 1973 at age 71. The grandmother and aunt were suspicious of Bill and Ruby Hall; in their house in the late 1950s percolated coffee instead of tea was prevalent – "perhaps the drink of the radical left!"

Fiona attended Oatley West Primary School, and then Penshurst Girls' High School, where she had a reputation for being quite self-assured.[24] While she was at high school, Fiona considered architecture as a career, a fascinating connection, similar to the interests of her mother.[25] But in the last year of high school, she decided to attend art school. In the early 1970s, Fiona studied at East Sydney Technical College, where she focused initially on painting and then photography.[26] She had started experimenting with a simple Kodak camera when she was 11. She and her brother competed for the use of the family laundry "darkroom" for the development of film. Between 1978 and 1982, Fiona Hall studied at the Visual Studies Workshop in Rochester, New York, with visits back to Australia before the death of her mother in 1981.[27]

Since the early 1980s, Fiona Hall has branched into many new art forms, often using everyday objects such as soap, TupperWare, video tapes and legal tender from numerous countries. Some of her most famous objects have consisted of finely

[23] Julie Ewington, *Fiona Hall* 2005, p. 180. In this book, Ewington has given a thorough description of Fiona's life.

[24] Telephone call from the late Marie Alewood, a neighbour of Payne-Scott's, 12 Febuary 2007.

[25] When Peter Hall went through the house in 1999 (after the death of his father), he found numerous architectural books. Though Payne-Scott's interests ranged from mathematics and physics through particularly to biology, a special interest of hers was architecture. She told Betty Hall that if she could have a second life she would return to life as an architect.

[26] In *Fiona Hall* (2005, p. 27), Ewington has written; "In 1974 it seemed that Fiona Hall emerged fully formed as an artist." At this time she began exhibiting photographs, which were collected and published.

[27] Based on extensive biographical details by Ewington (2005).

Fig. 12.12 Reconstructed painting, "Aurora Borealis/Aurora Australis" in the Antipodean Suite series from 1981, by Fiona Hall. The book's cover, "Starry Night – after van Gogh" is in the same series. This image may be related to solar radio emission and even the ionosphere (Ewington 2005, p. 48). Photo from and permission from Fiona Hall

cast metal sculptures inside sardine tins.[28] The book by Ewington and the accompanying volume from the 2008 exhibition by MacGregor, Savage, Webb, O' Brien and Hall, provide masterful oversights of the *oeuvre* of Fiona Hall.

Figures 12.12, 12.13, 12.14, 12.15 show some the striking examples of Fiona Hall's early work which illustrate the eclectic nature of her art. In Fig. 12.12 (painted in 1981 as was the cover of this volume), we see images that suggest a subconscious connection with her mother's astronomical research work from 1945 to 1951. These "Reconstructed Paintings" are "notable for the obvious pleasure Hall took in multiple slippages of meaning between historical templates and contemporary life. She substitutes contemporary artifacts for depicted objects ... "In Hall's 'Starry Night' [the cover of this book] electrical cords swirl in the night sky as in van Gogh's famous 'Starry Night'." (Note the black banana skins, suggesting "the precariousness of both representation and interpretation."[29]) In

[28] Ewington (2005, pp. 105–107).

[29] Ewington, p. 42.

Fig. 12.13 Fern Garden at the National Gallery of Australia in Canberra, completed in 1998. Overview of the garden. (Ewington 2005, p. 121) Photo from and permission from Fiona Hall

Fig. 12.14 Details of the Fern Garden. Fifty-eight *Dicksonia antarctica* large tree ferns are located in this garden which can be viewed from windows inside the museum (Ewington 2005, p. 118). Photo from and permission from Fiona Hall

Fig. 12.15 "A Folly for Mrs. Macquarie", completed in 2000. A public art work for the Sydney Sculpture Walk, commissioned by the City of Sydney. The "Folley" is located in the Royal Botanic Gardens, in front of Government House with a sensational view of Sydney Harbour (Ewington 2005, p. 123). Photo from and permission from Fiona Hall

Fig. 12.12, a related study shows the "Aurora Borealis/Aurora Australis", a study remarkably related to solar bursts and the ionosphere through magnetism (note the magnet at the bottom). With some imagination the Van Allen belts of the earth can be discerned in this inventive creation. These two works are from the "Antipodean Suite" of 1981, the year of Payne-Scott's death. Payne-Scott's achievements in the field of solar physics are possibly depicted by this image; magnetic fields and energetic particles are suggested. In addition the connection between Type II bursts, first detected at Dover Heights in March 1947, and the Aurora Australis is striking.

Fig. 12.16 "Give a Dog a Bone." The photograph of Bill Hall was the last photo taken by Fiona Hall for public display. The fabric of his cape, woven by Fiona from coke can strips, is also seen in Fig. 12.11 (Ewington 2005, p. 139). Photo from and permission from Fiona Hall

"Fern Garden" was a commissioned work completed in 1998; the location is the "inhospitable"[30] courtyard of the National Gallery of Australia in Canberra. Fifty-eight *Dicksonia antarctica* large tree ferns are located in this garden which can be viewed from windows inside the museum. The shape of the pathways suggests a diagram of a woman's uterus.[31] In Fig. 12.13, an overview of this garden is shown; in Fig. 12.14 the details of the garden are visible.

In 2000, Fiona Hall designed "A Folly for Mrs. Macquarie" (Fig. 12.15), located in the Royal Botanic Gardens in Sydney; the site provides a remarkable view of Sydney Harbour. The death of her father in 1999 is commemorated in the iron work.[32] The challenging installation shown in Fig. 12.16 ("Give a Dog a Bone")

[30]Ewington (2005, p. 118): "In a seemingly inhospitable courtyard, formed by the junction of the original National Gallery building of 1982 and the extensions opened in 1998, Hall sited 58 mature ... giant tree ferns that are amongst the most ancient plants in Australia."

[31]Ewington, p. 118.

[32]Her father's name is inscribed on the scythe. Fiona has told Goss (18 April 2009) that she often walked through the Royal Botanic Garden (the location of the sculpture) with her father in the early 1990s; she would meet him near Sydney Harbour and they would walk through the Domain to Kings Cross railway station. Bill Hall would then take the train back to his home in Oatley.

was shown in three versions during 1996–1997. This photograph of her father was the last photograph taken by Fiona Hall for public display; Bill Hall died about 3 years later. The ironic nature of the installation is obvious:

> Hall required a king for this castle of worthless wealth [supermarket objects carved from soap]. The children's song which gave the works its title mentions "this old man comes rolling home" ... and Hall commissioned a larger than life-size photograph of her elderly father, William Holman Hall, naked except for a fabulous full-length cloak knitted from aluminum Coca-Cola tins.[33]

The children of Ruby Payne-Scott and W.H. Hall continue to have remarkable careers, clearly influenced by two gifted and loving parents.

Fiona thinks that her father was attracted to this location as it was close to Garden Island where he worked at the naval dockyard during World War II.

[33] Ewington, p. 138. The knitted cloak photograph has the title "The Social Fabric" (1996); Greg Weight was the photographer. Note the same fabric in Fig. 12.10. Fiona Hall has told Goss that she was apprehensive about asking her father to pose in this semi-nude fashion. She was pleased that he was totally at ease during the photographic session.

Chapter 13
Payne-Scott, Communist Party of Australia, Commonwealth Investigation Service and Australian Security Intelligence Organisation

Payne-Scott's left-wing political orientation was a well known characteristic of her life, often described by her colleagues in contemporary accounts. Through the years a question about her has frequently been asked: was she, in fact, a member of the Communist Party of Australia (CPA)? In recent years, we have discovered that she was indeed a member for a period during the 1940s and perhaps early 1950s. An example of contemporary opinion was voiced by Joan Freeman in her autobiography, (see Chap. 11) in reference to her experiences of 1941–945:

> She was a vociferous member of the pro-communist group, which was fashionable at that time, and tried to influence me in that direction, though without success; I was not at all politically minded.[1]

Since the CPA was a proscribed organisation under National Security Wartime Regulations between 15 June 1940 and mid-1943, Payne-Scott was highly unlikely to divulge her CPA membership to anyone outside her party branch.[2]

Harry Minnett (also Chap. 11 – and certainly not a left wing sympathizer) remarked that Payne-Scott was quite left-wing in her politics, this being a common thread at RPL during the War. Many of the scientists were sensitive to:

> ... the blemishes and injustices of the capitalist system. Some embraced Marxist ideas with enthusiasm particularly when Russia, attacked by Germany, became an ally of the west. However, even among the most fervent of these, I cannot recall anyone admitting to membership in the CPA. Some of us felt that such notions [the superiority of the Marxist dogma], when translated into practical politics, would lead inevitability to authoritarianism and the repression of dissent ... She [Payne-Scott] was certainly a vociferous debater of

[1] *A Passion for Physics*, 1991, p. 76.

[2] Mills (letter, 14 September 1997) wrote in connection with Payne-Scott's membership in the CPA: "This is not something which could be bandied about as any known communist would have faced instant dismissal."

W.M. Goss and R. McGee, *Under the Radar*, Astrophysics and Space Science Library 363,
DOI 10.1007/978-3-642-03141-0_13,

> communist ideas, but was she really a communist enthusiast and if so for how long? ... She was uncompromisingly logical and down to earth and I think she would eventually spot the flaws in the dogma.[3]

With this political orientation, it is not surprising that the Commonwealth Investigation Service (CIS) and after March 1949, the Australian Security Intelligence Organization (ASIO), opened a file on Payne-Scott. The Australian Archive has obtained a portion of the ASIO file. Of the 31 page report, all but pages 16–26 were released to the Australian Archive.[4] (A report dated 21 August 1950 quotes at length from a June 1947 report on CPA activities.[5]) The time frame for these CIS/ASIO reports was 3 August 1948 to 25 August 1959. The CIS and ASIO file on Payne-Scott has characteristics that are often present in these types of files[6] – a reliance on informers who appear to be highly ideological and who pay little attention to facts, as well as on CIS/ASIO officers who are pedantic about their files and have an obsession with anomalous details. The file indicates that CIS/ASIO paid a lot of attention to scientists within the CSIR and (after May 1949) CSIRO. The intelligence agencies were fearful of possible espionage related to military secrets and the Soviet Union. In addition, these organisations were generally opposed to any left-wing or communist influence in Australian society. Thus possible members of the CPA within the CSIR were of concern. CIS and ASIO were continually concerned that any communist or suspected communist was a serious threat to Australian national security. Scientists such as Payne-Scott, who had been prominent in radar research during World War II, were clearly of some concern because of the possible military value of their knowledge. There is absolutely no evidence that Payne-Scott was involved in passing along any information about radar; in any case, the research that started after 1945 was completely non-classified and was usually published in the "open" literature. None of the names in the Payne-Scott file have ever appeared in communications from US intelligence as being involved in passing secrets to the Soviet Union.

There was apparently an informant at the National Standards Laboratory or even RPL, who, in the earlier years from 1948 to 1951 reported on a number of activities.

---

[3]Letter to Goss, 21 January 1998.

[4]NAA: A6119/83, 679: Payne-Scott, Ruby Violet – aka Hall. Her ASIO file follows the pattern described in the *Wikipedia* article on ASIO: "People who have obtained ... files have said that their files contain notable inaccuracies and have noted considerable censured content."

[5]This June 1947 document is called "Australian Communist Party" (ACP) and originated from the CIS in Hobart, Tasmania; the text warns of the danger of possible leakage of information of secret information. A list of 30 public servants is given "who are either definitely known to be A.C.P. members or whose association with Communist activities is so close as to justify the conclusion that they are convinced Communists. It is known that the A.C.P. in certain cases has advised sympathizers to remain outside the party." Four individuals at 'Radio Physics (sic) CSIR' were named: Leonard Hibbard, Edward Inall, Thomas Kaiser and Miss R. Payne-Scott.

[6]We are indebted to Peter Murphy, Secretary of the SEARCH Foundation, Sydney, for extensive comments about the Payne-Scott CIS/ASIO file. At the time of the dissolution of the CPA in 1991, the assets of the Party were transferred to the SEARCH (Social Education and Research Concerning Humanity) Foundation.

Possibly the informer was at RPL, but not in the radio astronomy group (see below). In addition, it is possible that more than one person was involved in collecting information. The CIS/ASIO reports contained a few factual errors, e.g., (1) Payne-Scott's year of birth was listed as 1919 instead of 1912 in two of the documents and (2) her married name was listed as Edwards in a handwritten document of 9 December 1958, with the name of the husband as Henry Napier "Edwards" (or "Edwardes"; there was confusion about the spelling).[7] After further discussions during 1959, the conclusion was reached that Payne-Scott's married name was in fact "Hall" and not "Edwards".

The first ASIO document, dated 3 August 1948, consisted of a list of five individuals that the informer considered to be Communist supporters within the CSIR Division of Radiophysics; these included Ruby Payne-Scott, Rachel Makinson and John (Jack) Warner (1917–2008), all listed with their home addresses, except for Makinson. (The other two names on the list were machine shop employees.) A few months later a memo from the Deputy Director of the CIS in Sydney to the Director in Canberra summarised some details about the suspected communist, Payne-Scott. The memo of 29 November 1948 is shown in Fig. 13.1 with substantial redaction, carried out before the file was transferred from ASIO to the National Archives of Australia; the blacked out text must certainly contain the name(s) of the informant(s). The letter referred to in this document was a "letter to the editor", submitted to *The Sydney Morning Herald* and *The Daily Telegraph* in July 1948, during the controversy regarding secrecy within the CSIR, focused on the Chairman of the CSIR, Sir David Rivett. The letter is summarised in Appendix H; signatories from RPL and the National Standards Laboratory included, among 34 others, Payne-Scott, H. Minnett (a future Chief of RP), J.P. Wild (a future Chief of RP and future Chairman of CSIRO from 1978 to 1985), Warner, McCready, and R. Giovanelli. The letter to the newspaper made a strong case for the proposition that fundamental research in Australia would be better served by ensuring that all defence (hence secret) projects in Australia were carried out in organisations outside the CSIR. This issue was a major component of the national debate that led to the formation of CSIRO the following year.

[7]The confusion must have been with Henry (Harry) Napier Edwardes, a well known left-winger who joined RPL as an Assistant Research Officer in late 1942 or early 1943, at approximately the same time as Payne-Scott. Edwardes was described by Ron Bracewell (interview, January 2007) as a "radiophysics oldtimer, a practicing left-leaner. For 5 years or more after World War II, the US refused entry to visitors who had been members of the Melbourne University Labor Club ... [they were] eventually allowed to attend international scientific meetings in the US". After World War II, Edwardes worked on DME (distance measuring equipment) at RPL. Deery (1999, 2000) has described the plight of Richard Makinson and Thomas Kaiser, both caught up in conflicts involving scientific freedom in post World War II Australia. He has pointed out that Thomas Kaiser (1924–1998), a well known member of the CPA, was also a member of this University Labor Club during World War II. Kaiser worked at RPL, after gaining a First Class Honours degree in Physics at Melbourne in 1943. We know of no connection between Edwardes and Payne-Scott. It appears that ASIO was quite confused; four memos were written in 1958–1959, before ASIO succeeded in sorting out her correct surname. See footnote 17 Chap. 4 and footnote 78 Appendix H.

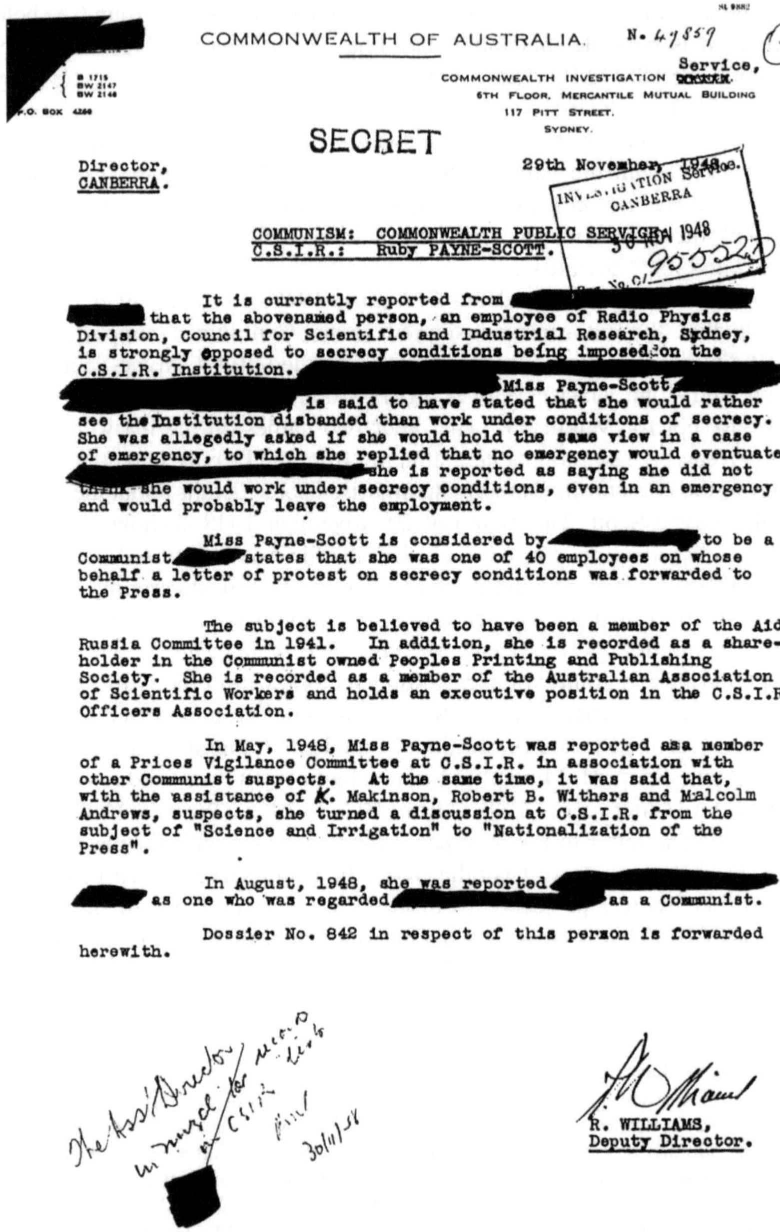

[redacted] B 1715
BW 2147
BW 2148
P.O. Box 4260

COMMONWEALTH OF AUSTRALIA. N. 47859 (5)

COMMONWEALTH INVESTIGATION Service,
6TH FLOOR, MERCANTILE MUTUAL BUILDING
117 PITT STREET.
SYDNEY.

SECRET

Director,
CANBERRA.

29th November, 1948.

INVESTIGATION SERVICE
CANBERRA
30 NOV 1948

COMMUNISM: COMMONWEALTH PUBLIC SERVICE.
C.S.I.R.: Ruby PAYNE-SCOTT.

It is currently reported from [redacted] that the abovenamed person, an employee of Radio Physics Division, Council for Scientific and Industrial Research, Sydney, is strongly opposed to secrecy conditions being imposed on the C.S.I.R. Institution. [redacted] Miss Payne-Scott [redacted] is said to have stated that she would rather see the Institution disbanded than work under conditions of secrecy. She was allegedly asked if she would hold the same view in a case of emergency, to which she replied that no emergency would eventuate. [redacted] she is reported as saying she did not [redacted] she would work under secrecy conditions, even in an emergency and would probably leave the employment.

Miss Payne-Scott is considered by [redacted] to be a Communist. [redacted] states that she was one of 40 employees on whose behalf a letter of protest on secrecy conditions was forwarded to the Press.

The subject is believed to have been a member of the Aid Russia Committee in 1941. In addition, she is recorded as a shareholder in the Communist owned Peoples Printing and Publishing Society. She is recorded as a member of the Australian Association of Scientific Workers and holds an executive position in the C.S.I.R. Officers Association.

In May, 1948, Miss Payne-Scott was reported as a member of a Prices Vigilance Committee at C.S.I.R. in association with other Communist suspects. At the same time, it was said that, with the assistance of K. Makinson, Robert B. Withers and Malcolm Andrews, suspects, she turned a discussion at C.S.I.R. from the subject of "Science and Irrigation" to "Nationalization of the Press".

In August, 1948, she was reported [redacted] as one who was regarded [redacted] as a Communist.

Dossier No. 842 in respect of this person is forwarded herewith.

[redacted]

R. WILLIAMS,
Deputy Director.

Fig. 13.1 A Commonwealth Investigation Service (later Australian Security Intelligence Organisation) report from 29 November 1950 about Ruby Payne-Scott. The informer may be a woman as the pronoun might be a three letter word. Rachel Makinson (K.R. Makinson) is also mentioned. NAA: A6119/83, 1679. From the collection of the National Archives of Australia

In addition in September 1948, Payne-Scott wrote a detailed letter to the CSIR Officers' Association (CSIROA) *Bulletin* about the issue of secret research within CSIR (see Appendix H); this complete document was in the ASIO file. Her impassioned letter contained a summary of the attacks in the US and ended:

> Are we to have similar attacks on Australian scientists who object to military control of research or hold "old fashioned" views on the international nature of science? They have already begun in our local press. The American examples show how easy it is to work up such attacks . . . The only defence lies in concerted action by all the scientists concerned to expose the ridiculous nature of the scare stories and the damage they will do to Australian science. Frightened men do not produce great research. Nor, I believe, do men behind barbed-wire [presumably researchers in secret laboratories]. Now is the time for the CSIROA to attack those who attack us, and insist that, if research is expected of us, we work under conditions that make research possible.

In a memo of 29 November 1948, there appears to be a mistake made by the censor. In the sentence: "Miss Payne-Scott is considered by BLANK [redacted] to be a Communist, BLANK[redacted pronoun] states that she was one of 40 employees on whose behalf a letter of protest on secrecy conditions was forwarded to the Press." Almost certainly the pronoun, referring to the informant is a three letter word, hence *she*. We have been informed by staff at the National Archives of Australia in Canberra that this mistake is a common occurrence for redacted text; i.e., the censor has neglected to disguise whether the singular pronoun has two or three letters. It is therefore possible that the informer was a woman, still unknown to us.

The next significant document was dated 26 April 1950, containing discussions of five prominent individuals at RPL: J.L. Pawsey[8] (the Head of the radio astronomy group and the Assistant Chief of the Division of Radiophysics), J.H. Piddington (an eminent radiophysicist who played a major role in the commissioning of the RPL radar near Darwin in early 1942, during the Japanese attacks), Payne-Scott, T. Pearcey (a mathematical physicist who worked on the first CSIRO computer after

[8]The assertion of Pawsey's membership in the AASW. (Australian Association of Scientific Workers) was plausible. There have been suggestions that this group was a Communist "front." But a thorough study by Jean Buckley-Moran ("Australian Scientists and the Cold War," in Martin et al. 1986) indicates that this claim is unlikely. The AASW existed only from 1939 to 1949. At one point, the group could claim about one-third of the total population of "all scientists [in Australia] irrespective of professional standing." During the War, AASW was highly regarded by the Australian government. Towards the end of the War, conflicts emerged between "progressive-liberals" and a smaller but influential radical group. On the issue of unionisation, the former group won out over the anti-capitalist radicals. In the post War period, W.C. Wentworth attacked the AASW as the anti-communist movement gained ground in Australia. Buckley-Moran wrote: "After several of its members were named in Parliament in March 1947, to be a member of the AASW *implied* Communist Party membership." The controversies around the creation of the CSIRO in 1949 contributed to the confusion. Finally Jack Lang, the prominent right-wing Labor parliamentarian, attacked the AASW. With all this pressure, AASW ceased to exist on 31 July 1949. Buckley-Moran wrote: "The expedient attack effectively put an end to the public articulation of social responsibility in science for a generation of scientists in Australia. ... AASW's premature demise marked a watershed in the attempt to negotiate a central relevance for science in society and to break down the isolationist mould of scientific production in Australia."

World War II, singled out as a "wild looking type with long locks") and V.D. Burgmann (later Chief of Textile Physics, CSIRO, 1958–1969, and Chairman of CSIRO from early 1977 to late 1978). This document is shown as Fig. 13.2. The text about Payne-Scott was remarkable: "She is a queer girl; a bright student but very erratic". Was a member of the University Christian Union which seems to be the forerunner of activity in Leftist Groups. It is thought that she is in a Feminist Group and that she may be a supporter of Jessie Street[9] but it is not known whether that is in the political or feminist field. "I would not put anything beyond her."

There is a strange document, dated 31 August 1950, concerning a visit to Camden, NSW (about 40 km to the south west of the centre of Sydney). The claim was made that a USSR Embassy car had been seen near this site and that A.L. Green (1905–1951, at the time Head of the Ionospheric Prediction Service) had been at this location.[10] The site was undoubtedly the Radio Research Board site of the CSIRO in Camden. In the mid 1950s this became the Radio Research Laboratory and later the Upper Atmosphere Section of the CSIRO with D.F. Martyn as the Chief Scientific Officer (past Chief of RPL – see Chap. 4). The document mentions the "Radio Research Board W.T. Transmitting Station which is ... part of the Ionospheric research activity." The text continues:

> I believe that B.Y. Mills and R. Payne-Scott, two persons of interest, are connected with this work. I am not aware of the classification of the work being carried out at Camden and if it would interest USSR but enquiries could be instituted.

Mills and Payne-Scott were working at Potts Hill at this time, sharing the 97 MHz interferometer (see Chap. 9). She carried out solar interferometry using the swept-lobe interferometer (Chap. 9); Mills was busy determining the position of Cygnus A at night.[11] Mills was in fact involved in some peripheral ionospheric research (related to the nature of scintillations at 97 MHz based on these Cygnus A observations) while Payne-Scott was not involved in any ionospheric investigations. The relevance of this document remains a mystery and shows how an unrelated event could be connected to individual scientists who were suspected of being Communists.

---

[9]Well known activist for women's rights, social justice and peace, 1889–1970. Her father in law, her husband , Sir Kenneth Street, and her son, Sir Laurence Street, were all Chief Justices of the Supreme Court of New South Wales. She was the only woman Australian delegate at the Founding of the United Nations in 1945.

[10]Green (Chap. 3) had worked with Payne-Scott at AWA in the years just before World War II.

[11]Mills (6 February 1999) has written that he never visited the Camden site but did visit George Munro at the Radio Research Board (RRB) office at Sydney University (Electrical Engineering Building) circa 1950 to discuss the ionospheric problems (scintillations arising in the F layer) that the Cygnus A interferometer at Potts Hill uncovered. Since the informer was ignorant of the work at Camden, the informer was not likely to have been from the RRB, but possibly from RPL, since the individual knew that Payne-Scott and Mills shared an interferometer in 1950. Based on this fact, Mills suggested that the informer might have been from outside the radio astronomy group at RPL; this person might have been responding to an ASIO request about possible links to the Camden RRB station and RPL.

[redacted]

SECRET

(8)

Director Sydney.

C..S.I.R.O. - Radio Physicists

A person [redacted]
says in regard to the following persons:-

N.T. x J.L. PAWSEY: It is thought that he did belong to the A.A.So.W. but it is not known whether he still does. Nothing has been heard of his association with political matters. He holds a very high place.

N.T. x J.H. PIDDINGTON: A graduate of Sydney University who is back now from abroad. He is one of the bright boys and is a good Physicist - "I should not be surprised if he is in the MAKINSON group". It is thought that he is a nephew of the late A.B. Piddington (Barrister).

283. x Miss Ruby PAYNE-SCOTT: She is a queer girl; a bright student but very erratic. Was a member of the University Christian Union which seems to be the forerunner of activity in Leftist Groups. It is thought that she is in a Feminist Group and that she may be a supporter of Jessie STREET but it is not known whether that is in the political or feminist field. "I would not put anything beyond her".

x T. PEARCEY: An Englishman; wild looking type with long locks. Expect he would be in this group, i.e. Leftist Intellectuals.

x V.D. BURGMANN: Son of Bishop BURGMANN; good man at his work; has been abroad; nothing else known.

[redacted]

26th April, 1950

[redacted] 7.50

File R. Payne Scott
1 R2 [redacted] /50

Fig. 13.2 A CIS report from 26 April 1950 about five possibly suspicious individuals, including Payne-Scott (described as a "queer girl... I would not put anything beyond her.") and other prominent CSIRO scientists, including a future Chairman of CSIRO, Victor Burgmann (Chairman, 1977–1978). NAA: A6119/83, 1679. From the collection of the National Archives of Australia

Two ASIO documents in 1958 and 1959 provided conflicting accounts of the cause of Payne-Scott's resignation from CSIRO in 1951. In the hand written note of 9 December 1958 (concerning the name "Edwards" or "Edwardes," see above), the question was also raised as to whether Payne-Scott was still employed by CSIRO. Another respondent also wrote a note by hand stating: "Is no longer with CSIRO and left because she had a baby," which was in fact correct.

A thorough report from 2 March 1959 mentioned that an informer had "identified a Ruby Payne-Scott, a Research Officer at Radio Physics NSW who BLANK had heard referred to as 'a Red' and who was dismissed from CSIRO for failing to give notification of her marriage ..." In the next item in this report, the resignation is listed as being on 20 July 1951 "when she resigned because of impending child birth". A critical remark followed about her earlier marriage in 1944 and failure to report this until 1950, together with her use of her maiden name until her resignation. As discussed in Chap. 4, the failure to report her marriage was not the cause of her resignation.

A summarizing document in the ASIO file appeared on 1 November 1950:

> The above named [Payne-Scott] is a Research Physicist employed at CSIRO since 18 August 1941. She is the holder of five one pound dividend paying shares in the People's Printing and Publishing Society. LARGE BLANK THREE LINES REDACTED ... there is no evidence at CSIRO that Payne-Scott is a member of the Communist Party. The subject person is reported to be married to William Hall, and they are temporarily residing at 16 Nymbodia Street, Coogee, the residence of Dr. Makinson [Richard and Rachel] who is overseas. The previous address of Payne-Scott was 5 Fairleigh Street, Ashfield.

In 1998 and 1999, the irony of the above conclusion was revealed. With the assistance of the SEARCH Foundation Secretary, Peter Murphy (see footnote 6 this chapter), we were contacted by a previous member of the CPA, Mal Andrews.[12] Andrews indicated that Payne-Scott had in fact been a member in the 1940s and possibly in the 1950s. , Three additional former members of the RPL scientific staff subsequently confirmed this fact.[13] Len Hibbard wrote:[14]

[12]Andrews appeared three times in the Payne-Scott CIS/ASIO file. See Fig. 13.1 and also a document from the CSIROOA *Bulletin* of April 1950, showing that Payne-Scott was the Branch correspondent for New South Wales. A.M. Andrews from the "Information" branch of CSIRO was on the *Bulletin* editorial committee while John ("Jack") Warner was the President of the Officers' Association at this time. A third instance is from an undated report about Payne-Scott, possibly in late 1948. Andrews, Payne-Scott, Rachel Makinson and a third individual "turned subject of lecture [details of the lecture were not included] from 'Science and Irrigation' to 'Nationalisation of Press.'"

[13]These include Rachel Makinson (K.R. Makinson), B.Y. Mills and Leonard Hibbard. In addition, Bob and Christa Younger (see Chap. 12, footnote 12) have reported on events during World War II when they were in their late teens. They were "reluctant to inquire into the lives of Bill and Ruby [Hall], being aware of the reasons for their secrecy in relation to their employment and association with the [CPA]. However, we did join them and others at performances at the left wing New Theatre Company from time to time." See Pfisterer, "Brave Red Witches: Communist Women Playwrights and the Sydney New Theatre", in Pons 2002, p. 168. Australian plays were performed as early as 1933 in the New Theatre.

[14]Letter to McGee, 14 October 1998.

> As regards Ruby, she was a person that I admired as a "scientist" and as a politically conscious person. She was not blessed with a robust larynx but, despite a thin tremulous voice, she could make her views felt in a very forthright manner and didn't hesitate to do so. I knew her as a fellow member of the "Lab" communist party branch, and a powerful bush walker. She could outpace me both in the city and the bush . . . I do remember an occasion when she shared my two-man tent on such a walk and that she solved the problem of my snoring by the simple procedure of capping my mouth and nose with her hand until I woke and rolled over.

Few of Payne-Scott's family or friends knew about her membership in the CPA.[15] Her activity in the party in later years remains unknown. We assume that she broke with the CPA; a possible time was the period in 1956 when the CPA leadership repressed any discussion of Khrushchev's "secret speech" on 24–25 February 1956 to the Twentieth Party Congress in Moscow. In the speech, he denounced the cult of Stalin and the resultant consequences; many CPA intellectuals were strongly influenced by this historic event. There is no doubt that Payne-Scott showed bravery in following her convictions; if her membership in the CPA had been known in the 1940s, she might well have faced dismissal from the CSIR/CSIRO.

[15]Her children did not know about this aspect of her life. Peter Hall did remember the magazine, *Soviet Life* , at their home in Oatley. Peter also remembers that his mother once told him that she had voted for the CPA candidate in an election in the 1960s; she told Peter; "the CPA has the best educational policy."

Chapter 14
Danebank School 1963–1974; the Ruby Payne-Scott Lectures

In 1963, after her children had reached the age of 12 (Peter) and 10 (Fiona), Payne-Scott took a position as a part-time science and senior mathematics teacher at the Danebank Church of England School for Girls in Hurstsville,[1] only 5 km distant from her home in Oatley. Hurstville was easily accessible by train from Oatley, and Ruby's children were, of course, at local schools (Chap. 12). At Danebank School, she was known as Mrs. Ruby Hall. Two photographs of the school in early 2007 are shown in Fig. 14.1a, b. She had already taught at Woodlands School (Chap. 3) for a year and a half in 1938–1939 and, though this new experience started off in a promising fashion, it became quite disappointing after a few years. A major reason was the debilitating effect of possible Alzheimer's disease (Chap. 15).

In 1963, Danebank was a small school and had no specialist scientist teacher. The Wyndham Scheme had been introduced into NSW schools in 1962, and science in Years 7–10 was compulsory for all students, leading to optional multidisciplinary sciences in Years 11 and 12. Biology was the only science taught at Danebank. Mrs Joyce Cowell (Principal) was impressed by Ruby's academic qualifications, but there was little or no discussion of Ruby's earlier achievements. At this time, teaching was viewed by women with young families as an ideal profession, as it was possible to combine school hours and holidays with the duties of raising a family. Teachers at private schools did not have to be academically qualified, and their salaries were lower than those of teachers in state schools. Private schools therefore often had a reputation of offering an inferior standard of education.[2] Some parents would send difficult or poorly achieving children to a private school in the hope that smaller classes and a different philosophy might bring about improvements in attitude and academic progress. Facilities at Danebank were basic, with only one small room for science. Payne-Scott was responsible for organising the new laboratories and equipment essential for the new State and Commonwealth

[1]We are indebted to Carolyn Little (retired Science Coordinator, Danebank School) for details of Payne-Scott's career at Danebank; much of the text in this Chapter has been contributed by Little.

[2]Danebank School had this reputation in 1963.

W.M. Goss and R. McGee, *Under the Radar*, Astrophysics and Space Science Library 363,
DOI 10.1007/978-3-642-03141-0_14,

Fig. 14.1 (**a**) and (**b**) Two views of Danebank School in February 2007. Images obtained by Goss with the assistance of Jan Christiansen and Sue Brian, Betty Hall's daughters

requirements for Science education. She remained effectively in charge of science and senior mathematics until her retirement in 1974.

Payne-Scott never achieved any real intimacy with other members of the school staff. Relationships at the school were formal – first names were never used. Payne-Scott never mentioned her previous career, other than to mention that she had spent time at the CSIRO. However, she spoke with great love and affection of her family, and came across as a typical mother with domestic concerns which were shared with other staff members. She made generous offers of clothing to other teachers' children. Despite her frustration with the progress of many students, she expressed a deep concern for each of them individually, displaying a surprising knowledge of their backgrounds as a source of their problems. However, as she aged and was probably affected by the onset of Alzheimer's disease, her stories deteriorated into longwindedness and her ability to cope also declined. At her retirement in 1974, her farewell speech occupied more time than that of Cowell, who was retiring at the same time as Principal.

The history of Payne-Scott's relations with students is troubling. Past students have commented that she did not engage well with many of them, even though the classes were small. She appeared remote and was unduly stern in her treatment of the pupils, and could be dismissive of their academic standards. From the mid-1960s students found her behaviour eccentric. Her class management skills were poor, with girls walking in and out of the class without being noticed.[3] On occasion she called girls by the wrong names. She would ring a handbell to attract attention, leading to the nickname "Tinker Bell". This was in contrast to her teaching at Woodlands School in 1938–1939; Kate Foy, at this time a mathematics teacher at Danebank, had been her student at Woodlands in Adelaide in the late 1930s and had memories of an effective and tolerant teacher (see Chap. 3).

The brighter students, however, fared well under her instruction. She clearly "knew her stuff" according to one student whom Ruby taught in the early 1970s; she brought books from home for this girl and favoured her with a gift at Christmas. Some students understood that she was clever and had had a prior career in physics, but the extent of that career was never discussed.[4]

These contradictory impressions were probably caused by her illness. Because of her deteriorating condition in the early 1970s, the School Council introduced a policy of compulsory retirement in order to force her to leave. Payne-Scott did not maintain contact with the School after 1974, and the lingering memory of her there is that of an eccentric individual.

When we renewed our interest in Ruby Payne-Scott in 1997, we contacted the Principal of Danebank School who was surprised to hear about her past

---

[3]Her poor eyesight could have contributed to this problem.

[4]A striking positive comment about Payne-Scott appears in a school history apparently from the late 1970s or early 1980s. "When she commenced at Danebank there were no real laboratory facilities ... she planned the new laboratories and ordered and installed the new equipment needed for the new science syllabus. She left behind ... a very fine Science Department." Mrs. Cowell wrote to us in 1997, "She was a dedicated and caring teacher for 12 years of her service".

achievements. The Science Coordinator, Carolyn Little, was herself a physics teacher and had an interest in radio astronomy and the history of science. Little decided that the life of such a significant former member of the Danebank staff should be celebrated by the School and even by the wider community. The School initiated an annual lecture, the Ruby Payne-Scott Lecture, to be presented at the School by successful women scientists. The first lecture in 1999 was given, appropriately, by Dr. Anne Green,[5] a radio astronomer at the University of Sydney, on the topic, "Microwaves in Space". Peter and Fiona Hall have both supported the lecture series; they continue to do so, attending a number of them.

During the 10 years of the Ruby Payne-Scott Lecture series, they have been presented by an outstanding women scientists in a variety of fields. The guest speaker has generally given an account of her area of research and achievement, together with some insight into difficulties she might have encountered along the way. Each lecture is preceded by biographical information by students and visitors who had some connection with Ruby. On occasions, this has included a short dramatisation of events in Ruby's life, interviews with former students, and excerpts from radio and TV programmes. In 1999, for example, a theatrical presentation, entitled "Who Was Ruby Payne-Scott? A Reflection on the Life and Work of Ruby Payne-Scott" was presented by the Year 11 Drama and Physics classes with the role of Ruby played by Catherine Bond. The Danebank audience has been joined by students from local schools such as Penshurst Girls' High School (Fiona Hall's old school), St. George Girls' High School, Georges River College, Oatley and St. George Christian School, among others. In Appendix K, we summarise the Ruby Payne-Scott lectures from 1999 to 2009.

Sadly, Payne-Scott's experience at Danebank School was not a happy one. But as her daughter, Fiona, remarked: "Danebank School is a minor part of her life. Radar, radio astronomy and our family were the major events. Mum's constant battling against adversity and her lucky breaks, such as coming to Sydney from the country as a young teenager, characterise her life."[6]

---

[5]In 2008, Professor Anne Green (*née* Barwick) was the Head of the School of Physics at the University of Sydney. She is probably the third woman radio astronomer in Australia after Payne-Scott and Professor Beverly Wills (*née* Harris) of the University of Texas, Austin. McGee was in the audience for this inaugural lecture.

[6]Interview March 2007, Adelaide.

Chapter 15
The End of Payne-Scott's Life: A Retrospective

Last Years

When Payne-Scott left her position at Danebank School in late 1974, her mental condition was deteriorating. Her conflicts with colleagues at Danebank were possibly due to the first signs of Alzheimer's disease, which accelerated until her death only 6 years later (1981). She aged rapidly as indicated in the passport photograph taken of her in 1976 before the trip to Europe (see Fig. 15.1). This photo shows an aged individual; her daughter, however, has little memory of her appearing as an old person since the transition occurred rapidly.[1]

The mental deterioration can also be judged by a letter she wrote to Joan Freeman Jelly (Chap. 11) on 16 February 1976. The letter was to congratulate Freeman on the receipt of the Rutherford Prize. She explained to Freeman that her name was now Ruby Hall, because she had adopted Bill's surname when the children were born in the 1950s. She wrote about her teaching and was effusive about her children's successes. At this time Peter was at Oxford working on a D. Phil. in probability theory. Payne-Scott wrote about Fiona:

> Fiona is our surprise. She is a very good artist and her favourite medium is photography ... [she] is now using her savings to explore Europe.

The striking aspect of this letter is the poor quality of the handwriting when compared to her handwritten letters and documents of the 1940s and early 1950s. Probably the poor quality of her penmanship can be associated with the mental degradation that was beginning at the age of only 64.

An ambitious overseas trip was undertaken by Bill and Ruby Hall in 1976. Peter was finishing his degree at Oxford, which he had begun in 1974. Peter remembers that she seemed in a reasonable mental state in 1974 although he later remembered that she had occasionally been confused before he left Australia. Bill and Ruby Hall

[1] Fiona Hall had been told by her mother earlier that as she aged she would be able to accept the loss of physical mobility but could not accept any loss of mental faculties.

W.M. Goss and R. McGee, *Under the Radar*, Astrophysics and Space Science Library 363,
DOI 10.1007/978-3-642-03141-0_15,

Fig. 15.1 Ruby Hall's passport photo before the trip she and Bill took to Europe via Japan and then Vladivostok (Trans Siberian Express). Her aging was apparent in this photo from 1976, 5 years before her death. Bill Hall family collection, used by permission of Peter Hall

flew to Japan and then took a boat to Vladivostok (Figure 15.2 shows the Australian group that undertook this trip.) for the start of the Trans-Siberian Express to Europe via Moscow.[2] After some sightseeing in Moscow and a visit to Paris, they met Peter in London in the summer of 1976. He noticed immediately that his mother was confused and showed signs of physical deterioration; his father had written to him earlier that his mother was quite unwell and that Peter should not be surprised by her worsening state. In fact, Bill Hall took his wife to a medical specialist in Harley Street (the area with many medical clinics), but this consultation was not at all helpful. Peter returned to Australia to begin his University of Melbourne appointment and met his parents when they returned to Australia by air at the end of 1976;

[2]Peter and his mother had discussed the possibility of them taking this train trip together; however, he did the trip by himself on his way to Oxford in 1974.

Fig. 15.2 The "Trans Siberian Group" at a hotel in Japan; Bill and Ruby Hall are in the front row, extreme right. Bill Hall family collection, used by permission of Peter Hall

he spent Christmas with his parents in Oatley. Clearly the cruel nature of Alzheimer's disease was beginning to take a toll.

As we have described in the Preface, Woody Sullivan talked to Payne-Scott on the telephone on 3 March 1978; she was rather confused with failing memory. On 30 March 1981 (about 2 months before her death), Sullivan phoned her house and likely spoke with Bill Hall; he told Sullivan that Payne-Scott was in a nursing home with no memory of the past.

As Payne-Scott's condition worsened, Bill Hall looked after her with admirable dedication. They had been members of the Thea Hughes League of Health for many years; even when her illness was well advanced, Bill Hall took her to classes where she was able to join in the discussions in a limited fashion. He also took her on short bush walks near their home. As she became more handicapped, for the last year or so of her life she was moved to a nursing home in Mortdale (a suburb immediately to the north of Oatley). A few years earlier she had taught her husband to cook some of her favourite foods, which he took to the nursing home.

After her death on 25 May 1981 (3 days before her 69th birthday), Bill Hall distributed her ashes in the grounds of their house in Oatley which they had designed and built in 1949–950 at 120 Woronora Parade. He told Peter that his mother was now "at peace". In a discussion with a representative of the funeral company, Bill and Peter were faced with a dilemma as to what "religion" to put on a form. Bill Hall said that Ruby Hall would have preferred "atheist"; in the end the rubric was left blank.

The Character of Ruby Payne-Scott

Payne-Scott was well known as a decisive individual with strong opinions. Her son described her as a fair person who was not always logical. Joan Freeman wrote:[3]

> [She made] – a strong impact with her distinctive, very positive personality. ... She was tall, solidly built, with straight fair hair, a strong-minded, no nonsense disposition, and a shrill voice which she could use very effectively in an argument. ... she had a sincere, kindly, and generous nature, to which I instinctively warmed.[4]

Payne-Scott had both friends and antagonists at RPL. She was influenced by Pawsey throughout her career at RPL from 1941 to 1951; he chose her to become the science leader of the new radio astronomy group in 1945. She was also closely associated with Alec Little (Potts Hill, Chap. 9), Bernard Mills (World War II radar, LW/AWH and Potts Hill) and Chris Christiansen (AWA and Potts Hill). Her scientific leadership was, however, challenged by John Bolton in 1947, only a year after his arrival at RPL and he did succeed in usurping that leadership. The challenge led to the antagonism of 1947–1948 (Chap. 7 and 8, the "bust-up" between the two was described to Pawsey by McCready). The conflict was partially resolved when Payne-Scott was exiled to the Hornsby site (Chap. 8) where the nature of Type III bursts was determined in an elegant manner. The Potts Hill interferometer was the highlight of her astronomical career, combining innovative engineering with the direct determination of the motions of Type IVM bursts for the first time.

As pointed out in the Epilogue (Chap. 16), however, the ironic fact is that this book would not have been written had John Bolton not changed his view of Payne-Scott in the 1970s.

---

[3]In *A Passion for Physics*, p. 76.

[4]A dissenting point of view has been expressed by Robert Coulson, a War time colleague of Payne-Scott, and a friend of Joan Freeman and Ron Bracewell. He was a managing director of the English Electric Valve Company for some years, responsible for establishing the traveling wave tube section of EEV. Coulson has written that Payne-Scott was a "uniquely difficult person to get close to and rather daunting". He was impressed by her low tolerance for "non-adherents to her faith [communism]" (email, 21 February 2007). Bracewell and Freeman did not share this negative assessment. B.Y. Mills (email, 23 August 2008) has pointed out that most conservatives at RPL shared Coulson's view; those who worked with Payne-Scott or knew of her achievements were quite positive in their assessment. On 2 August 1947, Coulson wrote Bracewell a letter to Cambridge (UK) with news from RPL back in Sydney. He wrote about the "R.P. Comrades [communists]" being "constantly vigilant" and was amazed to hear that Payne-Scott was married. He wrote Ron: "Did you know that Ruby is married? Has been for some time? I didn't." (From Bracewell archive July 2009.)

Fig. 15.3 Cat photographs by the Hall family. 15-3 is a well known Fiona Hall photo from 1974 when she was 21. The neighbour's large cat was in an adjoining garden. (from the Fiona Hall collection in *Fiona Hall* by Julie Ewington, 2005, p. 26)

Ruby Payne-Scott's Passions

As described in Chap. 12 and Appendix J, Payne-Scott's passion in life was bush walking. She found her husband in the SBW in the early 1940s and was an accomplished bushwalker for many years.

She also had a passion for cats.[5] A famous photograph, taken by Fiona Hall in January 1974, of a large cat in a neighbour's yard is shown in Fig. 15.3. This is Fiona's first photograph to appear in a public collection. Her brother's early affinity for cats is shown in Fig. 15.4, with the young boy holding a much smaller cat.

Payne-Scott also had a long association with another famous cat fan, John Jaeger (1907–1979), the well known Australian applied mathematician and geophysicist. Jaeger was a student of H.C. Carslaw (1870–954), with whom Payne-Scott had studied mathematics at the University of Sydney in the 1930s (Chap. 2).[6] In January 1949, a meeting of the Australian and New Zealand Association for the

[5]This feline interest was also shared by the next generation. See Ewington, p. 180.

[6]In Appendix C, Jaeger's contributions to the calibration of the lobe patterns of the LW/AW (later sea-cliff interferometer) system are described. Toward the end of World War II, Jaeger gave a series of lectures on the Laplace transform (usually a transform from the time domain to the

Fig. 15.4 A photo taken by Ruby Hall of a young Peter and the family cat. Bill Hall family collection, used by permission of Peter Hall and Fiona Hall

Advancement of Science (ANZAAS)[7] was held in Hobart, Tasmania, at which a number of radio astronomy papers were probably presented. Jaeger, who was one of the participants, invited Payne-Scott and others to his house in Hobart where he was a Professor of Mathematics. Payne-Scott was impressed by the fact that he had several cats and even a number of books containing poems by T.S. Eliot about cats. This mutual love of cats was a common bond; Jaeger spent a portion of his time at RPL in Sydney during these years.[8]

Ruby Payne-Scott's Legacy; Why Did She Not Return to RPL in the 1960s?

Payne-Scott certainly played a key role in the development of Australian radio astronomy in the decisive period of March 1944–1951. Pawsey relied on her for many aspects of the new research program that was to set the course of Australian astronomy until the end of the twentieth century. She had the requisite engineering experience based on the radar work of 1941–1945 and especially the mathematical

frequency domain); this integral transform was commonly used in radar systems. Payne-Scott and Bracewell were enthusiastic attendees at these lectures.

[7] During an excursion in the rain, the photo shown in Fig. 15.5 was taken; Payne-Scott and Pawsey stand next to Jaeger, the third from the right and, as was often the case, Payne-Scott is eating.

[8] Jaeger's affinity for cats is described by Paterson (1982); see Chap. 2.

Fig. 15.5 Photograph of a group visiting Mt Field National Park near Hobart during the January 1949 ANZAAS conference. From left : Frank Kerr (RPL), Kathleen Kerr, H.L. Humphries (then in the Cloud Physics group at RPL), Don Yabsley (RPL, hat), John Jaeger (University of Tasmania, hat), Pawsey (RPL, rain shawl!) and Ruby Payne-Scott (RPL). She is eating, as is often the case in group photos. (Identifications provided by the late Don Yabsley.) In 2002, Jim Lovell of the University of Tasmania found the cabin near Wombat Moor near Lake Dobson. The cabin is "Telopea," the botanical name for the well known Australia flowering plant, the Waratah. In 2002 the cabin was being renovated. Bill Hall family collection, used by permission of Peter Hall

skills to formulate the concept of radio interferometry and Fourier synthesis in radio astronomy (Chap. 7). Her contributions to solar physics were also important. She was without doubt the discoverer of Type III bursts[9] at Dover Heights and Hornsby, participated in the discovery of the rare Type II bursts in 1947 (and ironically ceased to believe in the reality of this important classification in late 1947), participated in the discovery of Type I bursts at Dover Heights in 1945 and 1946, and even observed Type IV bursts in 1946 and 1949 without realizing the uniqueness of these events. These bursts were only recognized to be a new type of solar radio emission event in 1957 by Boishot and Denisse. They suggested that the Type IV emission was caused by synchrotron emission from relativistic electrons. Payne-Scott and Little observed the motions of these Type IVM events directly for the first time using the swept lobe interferometer at Potts Hill in 1949–1951; Payne-Scott played a major role in the design of this innovative instrument. Payne-Scott and Pawsey set the stage for the monumental solar research carried out by Wild and colleagues using the Dapto swept frequency spectrograph in the 1950s, leading to the construction and opening of the Culgoora Radio

[9]Paul Wild invented the name Type III in 1950; Payne-Scott used the term "unpolarized bursts;" Pawsey preferred "isolated bursts."

Heliograph in 1967, an instrument that dominated solar metre wave radio astronomy until the instrument was closed in 1984 to make way for the Compact Array of the Australia Telescope, opened during the Australian Bicentenary in 1988.

It is tempting to ask the question: What would Payne-Scott's role have been had she returned to RP in the early 1960s, at the time she began her teaching career again? Probably the transition would have been difficult for her; the solar group had made major advances and the large scale Radio Heliograph was under construction. Perhaps she would have joined the 21 cm hydrogen group of Frank Kerr and colleagues Jim Hindman, John Murray, Brian Robinson and Dick McGee. In the late 1940s until her resignation in July 1951, she had been quite interested in the HI line, urging a dedicated search for the line.[10] The excitement of the confirmation of the HI line in Australia by Christiansen and Hindman on 6 July preceded her resignation by exactly 2 weeks.[11]

How did she feel about the end of her professional career in 1951, when she was only 39? We will never know. We can only hazard a guess, based on two newspaper interviews she gave in 1952, close to the time of the URSI conference of 11–21 August 1952 (Chap. 10). The first interview occurred before the conference; the date from the archival newspaper article is missing (as well as the name of the paper!) A likely date was the week of 4 August with the title: "Scientist Gives Up Her Career." The scientist was described as a shy, blonde from Oatley, formerly Ruby Payne-Scott, M.Sc. She told the reporter that she might take part in the URSI conference but:

> Obviously, I can't do two things at once.... I can't look after a young child and do my job properly. Peter is just 8 months old and I would like to have another child.... I was in the physics department as a demonstrator for 2 years.... Then I decided there was not much opening for women in that field... I worked on high frequency radar during the war when I joined [RPL]. At our laboratory, after the war, we started research – concerning radio waves emitted from the sun and stars. Then I became interested in disturbances in the sun that seemed to be connected with flares – eruptions of bright light. These flares were associated with radio fadeouts and magnetic storms on the earth.

This short summary of her career was comprehensive. We detect little sense of regret that she had left this part of her life behind.[12]

---

[10]Reported by John Murray in a letter from 24 January 2004.

[11]The detection had been made at Harvard by Ewen and Purcell on 25 March 1951.

[12]The other interview was after the conference in the Sydney Morning Herald of 24 August 1952. Much of this interview was a confused discussion with a young woman who was an observing assistant (Pamela White of Glebe) under the direction of Alec Little at Potts Hill; the text about Payne-Scott (also called Mrs. W.H. Hall in the interview) points out that she had been in charge of the Potts Hill station from 1948 to 1951. She now had a 9 month old baby. She and her husband had just built their own house in Oatley, having moved in about August 1951. She and Bill had ceased being active in the Sydney Bush Walkers. She and her husband did discuss the PMG's 'new methods. But I know absolutely nothing about telephones."

However, it is clear that Payne-Scott had no real desire to return to RPL.[13] In her earlier career she had been motivated mainly by the engineering challenges of the early solar radio research. (She told friends in later years that if she could live her life again she would be an architect.) Astronomy was not the driving force, although she quickly learned many aspects of astronomy in the years following the end of World War II. In the 1960s, she realised that she would be returning to a new RPL, a research establishment made up of teams working in the sphere of "big science". By then, RPL had become an *astronomy* institute, consisting of the new Parkes 64 m telescope (since 1961) and the new Radio Heliograph (since 1967). Her role, as a member of a small team working on isolated experiments with a handful of individuals designing, building, and then using the apparatus, as had occurred in 1950, was no longer feasible. Also her main collaborators (Mills, Christiansen and Little) had left to start new departments at the University of Sydney. Payne-Scott decided that she would be happier spending a few years as a school teacher. As Claire Hooker has so aptly said,[14]

> By her own desire, in the second half of her life [Payne-Scott] stuck to the women's roles of mother, wife and teacher. Yet if she was a brief feminine flare in radio astronomy, she was a bright one, both as a physicist and as a woman.

[13]These conclusions are based on discussions with Peter Hall, 12 February 2007.

[14]Hooker, p. 166.

Chapter 16
Epilogue: Why Did We Write This Book?

The rationale of our writing this description of the life of Ruby Payne-Scott has an ironic origin. John Bolton and Ruby Payne-Scott were viewed by colleagues at RPL as antagonists (Chap. 8); the "bust up" at Dover Heights between the two in late 1947 had been announced to Pawsey by McCready a few months after Pawsey left for his trip to the US and Europe. Payne-Scott had been "exiled" to the Hornsby site, a result that had positive scientific results (Chap. 8). We were surprised to learn of these antipathies in 1999.

In the 1940s and 1950s, RPL was characterised by strong personalities. The characterisation of the "triangle of antagonism at RPL" between Bolton, Payne-Scott and Piddington has been described in Chap. 1.[1] Indeed, there were turbulent relationships among some of the senior scientific staff.

In 2007, Mills[2] suggested a fascinating theory that provides a whimsical view of the political alignment of the early years of radio astronomy at RPL. This suggestion is consistent with the dynamic that led to the partial rupture of RPL in the early 1960s as Mills and Christiansen left for the University of Sydney (School of Physics and Electrical Engineering, respectively). Shortly afterwards, in 1962, Pawsey left RPL for the National Radio Astronomy Observatory (Green Bank, West Virginia, USA), just before his premature death on 30 November 1962, at the age of 54. Mills postulates that two camps were already taking shape in the late 1940s and early 1950s: the "English" faction of Taffy Bowen (the Chief of the Division), and the recently recruited scientific staff member, John Bolton.[3] This faction was in contrast to the "Australian" camp of Ruby Payne-Scott, Chris Christiansen and

[1] See footnote 5 in Chap. 1. There were other well known conflicts. The following sentence in the memoir continued: "One well known antagonism was that between the Chief of the Division of Radiophysics, Taffy [E. G.] Bowen, and an earlier Chief, D.F. Martyn, with whom Piddington was close friends."

[2] Interview with Goss, 1 April 2007, Roseville (Sydney), NSW. Mills suggested that the two camps had a different set of standards regarding rigour in science. In addition, the "English" camp had a fascination for cricket; the "Australians" were not cricket fanatics.

[3] Both were born in the UK.

W.M. Goss and R. McGee, *Under the Radar*, Astrophysics and Space Science Library 363,
DOI 10.1007/978-3-642-03141-0_16,

Bernie Mills. According to Mills, Pawsey was in neither camp as "he tried to keep things on an even keel. He knew there would be trouble; he hoped for simple and healthy competition which was not the way it turned out." Paul Wild, an Englishman, was "in the middle with a foot in both camps." In the end, Australian radio astronomy flourished. After 1960, Mills, Christiansen, Wild, Bowen and Bolton remained leaders in this rapidly expanding field.

In the early 1970s, Goss[4] was a young staff member working at CSIRO RPL in the "cosmic research group", under the direction of Brian J. Robinson (1930–2004) in Sydney and John G. Bolton (1922–1993) at the Parkes Radio Telescope in New South Wales, Australia. One day during an informal conversation there, Bolton said to Goss: "You should have been here at RPL 20 years ago. The brightest staff scientist was a woman, Ruby Payne-Scott. She was the first person in Australia to recognise the importance of radio 'confusion'."[5]

Unfortunately, Goss had never heard of Payne-Scott at the time and the importance of this off-hand remark was lost on him. He forgot this conversation for several decades. In 1996, three years after Bolton's death on 6 July 1993, the memory resurfaced and played a decisive role in the impetus leading to the beginning of the collaboration in 1997 that ultimately culminated in this book.

It was only 25 years later that the ironic nature of this praise from Bolton became obvious. Throughout this volume numerous examples of their mutual antipathy have been described. Pawsey tried in vain to mediate from a distance in 1947–1948; he only had limited success when he returned to RPL in October 1948.

To our surprise, we found that in later years, well after Payne-Scott had left RPL in mid 1951, Bolton had praised her highly to a number of colleagues. He told Woody Sullivan on 13 August 1976 (during an interview recorded at Jodrell Bank Observatory) that "... one of the most outstanding people was Ruby Payne-Scott and unfortunately women in public service ... had ... well, she was discriminated

[4] Goss had been a NATO Postdoctoral Fellow at RPL under the direction of Bolton from 1967 to 1970. From 1974 to 1977, he was a Principal Research Scientist at RPL.

[5] Confusion is the limitation in radio astronomy which occurs when the density of radio sources in the sky is so large that individual sources blend with neighbours. In the presence of confusion, even a *sensitive* radio telescope cannot be used to recognise weak sources. Pawsey described confusion to Reber on 20 March 1950 (NAA; A1/1/1 Part 5): "There is also a major difficulty in interpretation of records where several sources produce interfering patterns. We at present use common sense and low cunning but there may be better methods." Sadly, we can find no-one who can confirm the assertion by Bolton that Payne-Scott initiated discussions about the relevance of confusion in radio astronomy among the RPL group. [Grote Reber (1911–2002) was a prominent pre-World War II radio astronomer; he constructed a radio telescope in his backyard in Wheaton, Illinois (near Chicago) in 1937. The radio telescope (now at the NRAO), was 31.4 ft or 9.6 m in diameter. In 1938 he confirmed Jansky's 1932 detection of radio emission from the Milky Way. In the 1950s he moved to Bothwell, Tasmania, where he constructed low frequency arrays. Sullivan (*Cosmic Noise, A History of Early Radio Astronomy*, 2009) gives a comprehensive summary of Reber's early career.]

against for basic reasons of sex and other reasons – she was a very aggressive person."[6]

John D. Murray (a scientific staff member at RPL from 1948 to 1989) reported[7] that in about 1970 Bolton received some US guests at the Parkes Radio Telescope. These guests were friends of John and Letty Bolton from Bishop, California (location of the Owens Valley Radio Observatory, founded by Bolton in the mid 1950s, and run by the California Institute of Technology); it is likely that the friends were not astronomers. According to Murray, John Bolton told the guests during a lunch time conversation that there had been a woman research officer at RPL [clearly Payne-Scott], who had worked on radar in World War II and had then become a radio astronomer. Bolton said that he considered her to have been "one of the best physicists at Radiophysics – no, one of the best physicists in Australia." Bolton then said that she had to resign when she became pregnant. These opinions are consistent with discussions that Goss had with Bolton in the early 1970s at Parkes, as well as conversations during visits to the Bolton home in Queensland in 1988 and 1992.

In a real sense, John Bolton has left us a valuable legacy: his praise led to an increased understanding of this remarkable scientist and mother, Ruby Violet Payne-Scott. As her son, Peter Hall, has said[8] about this change of heart:

> I [think] this is not really inconsistent . . . it's often true in human relations that people underneath really do recognize the strengths of colleagues ... [even when] other things get in the way for a long time.

[6] Used with Sullivan's permission (October 2008).

[7] Letter to Goss, 24 January 2004.

[8] Interview, February 2007.

Appendix

Appendix A

Status of RPL in 1941. Propagation Committee (PC) at RP, 1944–1949 and Radio Astronomy Committee, 1949–1954

Status of the Radiophysics Laboratory (RPL) of the CSIR in 1941

In order to understand the role played by Payne-Scott in these years of explosive growth in radio astronomy starting in 1944, it is necessary to describe the circumstances under which the CSIR Radiophysics Laboratory (RPL) was founded in August 1939 just before the beginning of World War II (Australia followed the United Kingdom into war with Germany on 3 September 1939). It was based on the activities of the Radio Research Board, under the leadership of Sir John Percival Vissing Madsen (1879–1969).[1] The purpose of the new organisation in Australia (as part of the existing CSIR) was to coordinate War time research into radio direction finding (RDF) or radar.[2] The exciting story of the details of the formation of the RPL has been summarised by Sullivan (2009), Schedvin (1987), and Mellor

[1]Madsen was Professor of Electrical Engineering at the University of Sydney for much of his life, from 1909 to 1949. The history of the founding of RPL has been described in an impressive manuscript by W.F. Evans in two volumes, *History of the Radiophysics Advisory Board 1939–1945* (1970) and *History of the Radio Research Board* (1973). Both were published in limited typewritten manuscript form by CSIRO.

[2]RDF was the British term which evolved into the US acronym RADAR (radar detection – or direction finding – and ranging). This term was invented by S.M. Tucker of the US Navy in 1940 (Louis Brown 1999). As Hanbury Brown (1991) wrote: "Some years later we [the British] adopted the name Radar which, as Watson-Watt [one of the UK inventors of Radar] used to say in mildly disparaging tones was a 'synthetic palindrome' invented by our friends the Americans." The British adopted the new term RADAR after 1 July 1943; probably the Australians also changed their terminology at about this time.

(1958). David F. Martyn (1906–1970) had been visiting the UK since March 1939 to study the secret developments in radar that were to play such an important role in the defence of the UK during the massive onslaughts of the German Luftwaffe in the years to come. The actual formation of RPL occurred on 22 August 1939.

RPL was first housed in the Department of Electrical Engineering at the University of Sydney. The move to the newly constructed building that housed the National Standards Laboratory (with later additions for the RPL dating from the last days of 1939) was complete by April 1940. The name "Radiophysics" was deliberately invented to draw as little attention as possible to the real purpose of the new institution. As Evans (1970) has written, "... an innocuous scientific nomenclature [was made up], which masked the immediate wartime concentration on radio direction finding." Even the original "National Security Research Board" was changed to the "Radiophysics Advisory Board", which had its first meeting at Victoria Barracks, Melbourne, on 29 December 1939.

Today the two buildings (the National Standards Laboratory and the RPL) have been fully integrated, and are now called the Madsen Building. This name is ironic since the common name used by staff and students in the 1940s in order to make fun of the Gothic towers was "Sir Percival's Fortress", after Sir John Madsen's second name.[3] The building in 2007–2008 is shown in Fig. A.1a, b.

Recruitment of staff was a major activity. In the first year of operation, RPL had a small staff of only five research officers and eight assistant research officers, plus the Chief of the Division, D. F. Martyn (Chief from 1940 to 1942). One of the most prominent early scientific appointments at RPL was Joseph L. Pawsey. Martyn had met this 31 year old Australian in London earlier in 1939. Pawsey had completed a PhD degree at Cambridge (UK) in 1934 with J.A. Ratcliffe, working on ionospheric research and was working at EMI (Electric and Musical Industries Inc). He was appointed on 23 October 1939 as a Research Officer and left the UK for Australia on 22 December 1939; he began work at RPL on 2 February 1940.[4]

By the end of the War, there were over 300 staff members at RPL with 91 research staff and 97 engineering staff (Evans 1970). The success of the RPL in providing radar designs for the SWPA (Southwest Pacific Area) allied forces was substantial; more than 20 major radar development projects were undertaken during World War II, with RPL being responsible for the manufacture of more than "2,000 complete items of Service radar equipment" (Evans 1970). The LW/AW (light weight aircraft warning) 200 MHz radar sets (see Appendix C) were particularly successful, with roughly 800 radar stations throughout the SWPA (e.g., in Papua New Guinea and the Solomon Islands).

> ... [This] led to the realization of a new conception of light-weight transportability essential for jungle, island-hopping warfare. So evolved the highly successful LW/AW, which were entirely Australian in development. Many LW/AW Mk.Ia (Aust.) sets were eventually to be supplied to the American forces, who found the Australian equipment

---

[3]Interview with Rachel Makinson, 21 February 2007.

[4]NAA: A8520, PH/PAW/1, Part 1.

Fig. A.1 (**a**) The Madsen Building of The University of Sydney, the Eastern portion was previously National Standards Laboratory of the CSIR and the CSIRO. This building was called "Sir Percival's Castle" in the 1940s after Sir John Madsen's middle name. Photo from February 2007 by Goss. (**b**) The Western portion of the Madsen Building, previously the Radiophysics Laboratory of the CSIR and the CSIRO. In January 1968, RPL moved to the CSIRO Laboratories in Eastwood, NSW (Sydney). Photo by Goss, September 2008

> more adaptable for the S.W.P.A. campaign than their own more cumbersome A.W. equipment. In fact the ultimate success of the LW/AW series alone would have justified . . . the entire wartime R.D.F. undertaking [in Australia].[5]

Figure A.2a, b shows what is probably the only surviving fully restored LW/AW radar (Australian War Memorial, Canberra).

---

[5] Evans (1970), page 95.

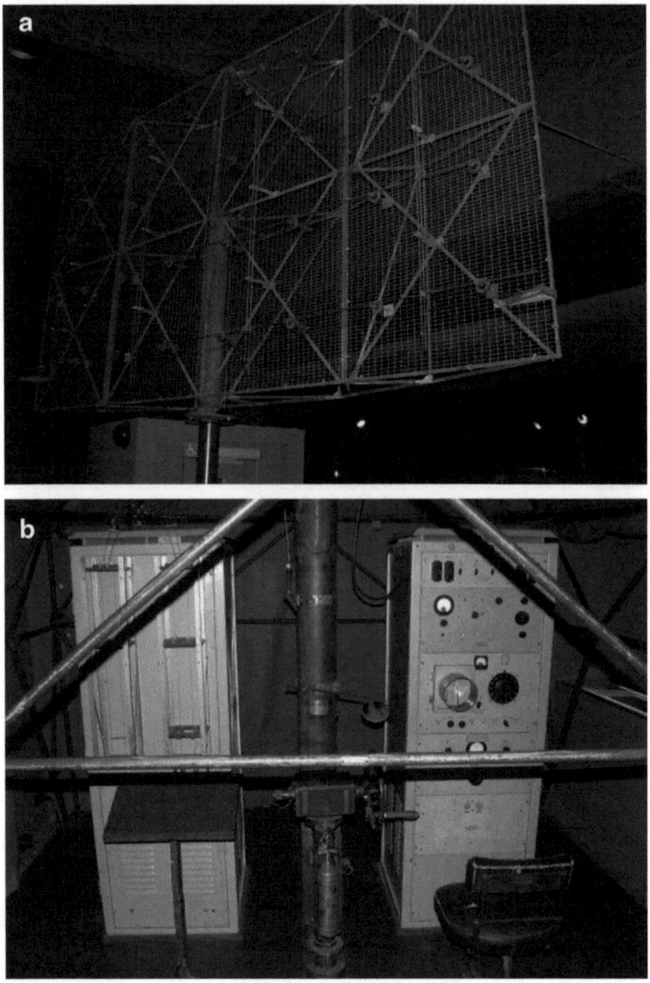

Fig. A.2 (**a**) and (**b**) RPL designed the LW/AW (light weight aircraft warning) during World War II. A beautifully restored unit at the Australian War Memorial in Canberra. (**a**) shows the top side, a 200 MHz broadside array designed by Pawsey, Minnett and colleagues. In (**b**), the lower control portion of the radar station is shown, power supplies (*left*) and the displays (*right*). The operator sat in the chair to the right; the azimuth control is visible in the middle of the room; the operator could adjust the azimuth of the radar antenna by turning the crank. Photo by Goss, September 2008

Propagation Committee (PC) and the Radio Astronomy Committee

The "Propagation Committee" (PC) Minutes of the CSIR Division of Radiophysics from 14 September 1944 to 7 March 1949, and of the renamed "Radio Astronomy Committee" from 11 April 1949 to 9 April 1954, have provided an invaluable source of information about the research program of RPL in these years. The Committee had probably been instigated during the later stages of World War II under the leadership of Pawsey (Chairman) for bi-weekly discussions (later the frequency was reduced to about once per month). Up to 1947 all RP research programs, such as super-refraction, ionosphere, Loran, mathematical physics (including the first valve [vacuum tube] computers), and cloud physics, were discussed in the PC. Members of the Committee in the first year included a number of Royal Australian Air Force (RAAF) personnel (e.g., Wing Commander Taylor and Flying Officer McDonald). Frank Kerr (1918–2000) was the Secretary from 1945 followed by Chris Christiansen in 1949. (Many other members of staff, including Ron Bracewell, served in this role in later years.)

Often the meetings were held on pay days (originally Friday and later Thursday); the various groups would come to the laboratory from the far flung field stations to be paid, usually in cash. The latest time for being paid was 3 p.m.; the PC meeting would start at 3.30 and would cover a wide range of topics such as progress reports on observations and equipment, papers in preparation, proposals for new projects and instruments, and extensions or closing of existing projects.[6]

The detailed minutes of the period 1945 to mid-1950 were provided by Don Yabsley (1923–2003) in 1999; the other copies were obtained from the National Archives of Australia.[7] These minutes (typically 2–3 pages per meeting) are a priceless resource since we can use them to track the progress of each project and its participants. The minutes frequently included the names of technicians as well as of the scientists; only scientists and engineers attended the meeting.

The minutes show that even reports of rumours were exchanged by the participants. Referees' reports on publications submitted and letters between colleagues, both within and outside Australia, were common topics, as were reports of false starts on new projects and even failures. Such information is seldom reported in the published papers, which sometimes even fail to list the dates of the observations as well as the names of the field stations. This contemporary record provides a myriad details of a major historical sequence of ground breaking progress in the early years of post War radio astronomy in Australia; many of these details may have been partially lost (or incomplete) without the PC minutes. The minutes were classified with a "Secret" designation through 1945, and became "Confidential" for only a few weeks in early 1946, before the classified nature of the reports was dropped.

Throughout this book we have made extensive use of the PC minutes to provide a time line for various activities of the different radio astronomy groups at RPL; these summaries also provide clues as to the nature of important discoveries and even conflicts between the scientific staff.

---

[6]Details of the format of the PC meetings provided by John Murray, letter from 24 January 2004.

[7]The time span of the Yabsley collection, 14 September 1945–27 July 1950, coincides with the time he was a member. The NAA collection is NAA: C3830, B2/2, Part 1 and Part 2, 332 pages.

PC Meetings in the Immediate Aftermath of World War II

Some surprises and novel facts appear in the minutes in 1945 and 1946. The role of "girls" (their term!) at RPL was discussed. These personnel were initially Women's Auxiliary Australian Air Force (WAAAF), who were carrying out the routine and tedious data reduction of super-refraction events reported from 112 radar stations (200 MHz, the LW/AW, Appendix C) around Australia from March 1944 to August 1945. These "over the horizon" echoes, due to unusual meteorological conditions, could lead to radar returns from distances up to 1,000 miles (Geraldton, WA to Ceylon) and the North West Cape of Australia to Java (900 miles); the implied military importance led to an intense study. Surprisingly the study was declassified quickly and published by Frank Kerr in 1948. Kerr found that coastal super-refraction was associated with the offshore movement of a warm and dry air mass. As the War ended, a report[8] in the 14 September 1945 PC meeting indicated that data from the 18 month period of 1944–1945 were being analysed. Flying Officer McDonald said that "more girls would be needed than were at present at RPL (19), if the work is to be completed in time . . . Further, no girls here are willing to stay after Christmas (1945), so it will be necessary to get extra girls for the Loran job, drawn from those WAAAF, who have agreed to stay in the Service some time after general demobilisation". A month later, this project was well under way with 30 girls, with the expectation that a report on super-refraction in Australia would be available by Christmas 1945. Most of the girls left RPL at the end of November, but a few remained for an additional week.

On 13 May 1946, a new type of "girl" was introduced in the PC minutes – a "computer". This term was common in describing women who did routine computing tasks in the days before digital computers. (A famous example was the Harvard College computer Henrietta Leavitt [1868–1921], who discovered the period luminosity relation of Cepheid variable stars during the first decade of the twentieth century.) This group of four computers was given a course in numerical methods. They worked in the mathematical physics group under the direction of T. Pearcey; see also Chap. 6, footnote 31.

For the last but one meeting of the PC in 1945 (12 November 1945), a two page report concerning the future of solar and cosmic (everything non-solar!) noise investigations was prepared, probably written by Pawsey with input from Yabsley and Payne-Scott (see also Chap. 6). The group planned to branch out into new wavelength bands other than 200 MHz (1.5 m). As an example, future solar noise investigations at the shorter wavelengths of 50, 25, 10, 3 and 1.2 cm was suggested. Dramatic success at 50 and 25 cm followed within a year and a half. Starting in mid-August, 1947 to late November, 1947, Yabsley and Fred Lehany used a 16×18 ft

[8]This report followed a report from Payne-Scott about the intention to start observing the sun with a radar receiver at 200 MHz, following up on COL (Chain Home Overseas Low) 200 MHz data from Norfolk Island and New Zealand. (See Chaps. 6 and 7.)

(4.9 × 5.5 m) paraboloid at Georges Heights on Middle Head;[9] an intense solar event was observed by this group on 4 October 1947 (in excess 10^7 Jy at 50 and 25 cm) and published in *Nature*, on 24 April 1948 (Lehany and Yabsley 1948).

Propagation Committee and Solar Noise Group Meetings in 1946–1947

In Fig. A.3, we show examples of two pages from the PC minutes of 19 March 1946. Payne-Scott gave a detailed report but was not an official member of the committee! Also Bruce Slee (see Chap. 11) was mentioned for the first time; his solar detection at the radar station near Darwin, while he was still serving in the RAAF as a Radar Operator, was reported. Finally, there is a fascinating report about an attempt to detect "stellar noise", presumably non-solar emission. No details of this observation have survived; the first successful attempted non-solar observation at Dover Heights was carried out by Bolton and Stanley in June 1947.[10]

Occasionally longer meetings were held for major reviews of existing projects and planning for the future. The most striking case was the long meeting ("meeting of Dr. Pawsey's Solar Noise Group", the printed minutes ran to seven foolscap pages) on 23 September 1947, 2 days before Pawsey's departure with his wife on the *Marine Phoenix*. He was absent for a year, visiting scientific institutes in the US, Canada, the UK and Europe.[11] The minutes of this PC meeting provide a fascinating snapshot of the status of RPL at this crucial period in the post War era, with a complete summary of the status in mid-1947, as well as plans for the following years.

[9] See footnote 34, Chap. 6.

[10] Bolton (1982), in a retrospective about these early days at Dover Heights, wrote: "Shortly after I joined Radiophysics in September 1946 Pawsey had attempted to confirm J.S. Hey's newly reported discovery of fluctuations in the cosmic background from the constellation of Cygnus. He was unable to repeat Hey's result." Perhaps, Bolton was referring to this earlier attempt, which apparently had been carried out some months earlier.

[11] See Chaps. 7 and 8. This visit was crucial in Pawsey's career; for the first time he met famous optical astronomers, radio astronomers and saw again many physicists he knew from his period at the Cavendish and from a War-time visit to the US in 1941 (his previous thesis advisor, Ratcliffe, as well as Appleton, L.A. DuBridge and I.I. Rabi). His introduction to the world of astronomy began with this visit; the contacts with the RPL were intensified, aided by a constant flow of correspondence.

B.51/2
FJK:DBN

MINUTES OF PROPAGATION COMMITTEE MEETING

Held on 19th March, 1946

PRESENT: Dr. Pawsey (Chairman), Drs. Bowen, Pulley, Piddington, Messrs. Pearcey, McCready, Yabsley, Parker, Heffernan, Miss Payne-Scott. Mr. Kerr (secretary)

Staff Changes

Mr. P. Squires will be joining the staff in a few weeks time, when his discharge from the R.A.A.F. has been completed. Mrs. Mills has now left to return to the University.

Membership of Committee

The previous membership list is now obsolete. It was agreed that the formal membership of the Committee should comprise:- Dr. Pawsey (Chairman), Mr. Briton, Dr. Bowen, Mr. Higgs, Mr. Kerr (Secretary), Drs. Pulley, Piddington, Messrs. Pearcey, McCready, Parker, Yabsley, Price (extra-mural), Prior (representing R.R.B.), and (later) Squires.

PROGRESS REPORTS

1. Superrefraction

(a) Analysis of Service Data (Kerr)

Attention is being directed to the drawing up of a detailed plan of the papers to be published giving a survey of superrefraction in Australia. The present position is being assessed, to determine the still outstanding questions, to which answers are desired.

(b) Experimental Work (Heffernan)

One flight has been carried out with the field strength receiver in the aircraft. No interference from the camera was noticed. The flight was successful, except for minor troubles. A smoke bomb was tried out successfully. Another flight is planned shortly on under S.R. conditions.

An important object of the flight is a more detailed investigation of the field strength decrease near the line of sight.

During the flight a record was also obtained of conditions from Parramatta to Wentworth Falls, along the track of the Survey Group's link. Much turbulence was observed.

2. Loran (Yabsley)

The girls are still analysing results and are now almost up to date. 1½ months now remain to complete the year of observations. The preliminary report is now being typed.

3. H.F. Noise (Yabsley)

Nil.

4. Scattering from Clouds (Parker)

Daisy Mae (3 cm) is now working and should be fully operational shortly. The 10 cm. G.C.R.A. set is also being tested. The structural work for the 50 cm. set is now ready for erection.

Fig. A.3 (**a**) and (**b**) Two pages from the Propagation Committee (RPL) minutes from 19 March 1946. These examples shows the level of detail in the reporting of activities of the research groups at RPL. In 1949, the name was changed to Radio Astronomy Committee after Pawsey's return in late 1948 from a trip of over a year to North America and Europe. From the collection of Propagation and Radio Astronomy Committee minutes of the late Don Yabsley. Pawsey was one of the inventors of the word to describe the new discipline "radio astronomy" in early 1948 as pointed out by Woody Sullivan

b)

The report written some months ago is now being typed.

5. Solar and Cosmic Noise (McCready, Miss Payne-Scott)

Analysis of $1\frac{1}{2}$ metre data is proceeding. When this is complete, it is hoped to publish a paper covering observations to date.

$1\frac{1}{2}$ metre equipment has been installed at Mt. Stromlo, for use in conjunction with spectroheliograph observations, and for simultaneous fading records between Sydney and Canberra. There has been however, little solar activity lately. It is proposed to set up 3-4 metre equipment at Stromlo, as the level of solar noise is believed to be higher at these wavelengths.

Equipment is being developed for further investigations into the spectrum, polarisation and short period fluctuations. It is planned to make S and X band observations at the Laboratory. At Dover there are 75 and 200 mc/s aerials with recorders. The A.W.H array has also been moved there.

Some exploratory investigations of stellar noise have been conducted with inconclusive results.

An attempt is being made to determine the position on the sun of the source of the energy, from a study of the lobe patterns obtained in the February observations. The theoretical treatment of refraction right through the earth's atmosphere has been worked out with some approximations. A calculation in one case gave a result in which the source of energy lay somewhere on a horizontal band passing through a large sunspot.

Little activity was noticed during the second appearance of the large spot, except for some bursts of about $\frac{1}{2}$ hour duration, in which the noise increased 15% with rapid short term fluctuations.

Simultaneous fading records have shown the fading to be similar at Dover and Collaroy.

A letter has been received from LAC Slee, Darwin, saying that he has observed interference presumably due to the sun and offering to take systematic observations. A lengthy reply has been sent welcoming his interest and offer of co-operation.

6. Mathematical Physics (Pearcey)

The following problems are being tackled using wave theory: Numerical work on a system of functions to be used in work on ground-based trappling. Reflection by discs of diameter order of a wavelength (relevant to reflection by ice crystals at high altitudes). Field strength structure in the neighbourhood of the skip distance.

Fig. A.3 (Continued)

Appendix B

Ruby Payne-Scott: Bibliography (Format as typed by Sally Atkinson in the early 1980s)

Publications

R. Payne-Scott: "Relative intensity of spectral lines in indium and gallium", Nature, Vol. 131, pp. 365–366, 11 March 1933.

R. Payne-Scott and W. H. Love: "Tissue culture exposed to the influence of a magnetic field", Nature, Vol. 137, p. 277, 1936

R. Payne-Scott: "Notes on the use of photographic films as a means of measuring γ ray dosage", J. Cancer Res. Commun., Vol. 7, pp. 170–175, 1936.

R. Payne-Scott: "The wavelength distribution of the scattered radiation in a medium traversed by a beam of X or gamma rays", Br. J. Radiol., Vol. 10, pp. 850–870, 1937.

Ruby Payne-Scott and A.L. Green: "Superheterodyne Tracking Charts – II", A.W.A. [Amalgamated Wireless (Australia)] Technical Review, Vol. 5, No. 6, pp. 251–274; and Wireless Engineer, Vol. 19, pp. 290–302, July 1942.

Ruby Payne-Scott: "A Note on the Design of Iron-Cored Coils at Audio Frequencies", A.W.A. Technical Review, Vol. 6, No. 2, pp. 91–96, 1943.

Pawsey, J.L., Payne-Scott, Ruby and McCready, L.L.: "Radio-frequency energy from the Sun", Nature, Vol. 157, pp. 158–159, 9 February 1946.

Payne-Scott, Ruby. 1947. "A Study of Solar Radio Frequency Radiation on Several Frequencies during the Sunspot of July–August, 1946." CSIR Radiophysics Laboratory Report, RPL 9.

McCready, L.L., Pawsey, J.L. and Payne-Scott, Ruby: "Solar radiation at radio frequencies and its relation to sunspots", Proc. Roy. Soc. A., Vol. 190, pp. 357–375, 12 August 1947.

Payne-Scott, Ruby, Yabsley, D.E. & Bolton, J.G.: "Relative times of arrival of bursts of solar noise on different radio frequencies", Nature, Vol. 160, pp. 256–257, 23 August 1947.

Payne-Scott, Ruby: "The visibility of small echoes on radar PPI displays", Proc. I. R. E. (America), Vol. 36, No. 2, pp. 180–196, February 1948.

Payne-Scott, Ruby and McCready, L.L.: "Ionospheric effects noted during dawn observations on solar noise", Terrestrial Magnetism & Atmospheric Electricity, Vol. 53, No. 4, pp. 429–432, December 1948.

Payne-Scott, Ruby: "Bursts of solar radiation at metre wavelengths", Aust. J. Sci. Res. A, Vol. 2, No. 2, pp. 214–227, June 1949.

Payne Scott, Ruby: "The noise-like character of solar radiation at metre wavelengths", Aust. J. Sci. Res. A, Vol. 2, No. 2, pp. 228–231, June 1949.

Payne-Scott, Ruby: " Some characteristics of non-thermal solar radiation at metre wave-lengths." J. Geophys. Res., Vol. 55, No. 2, pp. 203–204, June, 1950.

Christiansen, W.N., Hindman, J.V., Little, A.G., Payne-Scott, R., Yabsley, D.E. and Allen, C.W.: "Radio observations of two large solar disturbances", Aust. J. Sci. Res. A., Vol. 4, No.1, pp. 51–61, March 1951.

Little, A.G. & Payne-Scott, Ruby: "The position and movement on the solar disk of sources of radiation at a frequency of 97 Mc/s. I – Equipment", Aust. J. Sci. Res. A., Vol. 4, No. 4, pp. 489–507, December 1951.

Payne-Scott, Ruby & Little, A.G.: "The position and movement on the solar disk of sources of radiation at a frequency of 97 Mc/s. II – Noise storms", Aust. J. Sci. Res. A., Vol. 4, No. 4, pp. 508–525, December 1951.

Payne-Scott, Ruby and Little, A.G.: "The position and movement on the solar disk of sources of radiation at a frequency of 97 Mc/s. III – Outbursts", Aust. J. Sci. Res. A., Vol. 5, No.1, pp. 32–49, March 1952.

In Addition Ruby Payne Scott wrote many unpublished reports, both in the wartime series TI, PD and RP and in the postwar series RPR and RPL.

Unpublished Reports by Ruby Payne-Scott (since joining Division of Radiophysics on 18/8/41)

	Report No.
T.N. Basnett & R. Payne-Scott: "Oscillator frequency required for Shore Defense and Gun Laying Calibrator" MOST SECRET, 10 February 1942, (with Bibliography), 3 pp.	RP.138/1
R. Payne-Scott: "S band signal generator", 8 November 1943, 10 pp., figs.	PD. 30
R. Payne-Scott: "Effect of summer temperatures on S. band crystals" SECRET, 23 February 1944, 3 pp., 1 table	TI. 80/2
R. Payne-Scott: "Measurements of the noise level picked up by an S-band aerial" CONFIDENTIAL, 11 April 1944 (with bibliography) 5pp.	RP. 209
R. Payne-Scott: "Thermal noise generator for absolute measurement of receiver noise factors at 10 cm" SECRET, 29 May 1944 (with bibliography), 7 pp., figs.	RP. 211
R. Payne-Scott: "The present position of low power S-band measurements in the Radiophysics Laboratory" SECRET, 6 June 1944, (with bibliography), 4 pp. figs	TI. 121/1
R. Payne-Scott: "Relative merits of intensity and deflection modulation for viewing small signals, and the effect of sweeping the intensity-modulated trace, 10 January 1945, 5 pp, 1 fig.	RP. 233
R. Payne Scott: "Ultimate visibility of signals on a PPI display, and the effect of electrical parameters on visibility", 20 May 1945, 59 pp.	RP. 252/1
R. Payne-Scott: "Charts for the calculation of the smallest signal visible on PPI displays", 6 July 1945, 6 pp.	RP. 252/2
R. Payne-Scott: "Present position of fundamental R.F. measurements in the Radiophysics Laboratory" RESTRICTED, 6 August 1945 (with bibliography), 17 pp, figs.	TI. 191/1
R. Payne-Scott: "Solar and cosmic radio frequency radiation; survey of knowledge available and measurements taken at Radiophysics Laboratory to December 1, 1945", 1945 (with bibliography), 22 pp, figs.	SRP. 501/47

(continued)

Unpublished Reports (Continued)

	Report No.
J.L. Pawsey, R. Payne-Scott & L.L. McCready: "Solar radiation at radio frequencies and its relation to sunspots", 16 June 1946 (with bibliography), 22 pp. figs.	RPR. 24
R. Payne-Scott: "Visibility of small echoes on radar PPI displays", 8 October 1946, 34 pp., figs	RPR. 40
R. Payne-Scott: "Study of solar radio frequency radiation of several frequencies during the sun-spot of July-August, 1946", August 1947, 33 pp, 15 figs.	RPL. 9
R. Payne-Scott: "Solar noise records taken during 1947 and 1948", December 1948, 2 pp, chart.	RPL. 30

Appendix C

The Sea-Cliff Interferometer: Lloyd's Mirror Radar from World War II

The key role that the sea-cliff interferometer played in the early solar physics research after World War II has been outlined in Chap. 7. Payne-Scott was a major player in the use of this early interferometer to determine both the positions and sizes of radio sources. The first interferometer in radio astronomy was the sea-cliff interferometer, a "Lloyd's mirror",[12] using a single antenna. The astronomical importance of the early 1946 data was the determination of the coincidence of Type I bursts with sunspots (Pawsey et al. 1947) and the detection of discrete radio sources by the RPL group of Bolton, Stanley and Slee (e.g., confirmation of Cygnus A, discoveries of Taurus A, Virgo A and Centaurus A).

Contrary to a common misconception in the history of early radio astronomy, the phenomenon of the Lloyd's mirror effect near the sea was first discovered by several radar groups, working independently during World War II. In this appendix we provide a brief description of these groups: the RPL scientists and several groups in the UK involving well known scientists, Fred Hoyle and J.A. Ratcliffe.

Sullivan (1991 and 2009) has provided an extensive history of the use of the sea-cliff interferometer in radio astronomy. As Sullivan (1991) has written:

> It is important to note that this phenomenon [radio Lloyd's mirror] was nothing new to those who had been developing radar systems, for during the war radar beams often pointed near the horizon . . . the reflected signal from a distant aircraft was well known to oscillate as it passed through the radar's lobes. This effect was both a blessing and a curse . . . for it could be used to gather precise information on the target's height, but on the other it meant that low-flying aircraft could sneak in "under" a radar, since the first lobe was *not* at the horizon, but above it by a considerable amount . . .

[12] After the Irish physicist and university administrator Humphrey Lloyd (1800–1881); the initial use of this method involved oblique interference at optical wavelengths, demonstrating the wave nature of light.

The Australian development of this interferometer was partially based on the work of J.C. Jaeger (1907–1979, Australian mathematician) during the War. In RP 174[13], "Theory of the Vertical Field Pattern for R.D.F. [Radio Direction Finding] Station", a classified report dated 17 March 1943, Jaeger described the use of the sea-cliff radar to determine *elevations* of incoming aircraft. The use of the sea-cliff interferometer for Fourier synthesis was proposed for the first time in a 1947 paper by McCready, Pawsey and Payne-Scott, which contained no reference to the Jaeger publication. Payne-Scott and Pawsey had carried out a substantial reformulation of the mathematics of the 1943 report.[14]

The Jaeger report of 17 March 1943 covered many of the effects that future radio astronomers would need between 1946 and 1952. The effects of the curvature of the earth, refraction, surface roughness and changing tides (sea surface height changes) were discussed.[15] In those pre-electronic computer days, calculations with mechanical calculators were tedious. A number of complex cases were treated by Jaeger for different geometries of transition from land to sea, e.g., sloping ground between the transmitter and the sea cliff. As discussed in Chap. 7 and especially in Appendix L, the most severe problem was the large and uncertain refraction close to the horizon. Jaeger had tried to estimate this in 1943; as an example, he calculated that due to refraction an aircraft would appear to increase in elevation by 340 ft at a range of 50 miles under common conditions close to the horizon.

J.N. Briton, the Chief of RPL at the end of the War, wrote a comprehensive report in 1947, "Lightweight Air Warning and G.C.I. [Ground Control Interception] Radar in Australia", in the *Journal of the Institution of Engineers*, Australia.[16]

---

[13]From the Sullivan archive of RPL publications. Based on interviews with Peter Hall, we know that Payne-Scott knew Jaeger well. See Fig. 15.5, Jaeger and Payne-Scott in Tasmania in 1949 with a number of colleagues.

[14]In the text book, *Optics*, by Bruno Rossi (First edition, 1957), a series of problem sets (questions for the student) concerning the Australian sea-cliff interferometer are presented in Chap. 7, "Electromagnetic Theory of Light". Accompanied by a figure of a sea-cliff (similar to Fig. 7.7), the student was asked to calculate the intensity of the radiation from the radio telescope as a function of the angle characterising the height of the radio star (elevation) above the horizon. Goss was given this problem as an undergraduate at Harvard College in late 1961; his fascination for this instrument has survived to the present.

[15]A few years later, more extensive treatments of these disturbing effects were presented by Stanley and Slee (1950) and Bolton and Slee (1953); the influence of the reflection coefficient of the sea, reduction of the amplitude of fringes from low elevation sources due to the divergence of the reflected radiation from a curved earth, and the reduction in amplitude due to finite bandwidth were treated in detail. Also these authors discussed the disturbing impact of sea waves; later this group found that the sea-cliff interferometer would not operate at frequencies as high as 1.2 GHz due to this effect, because of the roughness of the reflecting surface of the sea.

[16]Minnett (1999) has provided a through description of the LW/AW (Light Weight /Aircraft Warning) radar with a clear time line of modifications from 1941 to 1945; he has pointed out that Briton (1947) has confused the 1941–1942 prototype with the final LW/AW Mark I system as well as the relative roles of RPL and the RAAF. Additional historical documents are found in Alexander (RP 207/3), 11 January 1945, "History of the Development of the Australian LW/ AW Equipment". In addition, E.M. Bullock (1999) has written a history of the LW/AW system

Many operational details of RPL during World War II are included in Briton's publication, such as manufacturing procedures, tropic proofing and shipping procedures for the portable LW radars. A number of diagrams were included showing the range and height distributions as a function of distance from the transmitter. An example is shown in Fig. C.1, clearly based on the Jaeger report. The solid lines give the directions of strong echos while the broken lines show the directions of zero response. A design feature of the LW/AW (Light Weight/ Aircraft Warning) series of radars was obvious: due to the interference of the direct ray from the aircraft and the reflection from the sea, a series of nulls were obtained.[17]

A famous astronomer of the twentieth century was also involved in this problem of a radar "Lloyd's mirror", in this case for ship-borne radar. Fred Hoyle (1915–2001) solved a problem similarly to Jaeger; there was apparently no knowledge of the research in Sydney. Hoyle was working at the Admiralty Signal Establishment (ASE) near Portsmouth.[18] He realised that the ship-borne 7 m radars of the Royal Navy in early 1941 had no provision for the determination of the heights of incoming aircraft; a theoretical diagram of the response of the naval radar was carried out quite similar to the Australian effort. On board a ship, the low frequency radars produced lobes as an interference pattern of the direct and reflected returns from the incoming aircraft. The original design of these radars only incorporated methods to determine azimuth and range, not height. Mitton (2005) has written:

> Hoyle saw that the difference between the attackers being repelled and ships being sunk lay in knowing the height of the incoming aircraft.

---

from the point of view of the manufacture of the antennas by the New South Wales Railways in World War II, "Radar Manufacture by the NSWGR [New South Wales Government Railways] During World War II" (*The Bulletin of the Australian Railway Historical Society*, 1999); the aerial was called the "Worledge Aerial" after J.G.Q. Worledge, leader of the NSW Railways radar structures group.

[17] Briton uses the form in Fig. C.1 to illustrate how the lobe pattern could be used to determine not only the range (distance from the transmitter) and azimuth (the angle of arrival of the echo) but also a crude estimate of the height of the aircraft: "Thus if an aircraft approaches at say 6,000 ft, it would be first detected . . . at 97 miles and would be observed until it reached 75 miles. It would then be lost for a time until it reached 65 miles and held in view until 50 miles. As it flies in and out of lobes, it is alternatively seen and lost. . . . It has, however, the compensating advantage that an observation of the distances at which it is found and lost gives a means of estimating the height of the aircraft. This is an extremely useful item of information, if interception by fighters is involved.' The assumption had to be made that the aircraft flew at a constant height during this period. Later radars developed in Australia had a more direct method of height finding (see Chap. 5).

[18] Mitton (*Fred Hoyle: Conflict in the Cosmos, A Life in Science,* 2005) gives a detailed account of this "graphical" solution implemented by Hoyle. Cyril Domb (2003), a colleague of Hoyle at ASE, has also written an account of his experiences in the height finding project. Domb has published a figure that is quite similar to Fig. C.1. In his autobiography (1994), *Home is Where the Wind Blows*, Hoyle also describes this radar experience.

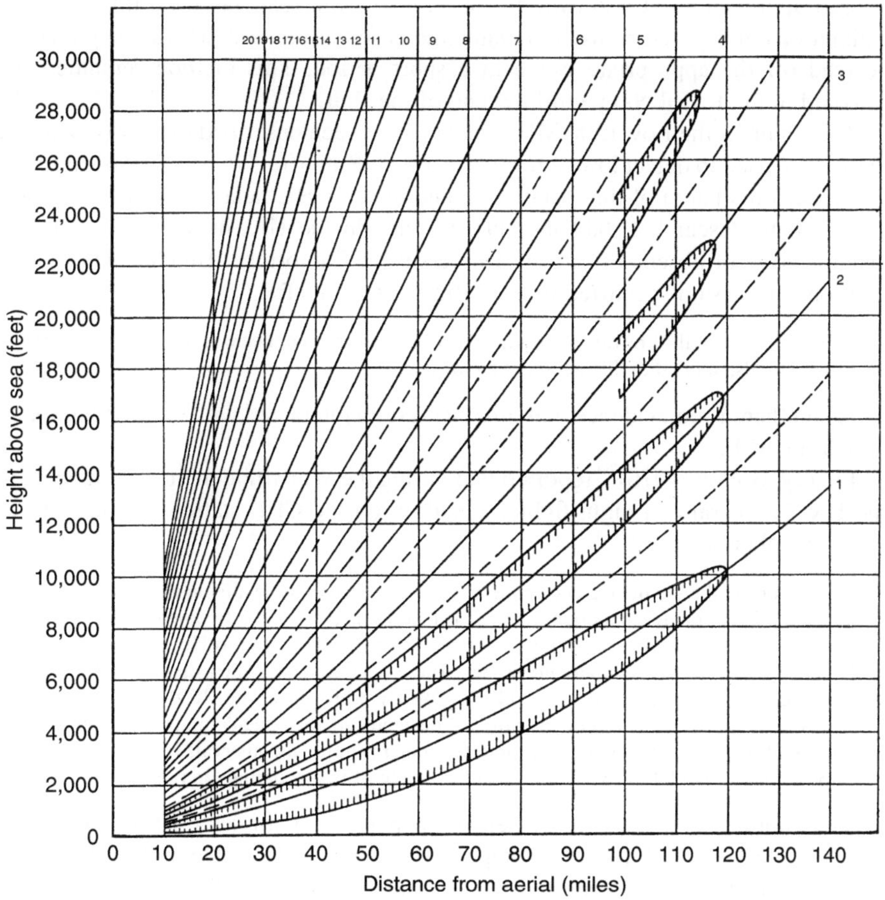

Fig. C.1 Based on a World War II era report by J.C. Jaeger, the response of a 200 MHz LW/AW radar was calculated. The x axis is the distance from the aerial to the incoming object (miles), while the y axis is the height above the sea (ft). (20,000 ft is 6,100 m.) Based on an assumed height of the radar above the sea of 76 m; the solid curve gives the response of maximum field strength of the radar echo and the dashed lines give the zero response. The post World War II use of this system as a sea-cliff interferometer was based on this type of calculation. Taken from Briton, his Fig. 1 (1947, *Journal of Institution of Engineers Australia*, vol. 18, p. 121). Figure from CSIRO, ATNF. RP 174. 'Theory of the Vertical Field Patterns for R.D.F. [radio direction finding] Stations.' J.C. Jaeger, 17 March 1943. Used by permission of ATNF, CSIRO

Hoyle carried out a theoretical calculation of the pattern in order to determine the height of the aircraft; the results were not satisfactory due to large systemic errors. The simple solution proposed was to calculate response curves for various values of the parameters of the radar. An elegant method was proposed in which each separate radar set could be calibrated with a trial aircraft flight, with the friendly aircraft approaching at a known elevation above the sea. Based on this trial, the calibration curve could be used to match the observations and the true height could be read off the appropriate curve in a short period. This method remained the standard in the Royal Navy until the end of the War.

The Chair Home Low (200 MHz) radars have been described in Chaps. 6 and 7; these were used to detect low flying aircraft. J.A. Ratcliffe[19] (Chap. 7, Appendix D) was placed in charge of the installation and use of CHL radars at TRE (Telecommunications Research Establishment), Dundee in the UK, in late 1939. He had taken a leave of absence from the University of Cambridge. Budden reports that vigorous discussions occurred in Ratcliffe's office at TRE,

> . . . reminiscent of the prewar Cambridge days using physical optics with Fresnel zones and Cornu spirals, applied to diffraction at a cliff edge.

Again there is no evidence that this TRE group interacted with their counterparts at RPL or ASE.

Finally, Bolton (1976)[20] reported that during the War while in the Royal Navy in the UK, he operated a night fighter squadron on the CHL station on the Firth of Forth (Scotland).

> And of course, the sea interference pattern was a nuisance to us. [Sullivan asked if he knew about this from practical experience.] Yes, on board ship the sea interferometer was nothing new. In fact, we had a short training course at Portsmouth [close to Hoyle's location at Nutbourne] before we went out into service. There is a Canadian radio astronomer, Carman Costain; well, his older brother . . . [Cecil] Costain, who was at the Herzberg Lab, and I were on the same course at Portsmouth. We designed an analogue computer to determine the height of an aircraft using the sea interferometer. Timing the intervals between passage through successive lobes [we determined the height of the incoming aircraft].

Thus numerous groups had extensive experience with sea-cliff antennas; but after the War only RPL took advantage of the nearby cliffs of Dover Heights to use this instrument as the first interferometer in radio astronomy. The sea-cliff interferometer certainly played an important role in these early years. But the inflexible observing times (only source rise or source set), with the large uncertain refraction corrections, were a major handicap and by the early 1950s, the sea-cliff interferometer had been abandoned at RPL. After the early 1950s, the Michelson interferometer, pioneered in radio astronomy by Martin Ryle and his colleagues (called a "spaced aerial interferometer" by this group Appendix D), became the high resolution radio telescope of choice.

---

[19]From Budden's 1988 memoir of Ratcliffe.

[20]Interview with W.T. Sullivan III, 13 August 1976.

Appendix D

Ryle, Payne-Scott, Bracewell and Bolton: "Solar Bursts" from Aircraft

At the end of 1948 (9 December) as Payne-Scott was completing the publication summarising the successful campaign to observe solar bursts at Hornsby (Chap. 8, footnote 46 and Chap. 9, footnote 26), at Pawsey's suggestion she wrote to Martin Ryle. It is likely that Pawsey had discussed some of these issues personally with Ryle during his visit to Europe in 1948. Payne-Scott included a copy of the paper on the Hornsby campaign (Payne-Scott 1949). She summarised the two types of bursts (the circularly polarised, "enhanced radiation" – Type I, and the "unpolarised bursts" – Type III) as described in Chap. 8. She pointed out that the latter bursts showed the delays at different frequencies, with higher frequencies arriving first.

Bracewell has reported[21] that he acted as a go-between for Ryle and Payne-Scott. Probably this interchange occurred after her December letter. He said that Ryle wrote to Payne-Scott rather sarcastically. "... [T]hese bursts did not occur in Cambridge, but any that did were certainly due to aeroplanes, as he knew from having a loudspeaker on line" [Bracewell's version]. Payne-Scott asked for actual pen recordings of these events. Ryle did not send any then, but did allow Bracewell (in his last year in Cambridge) to trace some of the recordings, which were sent on to Sydney in early 1949 (these are not in the archive). Payne-Scott asserted that these Cambridge data showed solar bursts similar to those she had observed at Dover Heights and Hornsby.

There was little direct correspondence throughout early 1949; most of the correspondence in this period between Ryle, Ratcliffe, Bolton and Pawsey was about the positional determination of Cygnus-A (at that time a "radio star") and the cause of the amplitude scintillations. But on 15 November 1949, Ryle chose another Australian colleague for his next message. J.C. Jaeger (Appendix C) had just visited his former university and must have met Ryle; clearly the subject of the RPL "unpolarised bursts" had been discussed. Jaeger (Chap. 8) had, in fact, worked on the theory of these events. In the letter to Jaeger, Ryle was again sarcastic in his criticism of the claimed reality of the "unpolarised bursts":

> The main point seems to be that Ruby Payne-Scott has misunderstood our earlier remarks. We do not claim that the "unpolarised bursts" are polarised; it is just we can't see them at all! This confusion arose because of the nomenclature, because we said that the only bursts we saw (meaning the fluctuations associated with sunspot activity) [the Type I "enhanced radiation"] were characteristically polarised.[22]

Ryle complained that in the northern hemisphere at 175 MHz, there were no "unpolarised bursts" (or "isolated bursts" in the Pawsey terminology), while these

---

[21] Bracewell interview, 10 November 1997.

[22] NAA: C3830, A1/1/1, Part 4.

were common in the southern hemisphere at lower frequencies. For the first time, he directly named the culprit: "... we have quite a large number of "unpolarised bursts" which correlate with the presence of light aircraft!" He does admit that the 80 MHz data is more or less useless due to the high level of interference. "We feel quite strongly that they do not occur on 175 Mc/s ... Please give our best wishes to our friends at Radiophysics". Within a few weeks, Jaeger was back in Australia and passed the letter on to RPL (probably to Bowen or Pawsey).

On 29 December 1949,[23] Pawsey wrote back to Ryle; for the first time a sense of exasperation appeared in his letter:

> It seems to me that this must be simply a question of differences in terminology [exactly a point that Ryle had made previously]. Bracewell brought out a number of tracings of your records and these show phenomena which are like the bursts which we measure here. I am enclosing some rough tracings of Bracewell's tracings [of the Cambridge 175 MHz data] which are good examples. The point of our observations is that we get occasional large solar bursts which look just like the bursts shown on your records at times when the general level [of solar radio emission] is quite low and on the occasions when we have measured the polarisation of these they have proved to be not circularly polarised.

He then directly challenged Ryle to compare his claim that there were no Type III bursts in the northern hemisphere with his own record no. 2 (from Bracewell's tracing) "which seems to us to be a normal sort of record". *That is, this was not an aircraft!*

But the controversy continued. After Bracewell had been back in Sydney for a few months, Payne-Scott showed him a letter which Ryle had written to Pawsey on 21 March 1950.[24] The letter concerned the occurrence of "unpolarised bursts" at 175 MHz, "the higher frequencies". Ryle wrote:

> We would agree that at first sight the bursts on July 6 (record no. 2) [probably 1949] ... which Ron Bracewell traced, give the impression of being the same thing as your unpolarised bursts ... However we are enclosing a photostat copy [Fig. D.1] ... it is clear that much of the fine structure was lost in the tracing. Bursts of this type are often received as light aircraft, having unsuppressed ignition, fly across near our aerials ... The "bursts" at 1,342 and 1,402 are very characteristically aircraft interference, as is the one at 1,508 [UT] ... but my guess is that in all probability all the "bursts" on this afternoon were caused by a light aircraft doing circuits from Cambridge aerodrome. I think the [solar activity] at the same time is coincidental. I wish we had known that Ron was tracing these records ... as we could, I think, have saved confusion. I realise that these arguments may not sound convincing to you, but the appearance of such interferences on our spaced aerial system does provide very characteristic records, and ones which are unlikely to be produced by a genuine solar disturbance ... the records selected by Ron were by no means typical (if only because of the bad flying weather here!) ... Bursts on 9 days occurred when the sun was not in the radiation pattern of the aerials. Of the remaining possible bursts, 20% were definitely identified as being due to aircraft (somebody having been present at the time, and having marked the record), while 62% had the characteristic "beating" appearance associated with a rapidly moving source. The remaining 18% may have been of solar origin, or may merely have been due to aircraft flying N–S, and thus not crossing the lobes ... I think that this

[23] NAA: C3830, A1/1/1, Part 4.

[24] NAA: C3830, A1/1/1, Part 5. Harry Wendt provided help in obtaining a preliminary version of this letter from the National Archives of Australia.

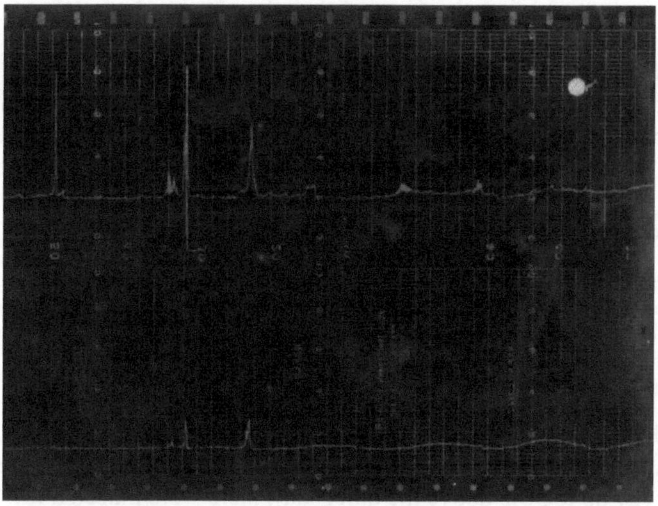

Fig. D.1 From a letter from Ryle to Pawsey from 21 March 1950. A copy of Cambridge radio noise observations at 175 MHz was enclosed in the letter. The figure was purported to indicate that aircraft were the major cause of solar bursts, leading Ryle to believe that at least the unpolarised bursts (Type III) did not arise from the sun. From the collection of the National Archives of Australia. NAA: C3830, A1/1/1 Part 5

> analysis shows that if bursts of radiation are received from the undisturbed sun on 175 Mc/s., they are extremely rare phenomena (compared with the figure of approximately one per day given by Ruby Payne-Scott for lower frequencies); they also do not have an intensity more than twice that from the undisturbed sun. Our evidence at 80 Mc/s is not very conclusive either way, because in addition to aircraft ignition bursts, there are additional sources of intermittent signals from various mobile services operating in this band, but we feel fairly convinced about our 175 Mc/s. results.[25]

Ten days later Pawsey wrote with some resignation, "I think I shall retire from the controversy concerning 'isolated' or 'unpolarised' bursts. I shall pass on your letter to Ruby Payne-Scott and others who have been taking actual observations recently".[26]

A month later Bolton was visiting the Cavendish Laboratory for a few weeks. On this European tour, he and Kevin Westfold visited Hey's AORG (Army Operational Research Group), Cambridge, Jodrell Bank, Paris, Leiden, Kiel, Copenhagen and the URSI meeting in Zurich.[27] At the end of the Cambridge visit, Bolton wrote a

[25]Bracewell's retelling of this event is more colourful. He quotes Ryle as saying "If I had known Bracewell was spying on me, I would not have given him these recordings to trace".

[26]NAA: C3830, A1/1/1, Part 5.

[27]He and Pawsey had an exchange of two letters (NAA: C3830, F1/4/BOL/1) about the possibility that one of them would present Paul Wild and Lindsay McCrady's new Penrith swept frequency radio data and Payne-Scott's Hornsby results at the 1950 URSI Assembly in Zurich. In the end Bolton and Westfold had to leave Zurich (by train for Copenhagen) on the last day of the

14 page report concerning the Cavendish Laboratory radio astronomy activities (date about 30 May 1950) (including four photographs made by him of the Cambridge radio telescopes).[28] There were three main topics: (1) solar noise, (2) discrete sources and (3) fluctuations in discrete sources (scintillations). Solar noise research was being carried out at 45, 80, 175 and 500 MHz, all with Michelson interferometer spacing of 10 wavelengths (about 6 deg compared to the solar diameter of 30 arc min). Polarisation data was also obtained. Most of the report concerned the "isolated burst controversy". Bolton wrote:

> Ryle's definition of an isolated burst is as follows: "According to Ruby Payne-Scott isolated bursts occur when the Sun is perfectly quiet – i.e., only thermal level present and not visible spots. They are on average about ten times the thermal value [quiet sun], last a few seconds, have a rapid rise and an exponential decay and occur at about one per day – but not in Cambridge".

Bolton looked at a large number of records himself and found that, if the sun were quiet, there were few if any bursts at 175 MHz. However, when the solar level was higher (by up to 50% above the quiet sun), there were indeed a few bursts with peaks up to three times the quiet level. These events had a sharp initial rise with an exponential tail (exactly as Payne-Scott claimed), lasting some minutes. Bolton showed these results to Ryle; his suggestion was rejected by Ryle since the claim was made that the sun was not absolutely quiet at the time and thus solar noise storms (Type I) would be expected. But as we have seen Type I bursts were usually circularly polarised. In the report, Bolton blamed the confusion on the slow rise time on the paper chart recorder in the presence of a fast signal increase. Ryle thought the "recovery time gave a false impression and that the response to a rapid increase was very good". Bolton wrote that he saw contradictory results when comparing some previously obtained solar data obtained simultaneously by Jodrell Bank and Cambridge ("spaced receiver" data). Finally, Bolton looked at the 80 MHz data and declared that the data was useless due to lots of external interference (perhaps real "aircraft") due to ignition or even C.W. (continuous wave from e.g., a broadcast station). "There is no listening watch kept on the character of the radiation". A whimsical sentence ended the report. "I do not think Ryle has sufficient evidence to deny the existence of isolated bursts; on the evidence of the Cavendish records one can only give the Scottish verdict – *NOT PROVEN*".[29] Perhaps real "solar bursts" were in fact masquerading as "aircraft" (Fig. D.1)![30]

---

conference and missed the solar session, after just enough time to present a paper at the galactic noise session!

[28]NAA: C3830, F1/BOL/1.

[29]A third possible verdict in a Scots trial, apart from guilty, or not guilty. Not proven is interpreted to signify that the suspect may not be innocent but there is insufficient evidence to the contrary.

[30]In contrast to the report which he published 32 years later (Bolton 1982), Bolton did accomplish a great deal at the Cavendish Laboratory during the visit. Although it began with a major misunderstanding based on an unwelcoming letter from the Cavendish Professor W.L. Bragg on 2 May 1950 (born in Australia in 1890), Bolton wrote 16 days later that the situation "has

As we now know, the Sydney achievements of Payne-Scott and subsequently of Wild and his colleagues have withstood the test of time; these results are now accepted as defining the characteristics of Type III bursts. Sullivan (2009, Chap. 14) has described the doubts that Ryle had in 1949 and 1950 about the reality of *any* solar radio bursts, when he became convinced of the local nature (in the ionosphere) of radio star scintillations. Based on the exchange with Payne-Scott and Pawsey, the aircraft "solar bursts" may well also have played a role in this conviction. As Sullivan has stated, Ryle never published any of these doubts. By the early 1950s, Ryle's interests in solar radio astronomy had diminished.

Surprisingly, Pawsey, Ryle and Ratcliffe continued these discussions about "avoiding possible clashes of interest" with correspondence on 3 November 1950 (Pawsey to Ratcliffe) and on 7 February 1951 (Ryle to Pawsey).[31] In the first letter, Pawsey outlined a number of major projects that RPL would start in the 1950s. In his response, Ryle gave the impression that solar work was to be given lower priority at the Cavendish. He described a program to follow the motion of sunspot sources (Type I bursts); no results were ever published. Ryle did emphasise the accurate position finding observations of F. Graham-Smith and also the work with the "Long Michelson" radio interferometer. Pawsey wanted to avoid the duplication of "expensive equipment". He wanted to know if the Cavendish group were about to start work at a new wavelength of 6.7 m (45 MHz): "This is a wavelength on which we held off by agreement with Ryle (in 1948) so he could have a clear field." Few concrete results were to come of these attempts at coordination, especially in the field of solar noise. After about 1950, these groups would no longer be rivals in the field of solar bursts. (The Cambridge group carried ground breaking radio synthesis observations of the quiet sun at 60 cm, 1.7m, 3.7m and 7.9m in the period 1950–1953.) The Australian group of Paul Wild and colleagues would dominate this important research area in the years to come.

In view of the later disturbed relations between the Sydney and Cambridge groups, a comment from Ryle to Pawsey in March 1950 (before the visit described above) is ironic: "We are very much looking forward to seeing Bolton, and to hearing about some of the more recent work at Radiophysics." Later in 1950, the strains between the Cambridge and Sydney radio astronomers were beginning to show. By the time the comparisons of the 2C survey (Shakeshaft et al. 1955) began in 1955,[32] the conflicts had reached their highest point. As Mills (1984) has pointed out, the

improved somewhat". Instead of a suggested stay of 1 day, Bragg had extended the welcome to 3 days after the first exchange of letters. After meeting Bolton personally, Bragg suggested a visit of 3 weeks! Bolton wrote a frank letter to Bowen on 18 May 1950, stating that "There is . . . a visitor problem [too many of them with the resultant loss of time] at the Cavendish but in spite of this they are both jealous and scared of R.P". NAA: C3830, F1/4/BOL/1. For another interpretation of this visit see Hoyle (1994); his claim that Bolton was ejected after only an hour from Ratcliffe's Cavendish office and ended up at Hoyle's rooms at St. Johns College, is unlikely to be totally correct. Based on Bolton's letters and the extensive report prepared by Bolton in 1950, he clearly spent substantial time at the Cavendish Laboratory, interacting with a number of colleagues.

[31] NAA: C3830, A1/1/1, Parts 5 and 6.

[32] Pawsey (1957); M. Ryle, "Radio Galaxies", *Scientific American*, 1956 correspondence from Mills in December 1956 (Mills 1984).

controversies of the 1950s seem surprising given the knowledge of the high red-shift universe more than 30 years later. As Mills (1984) aptly summarised the situation:

> The 1950s were stimulating and exciting for both radio and optical astronomers. For the first time it became clear to all of us that radio astronomy promised a great extension to the powers of optical astronomy for exploring the distant Universe. The Sydney-Cambridge controversy eventually concerned only the degree to which the radio data stood alone and could be interpreted by simple models.

The interchange between the Sydney and Cambridge radio astronomers in these years (1947–1951) was quite civil; Kenneth E. Machin (a graduate student of Ryle, working on radio aperture synthesis of the quiet sun at 81.5 MHz, the PhD being awarded in 1952,) was apparently frustrated by the length of time taken by Ratcliffe and Ryle to referee scientific papers. Machin wrote a ditty (found by Sullivan in Cambridge) that consists of eight stanzas. Here are the first and last, providing a clue to the competition between Sydney and Cambridge, with Ruby Payne-Scott in the place of honour:

> A paper on sunspots I promised to write,
>
> So I started one evening and stayed up all night,
>
> And by the next morning that paper was done
>
> Apart from the title and figure 21.
>
> [Then six stanzas which consisted of lengthy criticism of the manuscript by both Ryle and Ratcliffe.]
>
> At last it was finished, but then sad to say,
>
> I posted it off and then later that day
>
> I looked at *Nature*, I felt like such a clot,
>
> The whole thing had been published by Pawsey, Mills, Yabsley, Bracewell and Payne-Scott.

Appendix E

Pawsey's Review Paper of 1950: "Solar Radio-Frequency Radiation"[33]

In 1950 Pawsey (1950a) published one of the first solar radio noise review papers.[34] This publication is a vital resource for tracking the first years of both solar and

[33]This appendix is referred to in Chaps. 7 and 8; in particular, Payne-Scott's hesitation about using the figure from the March 1947 Type II event in the review paper (our Fig. 7.15) is discussed in Chap. 8.

[34]Other early radio astronomy review papers include Reber and Greenstein, "Radio-Frequency Investigations of Astronomical Interest" (1947), J.S. Hey, "Reports on the Progress of Astronomy, Radio Astronomy" (1949), and Ryle, "Radio Astronomy" (1950), a 61 page review.

cosmic radio astronomical research in its early years. "Solar Radio-Frequency Radiation" appeared in the British journal, *Proceedings of the Institute of Electrical Engineers*, 1950, Part III, vol. 97, page 290. The appearance of a major review paper by one of the leading researchers was an indication of the importance of the emerging field of radio astronomy. Strikingly the journal was far removed from traditional astronomy and closer to radio engineering. Later review articles were more accessible to the astronomical community. For example, Pawsey and Smerd published an extensive review article only 3 years later, "Solar Radio Emission", which was Chap. 7 of *The Sun*, in the series, *The Solar System*, edited by G. P. Kuiper and B. Middlehurst (University of Chicago Press). This review is discussed in some detail in Chap. 9.

Payne-Scott and Bolton played a major role in the preparation of the 1950 paper with a number of their unpublished investigations being summarised. In particular, essentially none of Bolton's individual solar research on the polarisation of "enhanced radiation" was ever published and only appears in this review paper.[35] A number of Payne-Scott's results (fading and details of the July 1946 solar campaign) were also only published in this review paper. Both colleagues provided substantial comments on the content of the publication by Pawsey.

A recounting of the tortuous history of the 1950 paper also reveals a great deal about the internal procedures within RPL, as this new radio astronomy group attempted to find their place in the world of astronomy. A prominent example is the evolution of the lengthy and often contradictory internal review process within RPL.

The fact that the RPL scientists were not astronomers, but physicists and engineers, meant that their acceptance in the world of astronomy was a complex process. The fact that this acceptance did occur within a decade is partially due to the contacts Pawsey made in the decade 1945–1955, both within Australia and overseas. Pawsey hoped that the review paper would showcase this new approach to solar physics.

The evolving relation of RPL to the rest of Australian science, in particular to the hierarchy of CSIR and later CSIRO, can also be partly traced by examining the controversies surrounding the publication of this paper and others in the 5 years after the end of World War II.

The genesis of the 1950 Pawsey publication was complex and probably began in early 1947. We can surmise that Pawsey wanted to write a review paper that would be ready before his departure in late September 1947 for his 13 month world tour. The PC Minutes of 6 June 1947 indicate that Pawsey was preparing to write a review paper; an outline was available but "work in abeyance while working on [the paper on thermal radiation with Yabsley]".

Pawsey must have regretted that the article was not complete before he left. We have found at least 27 letters, as well as an important telegram, to and from Pawsey in the following year, about the status of the paper, details about the writing of the

[35] Orchiston, Slee and Burman (2006) discuss the 1950 review paper.

paper, comments from numerous internal referees in Australia and the UK, and numerous contentious letters about the choice of the scientific journal to be used.

Even after the paper was submitted in September 1948, the publication was delayed until 1950. Pawsey made a great effort to keep the review up to date over this 3 year period. Payne-Scott, Bolton, Bowen, White, Ratcliffe, Minnett, Yabsley and McCready (who ended up doing much of the detailed editorial work in Sydney during Pawsey's absence, from September 1947 to October 1948) were all involved.

While Pawsey was on board ship in the Pacific in September–October 1947, McCready was already writing a letter[36] to him to San Francisco with details of the preparation of the figures; Bowen was passing the paper on to Harry Minnett (an "independent reader") for comments. The PC minutes of 14 November 1947 indicate that the internal referees were Minnett, White (of the CSIRO Executive) and Bowen.

A few weeks later (24 November 1947), Bowen wrote to Pawsey[37] describing the likelihood that the survey paper would be published in the newly formed *Australian Journal of Scientific Research*; Bowen wanted the complete manuscript so the survey paper could be published in the inaugural issue in March 1948. As will be seen, this did not happen. Bowen sent an edited manuscript by Harry Minnett:

> I think the best way of arriving at a final copy for submission to the Executive before the end of the year is for you to read the one carrying Harry's amendments ... and return it to us. Would you be sure to return the amended text to me well before December 31st ...

Bowen indicated that he had two additional suggestions, which were optional. (1) Bolton had discovered some new sources (after Cygnus A) which could be mentioned in some vague manner – "I think, too, it would be wise not to be too specific about them in the US and UK until Bolton has a chance of finalising his observations and getting them published". (2) Victor Bailey at the University of Sydney had a new theory of the origin of solar bursts which seemed to be more promising than the models of Martyn and Giovanelli.

The next step was a re-typing of the paper at the ASRLO (Australian Scientific Research Liaison Office) in Washington. Pawsey sent it back to RPL on 9 December 1947, including the corrections from Minnett; a number of additions were to be considered in Sydney, especially to the figures and figure captions. Pawsey certainly thought the paper (now the "Washington draft") was ready for submission to the new Australian journal.

But this was not to be. In the new year, Bowen became disaffected with the 'Washington draft"; he wrote that he "must admit to being a bit worried by your

---

[36]Unless otherwise noted, most of the correspondence during this period is NAA: C4659, 8.

[37]NAA: C3830, F1/PAW/1, Part1. This letter, which included manuscript and figures was formal – "Dear Pawsey" – and was sent to the ASRLO at the Australian Embassy in Washington. Bowen was very positive about the current health of the radio astronomy group during Pawsey's absence: "The way in which everyone is right on top of his job and continuing to do excellent work after your departure is a great tribute to your leadership." Pawsey arrived in Washington in the first days of December, 1947 after visits to San Francisco, Los Angeles and Chicago. A number of important letters were posted from the train from California to Chicago during a stop in Lincoln, Nebraska on 10 and 11 November 1947. NAA:C3830, F1/4/PAW/1, Part 1 and NAA:C4659, 8.

Survey paper and have asked Fred White to read it critically".[38] The Pawsey and McCready publication, "The Measurement of Solar Radio-Frequency Radiation" (called the "instrumental paper"), which was to be a companion paper, was held up (see below). Six days later, (same source as footnote 38 Appendix E), Bowen wrote an even more negative letter with details of his objections, along with the detailed comments from White and additional criticism from Jaeger and Martyn. He made several suggestions to break the log-jam including getting someone to re-write the paper or even wait until Pawsey's return to RPL. Bowen did not favour the latter due to advances in the field of solar noise. A third option was a major re-write by Pawsey.

Pawsey expressed mixed feelings in his reply on 9 March 1948 (footnote 38 Appendix E). 'I am naturally a little unhappy about it, particularly as I think the general criticism is right." His proposal, in order to get the paper published before he returned to Australia towards the end of 1948, was to ask that Bowen become a co-author! There were many issues that had to be resolved, in particular the intended audience. Was it intended for astronomers or ionospheric physicists? Bowen replied on 31 March 1948 (footnote 38 Appendix E) in a more conciliatory manner; he urged a completion of the paper before Pawsey's return later in the year (Pawsey arrived with his wife in the UK from New York on 1 April 1948 on the SS *Queen Elizabeth*), and he was clear that "you are the person to write it [i.e., Bowen was not to be a co-author] and I guess that a final version isn't too far off".

Buoyed by Bowen's renewed optimism, Pawsey rewrote the paper quickly in the UK and by 24 May 1948 Bowen was quite pleased with the new version (the "April draft"). He wanted to push ahead with submission to the *Australian Journal of Scientific Research.* In the meantime, Ratcliffe had read the paper while Pawsey was at the Cavendish; Ratcliffe was impressed and suggested only minor changes. On 9 July 1948 Pawsey sent Bowen yet another version of the survey paper, an updated "April draft".

Meanwhile, both Payne-Scott and Bolton had written to Pawsey with suggested changes. Bolton's suggestions (1 February 1948) were minor while Payne-Scott's were major (see Chap. 8). By now Payne-Scott no longer believed in the several minute time scale delay from the 8 March 1947 Type II event. Since Pawsey referred to the famous "Figure 4" in the review paper (Figs. 7–15 from Chap. 7) at least three times in his "April draft" of the survey paper, Payne-Scott insisted that the minute delays were not real. She objected to all references to these time delays. But Pawsey maintained his belief in the reality of the Type II bursts, as observed in the giant outburst from 8 March 1947.[39] Based on later experience and years of study of Type II bursts, there is now no doubt about the reality of the event observed in March 1947 by Payne-Scott, Yabsley and Bolton.

[38] 12 February 1948. NAA: C3830, F1/4/PAW/1, Part 1.

[39] In the NAA: C4659, 8 document there are two pages in Pawsey's handwriting in which he dealt with all Payne-Scott's points. In many he wrote "agreed". But in the case of Fig. 4, he stuck to his original point of view.

On 9 July 1948, Pawsey wrote Bowen with the "last copy" of the survey paper. He had detailed comments from Bowen, Ratcliffe and Payne-Scott. He had accepted all but one of the suggestions:

> There is one point on which I have departed from Ruby's ideas. She is suspicious of the long delay observed between frequencies in the famous letter to *Nature*, because the effect has failed to repeat. I wish to give the information . . . with the remark that the out burst [sic] was exceptional. I should only be prepared to delete if, on examination of the original records Bolton or Lindsay [McCready] were prepared to say that the records were definitely wrongly quoted. I am not prepared to alter anything on the grounds that other out bursts [sic] have been different.

There were two additional hurdles to overcome. By 9 August 1948 (letter from Bowen to Pawsey) it was clear that the new Australian journal would not, on principle, accept technical papers; thus the "instrumental" publication of Pawsey and McCready had been rejected.[40] The editorial board and the CSIRO Executive had decided that "survey" papers were also not acceptable. Bowen was worried that the Pawsey survey paper would certainly be rejected even though it contained a large amount of original material. Pawsey had anticipated this problem and had already been thinking of an English journal. "I believe that I could get the article accepted here [in the UK] straight away" (27 July 1948).

Then in a very surprising turn of events, Bowen was again displeased with the paper! On 30 July 1948, he wrote Pawsey – "I can't say that I am happy about the paper. Like a lot of others we have had recently, either it isn't very good or I have read it too many times in a rough form . . ." Pawsey was naturally surprised (9 August 1948) and wrote to Bowen:

> I also feel depressed about it and don't know what to do to break the bad sequence . . . but the two who have read it, yourself and Ratcliffe, are in disagreement. You are critical, Ratcliffe praised it . . . What a business this is!

On the other hand, in a letter dated the same day, Bowen wrote to Pawsey about how much he was missed:

> . . . have you made any plans for your return yet? I get the impression from your letters that your return may be imminent but I haven't seen any dates mentioned. [During the second week of September 1948, they would depart on the SS *Orontes*.] We are all becoming very anxious to see you back, particularly the Solar Noise people. There is no doubt that their work is suffering by your absence and I am looking forward to a great burst of activity and enthusiasm on your return.

Finally, in spite of Bowen's renewed pessimism, Pawsey was given permission by CSIR to submit the paper to a UK journal (20 August 1948);[41] clearly

[40] This paper was never published; in the end it was only issued as an internal report, RPR 74, "The Measurement of Solar Radio-Frequency Radiation", a paper of 24 pages with eight figures and three plates.

[41] On the same day, McCready sent Pawsey an updated version of the manuscript. He had made a few minor corrections suggested by Payne-Scott based on her 1948 data at Hornsby (Chap. 8). She had at this time established a clear observational distinction of the circular polarised bursts

Ratcliffe[42] had decided that a suitable journal would be the *Proceedings of the Institute of Electrical Engineers*. The paper was submitted on 20 September just as Pawsey left the UK for Australia. Ratcliffe wrote to Pawsey on 30 November 1948 with the news that the paper had been accepted.[43]

The paper was then corrected (probably for referees' comments) and re-submitted on 3 February 1949. The paper was "read" on 7 December 1949 before the "Radio Section" of the society by Stanley Hey, standing in for Pawsey, with extensive published discussion after the presentation. Ryle,[44] Millington, Hunter, Roberts and Stanesby posed questions with detailed written replies submitted at a later date by Pawsey and Hey.[45] The paper appeared 9 months later in 1950. The time from the original submission to publication was almost 2 years! Pawsey updated the paper numerous times after the original submission, and included a short addition describing the Wild and McCready swept-frequency results published in 1950. A small footnote was added in March 1950 about the detailed nature of "isolated bursts", based on the 1949 Payne-Scott publication of the properties of "unpolarised bursts".

The survey paper remains an important legacy of the early RPL solar work. Many of the figures were supplied by Payne-Scott, including a number from the unpublished July 1946 campaign at Dover Heights (RPL 9 – see Chap. 7). The invaluable publication of Bolton's solar work, especially the circular polarisation data from March to May 1947, has remained a lasting contribution.[46]

---

associated with "enhanced emission" (Type I noise storms) and the "unpolarised bursts" (Type III). These conclusions were added to the text of the survey paper at the last moment. NAA: C4659, 8.

[42] A cable from Pawsey to RPL on 8 September stated that Ratcliffe said that this journal had a new policy of "publishing integrating articles in scientific radio". He looked into submitting the publication to *Reports on Progress in Physics*, but was told that this journal was "booked solid until 1950".

[43] NAA: C3830: A1/1/,1 Part 3. Ratcliffe wrote: "I expect that you will have heard by now that the I.E.E. [Institute of Electrical Engineers] feels that is a very good one. I believe they are going to ask your permission to have it read in this country by Hey. They thought it was very important and would attract a good audience."

[44] As expected (see Appendix D), Ryle questioned the reality of the time delays for large outbursts. Pawsey gave a definitive answer by quoting the compelling evidence of Wild and McCready for the reality of Type II bursts; these were associated with optical solar flares with broad bands of radio emission, drifting to lower frequencies with a drift rate similar to the March 1947 giant outburst.

[45] In a slightly bizarre manner, there was confusion about the authorship of the replies to questions from the audience. Hey had told the Institution of Electrical Engineers in London that he would be "unwilling that any contribution to the discussion on [Pawsey's] paper should appear under his [Hey's ?] name .... [But Hey] was kind enough to prepare a statement suitable for attachment to [Pawsey's] own reply, to be published as though it emanated from [Pawsey]". In the Sullivan RPL archive (D5/4/25) we found the combined text, with the Pawsey text followed by the more lengthy Hey text. As Hey wished, all the published text is under the by-line of Pawsey. H. Stanesby made the interesting point that British TV stations in the VHF band would be susceptible to interference by strong solar outbursts; there was evidence that prominent interference had been detected in the British Post Office 80 MHz radio system at the time of the great sunspot of February 1946.

[46] Bolton has written that after the giant flare and Type II burst of 8 March 1947 (Chap. 7), he continued observations of the circular polarisation during the next appearance of the giant sunspots

TYPE	WAVELENGTH RANGE OBSERVED	DURATION	ASSOCIATED OPTICAL FEATURE	POLARISATION	SIZE OF SOURCE	REMARKS
Thermal	1cm–4m	—	—	Random	Whole sun	
Slowly varying component	3cm–60cm	—	Sunspots		Small	
Enhanced Radiation (steady or with bursts)	Above 1m	Days	Sunspots near central meridian	Circular	Small	Bursts selective
Isolated burst	Above 1m	Seconds	—	Not circular	—	Wide freq. range
Outburst	3cm–5m	Minutes	Flare	Not circular	—	Some of extreme intensity

TYPES OF SOLAR NOISE

Fig. E.1 A summary table from the Pawsey (1950a) review paper (Table 1), "Types of Solar Noise." This version from the ATNF historical photographic archive (B1924-2) appears to be a slightly updated summary of the characteristics of solar noise as compared to the published version. The published version in 1950 does not include the "slowly varying component," prominent at cm wavelengths. This cm component of solar radio emission arises from thermal emission of hot regions of high electron density and from magnetic fields existing in the vicinity of sunspots and chromospheric plages, Kundu 1965). (Plages are bright cloud-like features found around sunspots that represent regions of higher temperature and density within the chromosphere.) This radio emission varies from day to day and is closely correlated with sunspot areas

The paper has a valuable historical summary in the introduction, including a clever diagram showing the relation between flux density, mean effective temperature (called brightness temperature today) and brightness as a function of frequency. Sections were titled "Observed Characteristics" (including mainly data from RPL – see Fig. E.1[47]) and "High Frequency Characteristics of the Solar Atmosphere", which includes a very readable description of the Appleton magneto-ionic theory. The final sections were titled "Discussion and Hypothesis"[48] with a fascinating discussion of the early non-thermal theories of solar bursts. Most of these did not stand the test of time. The conclusion tries to bring solar and cosmic radio astronomy together.

> With the discovery of radio-frequency waves from the sun and the galaxy, radio science has provided astrophysics with a new tool ... The mode of origin of galactic noise is not yet established. While this remains so, the deduction of general astrophysical data from cosmic noise may derive from solar noise studies. Both cosmic and solar noise were discovered through the interference that they caused in radiocommunication and radar ... The full applications of a new science are seldom evident, however, and their discussion must be left to the future.

---

in April 1947. "I wrote up the observations for Martyn but the joint paper with his theoretical consideration never eventuated" (Bolton 1982).

[47]This figure dates from 1949 and is an update of Table 1 of the survey paper.

[48]Pawsey pointed out that the power in solar radio emission was apparently large, 10^{11}–10^{15} W; this power was trivial compared to the 10^{26} W of heat from the sun.

A highlight of the published discussion at the conclusion of the paper was the vexing issue of the comparison of solar and cosmic radio noise. Pawsey replied to a question: ". . .. the fundamental relations established between cosmic and solar noise are, first, the surprisingly high intensity of cosmic noise relative to solar nose,[49] and secondly, the existence of at least some dozens of discrete sources of cosmic noise such as radio stars." The newly published identification by Bolton, Stanley and Slee of Taurus A (1949; also Bolton and Stanley 1949) with the famous Crab Nebula, the remnant of the supernova of AD 1054, was at this point known to have a luminosity 10^{16} times greater than the sun:

> [Pawsey wrote in conclusion]. . . current opinion is that both interstellar gas and some forms of stars should be considered as possible sources of cosmic noise.

It would be some years before a more mature understanding developed of the roles of "interstellar gas" and "radio stars" (later on called "radio nebulae" or "radio sources") in the radio universe. During the next decades, a consensus developed within the astronomical community over the role of radio astronomy.

Appendix F

Payne-Scott and the Creation of a Facility for Joint Radio-Optical Observations of Solar Phenomena

Starting in 1947, Payne-Scott played a major role in planning a new optical facility for joint optical and radio observations of solar phenomena. The attempts to start this facility were painful and tedious; after 4 years the facility still did not exist. In September 1952, more than a year after Payne-Scott left RPL, the Commonwealth Solar Observatory lent a spectroheliograph[50] to their colleagues in Sydney, which was then installed at Potts Hill. With the long time collaboration of Ron Giovanelli (Division of Physics at CSIRO), this facility was the beginning of many years of active optical and radio collaboration in Australia.[51]

---

[49] Dr. A. Hunter had raised the issue in the discussion: "An area of the Milky Way apparently equal in size to the sun will produce something of the same order of intensity of radio radiation as that body [the sun], whereas in the matter of optical radiation the Milky Way is weaker by a factor of about 10^{12}."

[50] See Raymond N. Smartt, "Spectroheliograph", in AccessScience@McGraw-Hill, http://www.accessscience.com, DOI 10.1036/1097-8542.642400.

[51] This material is based on NAA: C3830, A1/1/1, Parts 2, 3, 4, 5, 6 and 7, NAA: 3830, C4659, 8, and NAA: C3830, F1/PAW/1, Parts 1 and 2.

During the period of her intense involvement, Payne-Scott was a strong advocate for the acquisition of a modest optical observatory for contemporaneous observations.[52] The goal was to observe solar flares with wide-field Hα[53] imaging, which would then be compared with the onset of solar bursts and outbursts. Payne-Scott wrote two detailed memos and spent considerable effort in learning the details of the optical techniques. The first suggestion was made in a memo of 17 November 1947, entitled "Possibility of Constructing a Spectrohelioscope or Spectroheliograph Suitable for Investigating Optical Correlation with Solar Radio Observations".[54] A probable rationale was the realisation that the correlations of prominent solar events (such as the 7 March 1947 event) using short-wave fadeouts[55] as a proxy for major flares, was an uncertain process.

Allen (1946) had described the Mount Stromlo procedures for the solar events of July–August 1946; he was rather negative about the reality of detailed correlations of radio noise with chromospheric or photospheric features on the sun. He also stated that "there are occasions when there appears to be a physical connection between an outburst of noise and the onset of a flare". He relied mainly on the gross properties of the solar surface such as sunspot sizes and numbers. Payne-Scott had visited Mount Stromlo earlier in 1947 and described the slow mechanical scanning of the sun taking about 10 s for a full image.[56] Rotating prisms were used with an instantaneous field-of-view with a strip of only about 5 arc min. The proposed solution was to obtain a single image of the Sun in about 3 s in the Hα line, using the monochromatic filter technique invented by Lyot. A telescope aperture of only 5 in. (13 cm) was sufficient. The memo indicates that a number of discussions had already taken place with W.H. Steel and R.G. Giovanelli of the CSIRO Division of Physics about the construction techniques for the required filters.

---

[52] Already on 13 May 1946 (PC minutes), Pawsey had visited Allen at Mount Stromlo and been impressed with the difficulty of rapid full solar imaging: "[Pawsey's] impression was that visual observations are very difficult." For some years the RPL group had also tried to rely on optical solar groups in Japan, India and New Zealand for common solar coverage; this process had been somewhat unsatisfactory.

[53] In physics and astronomy, H-alpha, also written Hα, is a specific red visible emission line created by hydrogen with a wavelength of 6562.8 Å.

[54] This report was discussed in the PC meeting of 14 November 1947; an earlier report on 16 October 1947 provided a detailed description of the rationale: "Some thought will shortly be given to the practicability of constructing a spectrohelioscope. It would be useful in quickly ascertaining whether a flare were [sic] in progress during intense noise storms." Clearly Payne-Scott had done substantial research on the issue of solar optical instrumentation. An additional motivation was a paper published by A.D. Thackeray of the Solar Physics Laboratory at Cambridge (UK) in the 27 September 1947 issue of *Nature;* he found a flocculus with a high radial velocity (about 100 km/s) in Hα. This event was correlated with an increase in radio noise at 175 MHz, as observed by Ryle at the Cavendish.

[55] As an example Payne-Scott used the disappearance of the radio frequency carrier of the 14.4 MHz short wave station VLQ3, about 700 km north-east of Sydney, as a proxy to determine the time of the flare.

[56] She does admit that her account may not be completely accurate due to "vague memory of what I saw there". R. Smartt pointed out to us that a full disk observation with a film type spectroheliograph would, in fact, require an observation of some minutes' duration.

A week later on 24 November 1947, Payne-Scott (as well as Bowen) wrote a letter to Pawsey during his US tour; the letter was addressed to the ASRLO in Washington, D.C. The report from the previous week was enclosed, with additional information about the possibility of obtaining calcite crystals (or a substitute), required for narrow-band filters, in the US. Quartz crystals are also required in Lyot-type filters, but such large, high-quality quartz crystals were readily available in Australia. She stressed that the monochromatic filter scheme (rather than the scanning prism or grating scheme) was preferable. She mentioned several possible sources for complete monochromatic filters in the US. Both Giovanelli and Allen were reported to be interested in future collaborations.

Pawsey began to acquire in depth information on this new topic. Probably he talked to Seth Nickolson of the Mount Wilson Observatories about optical solar instrumentation. The highlight of his trip to the west of the United States was a visit to the Climax, Colorado High Altitude Observatory of Harvard University and the University of Colorado. His host was John W. Evans (1909–1999), a well known solar physicist and instrumentalist (the "John W. Evans" Solar Facility is currently a major component of the National Solar Observatory at Sacramento Peak in New Mexico, USA). A short time later Pawsey met Walter Orr Roberts (1915–1990, founder of the National Center for Atmospheric Research) at Harvard. In a short period, Pawsey received an intense education in the techniques of optical solar astronomy; he also passed on to the US solar physics community an understanding of the importance of this new field of "solar radio noise".

Based on these new contacts, Pawsey wrote numerous letters to various optical component firms, inquiring about narrow-band Hα filters. The choice of a spectroheliograph with a monochromatic filter had been made; the bandwidth of the filter was to be at most 0.25 nm. There were even discussions about tuning the filter to detect large radial velocities in the motions of the solar events. Pawsey contacted, among others, Bruce Billings (described as a "shining technical light") of Baird Associates of Cambridge, Massachusetts; this group was engaged in building a set of narrow-band filters for Harvard to be used in optical solar research. He also wrote to Edwin Land of Polaroid, from whom he received a non-committal response. Apparently Pawsey also met Billings personally in Washington, D.C. and received an encouraging response.

Next, in April 1948, there was a progress report from Payne-Scott. From an unspecified date, a handwritten report[57] was prepared with a detailed scientific proposal for the optical observatory. This report was an update of the 1947 document, representing additional research in early 1948 on the feasibility of the project; clearly there had been further discussion with Pawsey while he was in the US (until late March 1948):

[57]The title page (from the RPL archive A1/3/1b from the Sullivan archive) has the typed text: "From Ruby Payne-Scott – to be held in file until decision made about spectroheliograph." Also there is a note in ink from someone else with a cryptic message: "Add more on Sun Echoes." This may refer to possible radar experiments from the Sun, an experiment never attempted by RPL, although discussed frequently by Kerr and Pawsey.

> A large proportion of the future solar noise programme is to be devoted to the study of bursts (e.g., application of interferometry to locate source of burst on Sun [the Potts Hill intererferometer], measurement of time differences between onset of bursts on different frequencies [the swept frequency observations at Penrith] and onset of any related visual phenomena, possible spectrum analysis of bursts). Interpretation of results will require fairly detailed data on the occurrence and location of visual solar phenomena that may be related to the radio frequency noise. However there is no observatory making anything like continuous observations on the Sun during daylight hours. At present we rely on inferring the occurrence of flares from the occurrence of radio fade-outs – an unsatisfactory method as the correlation is not simple. On other phenomena we have practically no information. There is no hope of obtaining the detailed information we would require from the Commonwealth [Mount Stromlo], as they have not [sic] facilities for continuous observation.

Payne-Scott then suggested that the obvious solution was the co-location of the "simple spectrohelioscope" with the relevant solar noise equipment. The visual observations were to obtain short time-scale exposures in Hα, to be triggered from the radio receiver.[58] The narrow-bandwidth filter was described again ("analogue of an electrical band-pass filter") with the suggestion that exposures of about 1 s were desired. The Mount Stromlo scanning system was again rejected due to the long period of exposure. A planned construction scheme was then described with construction of the telescope in the RPL workshops, the optical components supplied by the Division of Physics. The narrow-band filter was seen to be the major challenge, consisting of plates of doubly refracting material which were to be separated by polarisers. A possible scheme was described for a potential purchase from a US company of a complete filter system (fixed tuned filter with a bandwidth of about 0.1–0.2 nm); Pawsey had been in touch with this vendor – Baird Associates – in the US and the cost was expected to be $1000 to $2000 US.

The Propagation Committee report of 22 July 1948 then stated that it was hoped that a light filter from America for construction of a spectroheliograph for RPL use would be obtained.[59]

However, the optimism of April 1948 was not sustained. After many months, these contacts with Billings produced no response. Bowen wrote one last time to Baird Associates on 2 July 1948 with a request for a quote on the delivery of the sub 0.1 nm filters. No reply was received and Bowen was clearly frustrated. Pawsey even suggested that either Evans or Roberts should go to Australia to help build the new spectroheliograph.[60] When Pawsey left the US on 27 March 1948, his interest lagged until he returned to Australia at the end of the year.

---

[58] In a letter of 8 March 1948 to Pawsey, Evans suggested a fanciful method of alerting a sleepy observer to the presence of a new optical solar flare. "[The alarm] will be connected to an electrical buzzer, an electric shocking device in the seat of a favorite chair, a flashing red light, and a radio beamed to the National Bureau of Standards".

[59] Payne-Scott also wrote Pawsey on 11 June 1948 after he arrived in London about the necessity of associating flares with outbursts (e.g., Type II outbursts): ". . . but a lot more evidence is needed before we can be definite about this – spectrohelioscope badly needed."

[60] There had even been a suggestion in March 1948 that a gift might be obtained from the Perkin-Elmar Corporation of a coronagraph (a telescope with a circular disk that covers the Sun, allowing observations of the outer corona of the Sun). Evans and Roberts had proposed this to Pawsey. But

Fig. F.1 Ron Bracewell and Ron Giovanelli visit Meudon Observatory in Paris in 1949. From right , Bracewell, Giovanelli, Bernard Lyot and L. d'Azambuja. From Bracewell in Sullivan 1984 (page 170). Used by permission of Mark Bracewell. From the Sullivan collection

In early 1949, Pawsey and Bowen decided to attempt to restart the project; contact was made with B. Lyot (1897–1952, the inventor of the coronagraph) in Paris. C.W. Allen, from the Commonwealth Solar Observatory on Mount Stromlo, went to Paris to visit Lyot. The latter suggested a French firm that might be able to supply a "Lyot filter" for Hα observations. By now it was clear to Bowen and Pawsey that Woolley, of Mount Stromlo was only willing to provide advice; thus collaboration with the Division of Physics (Chief G.H. Briggs) was necessary. Luckily Ron Giovanelli was in Europe. Ron Bracewell was in the last year of his Ph.D. with Ratcliffe at Cambridge (UK) and was expected to return to RPL at the end of the year. He had expressed interest, in a letter to Bowen on 9 February 1949, in a new spectroheliograph in Sydney. Pawsey then proposed to Bracewell, on 9 June 1948, that both he and Giovanelli should visit Optique et Precision de Levallois (OPL, the optical firm that would deliver the filter, and to which Lyot was a consultant), as well as visiting prominent French solar optical astronomers. The two young Australians went to Paris in the first days of July 1948 (see Fig. F.1). Bracewell wrote a long report to Pawsey on 25 July on his return to the UK from

in quite definite terms, Richard Woolley, the Commonwealth Astronomer at the Commonwealth Observatory at Mount Stromlo, wrote that he had no interest (letter to Bowen on 3 March 1948): "Australia does not possess any high mountains nor any other conditions to make it suited to coronagraph observations."

France.[61] Finally, in late November, OPL sent a detailed quote for the set of three filters with additional equipment; the total price was US $5300, with a delivery time of only 5 months.

Back at RPL there was optimism; the Propagation Committee (now the Radio Astronomy Committee) minutes for 28 July 1949 reported that the efforts to obtain a narrow-band filter in the US were not successful, but that negotiations with a French company were underway. But this project ran into trouble. In early February 1950, Bowen wrote to the CSIRO in Melbourne, supporting the allocation of funds for construction of a spectroheliograph in the Division of Physics; this instrument would support the RPL work on solar noise. Bowen admitted that they had tried without success to obtain support from Mount Stromlo and had even thought of starting an RPL optical observatory. Given the expertise of the Division of Physics in optics and the presence of an experienced solar physicist like Giovanelli, a new project in the Division of Physics (in the same building as RPL on the University of Sydney campus) was quite timely. We do not know the exact outcome of the attempt to buy OPL filters; probably the request was denied and Giovanelli was given reduced support to construct narrow-band filters at the Division of Physics. By 14 November 1951 Pawsey reported to Woolley that Giovanelli was expected to have filters in early 1952 suitable for viewing the solar disk.

In spite of these setbacks, a new enthusiasm was exhibited during a meeting of the Solar Noise Group with the "optical group of the Division of Physics" on 24 January 1950; the meeting was entitled "Discussion of the Construction and Purchase of an Optical Filter for Solar Observing". Pawsey, Payne-Scott, Giovanelli, Bracewell, and Steel[62] were present at the meeting. Giovanelli and Steel (also of the CSIRO Division of Physics) discussed the detailed construction of the filters and the optics of the system. A major concern was the "slow" photographic film then available; a faster red film would be required for the red region, e.g., Hα. Bracewell discussed the details of the system that was under discussion with OPL in France, e.g., bandwidth in the range 0.075–0.3 nm. Giovanelli responded to a question by Pawsey that daily imaging would be the goal with particular attention paid to the development of prominences and flares on a large scale. Payne-Scott then added:

> That radio interferometer indicates where radio bursts originate on the solar disk. What is required is simultaneous optical observations of the sun, i.e., photographs taken at a moments notice. The optical equipment should be at the same place as the radio gear. Movements in the solar atmosphere ... are also of interest.

Apparently Giovanelli was not completely satisfied:

> Dr. Giovanelli said that the optical gear would probably not require continuous supervision. He regretted that the radio people wanted the whole solar disk to be under observation.

[61] Certainly the fact that Bracewell was fluent in French was a major asset (he had come second in the 1930s in the New South Wales high school Leaving Certificate Examination).

[62] It is remarkable that William Howard (Beattie) Steel's name is consistently misspelled in all RPL documents in this period. He was always referred to as W.H. STEELE. See his photograph in Chap. 6.

A consensus had developed that radio-optical comparison was useful; but the details of coordinating the observations at radio and optical wavelengths were still not clear in 1950.

In the end, however, an interim system was developed. The detailed development of this solar Hα project is not known.[63] But from the archival material we can trace the history of the instrument. On 14 November 1951, Pawsey wrote to Woolley at Mount Sromlo. Giovanelli was some months away from having a working filter (see above); Allen had left for London the previous month and the solar staff at Mount Stromlo was depleted.[64] Pawsey asked if RPL could borrow the spectrohelioscope from Mount Stromlo. "... my people would then be able to get some first-hand experience of optical observations, while Giovanelli would undertake the general programme of work and maintenance. I think this outweighs the disadvantage at this stage, of the poor seeing in Sydney". Pawsey could see that Woolley was more interested in the moves to take Mount Stromlo into stellar astronomy with the new "big telescope".

Nothing was heard for some months. In a letter to Woolley on 27 June 1952 Pawsey again asked for a loan of the instrument, suggesting that the location would be at Potts Hill. "I realize all Giovanelli's objections about dirty atmospheres apply, so it might be a foolish scheme. What particularly appeals to me is the possibility of our people who are working on solar physics at least trying to look at the Sun. I feel it would be most stimulating". Within a few days Woolley agreed. The condition was that the loan was for a year; the Ionospheric Prediction Service wanted the instrument to be returned before the solar maximum of 1958. Pawsey was thrilled (23 July 1952); the intention was an installation in a hut at Potts Hill. Joe Warburton and Alec Little drove to Mount Stromlo with a utility truck on 8 September 1952 (a few weeks after the end of the URSI Sydney meeting of 11–21 August); the instrument was brought to Potts Hill in Sydney and installed.

R.D. Davies has described the operation of this spectrohelioscope at Potts Hill in 1952, making Hα images of the Sun. Davies had joined RPL in January 1951 after finishing his first degree in physics at Adelaide. The details of his experiences at RPL in the years to 1953 are described by Davies (2005, 2009).[65] He thus overlapped with Payne-Scott for about 7 months in the first half of 1951. Of his experience with the optical imaging at Potts Hill he writes:

[63] R. Smartt contacted several former and current members of the Division of Physics staff in late 2006 in an attempt to reconstruct the history of the Potts Hill optical solar instrument. No details of this instrument can be reconstructed; Smartt joined the group in the late 1950s. Steel has given a retrospective on his career (*Optics in Australia*, 1987). A successful project described by Steel was the impressive 0.013 nm birefringent filter for Hα solar work constructed by R.N. Smartt (Steel et al. 1961); this was a Lyot style filter that was the world's biggest at the time and could be tuned over plus and minus 1.6 nm. The field of view was 2.25°. Later this filter was a key component of the optical instrumentation at the Narrabri Solar Observatory.

[64] Woolley wrote to Pawsey: "Now that Allen has left we are faced with a serious reduction in solar work. Miss Daly is carrying on, but she is going to take a year's leave, roughly the whole of 1952."

[65] The correct year (1952) for the move of the spectrohelioscope to Potts Hill is given in the 2009 publication.

The instrument had been installed by R.G. (Ron) Giovanelli from the Physics Division of CSIRO ... The purpose of the instrument was to locate activity on the Sun for comparison with maps generated by the array [the instrument built by Christiansen et al. for imaging the Sun at 21 cm with a resolution of 3 arc min]. The array came into operation in early 1952, and on clear days I made maps of the features of the Hα Sun ... The features included sunspots, bright plages, and prominences (which were visible in emission on the limb or in absorption as dark filaments on the disk). It turned out that the most interesting result of this early work was the association of the 21-cm emission with plage areas which emerged before sunspots and continued for up to one rotation after spots had disappeared ... While observing the Sun with the spectrohelioscope on the afternoon of 26 February, 1953, I noticed a large prominence on the NE limb. At 0450 UT it suddenly erupted to a height of 3–4 arc minutes. The radio records at 62, 98 and 200 MHz showed a simultaneous burst. However, an even larger burst occurred at 0630 UT, and this was associated with material streaming back towards the Sun. Such eruptive prominences are not infrequent, although not all gave strong radio emission.

This paper was published in *Nature* in 1953 (Davis 1953), while Davis was on board ship travelling to Britain to begin a career at the Jodrell Bank Observatory. The importance of this result was that no prominent sunspots were involved and no associated flares were implicated.[66] In 2009, Davies wrote: “the discovery that radio emission was closely linked to the chromospheric plages rather than photospheric sunspots was my great excitement at Potts Hill.” He also remembers that Pawsey was very upset that he did not stay for the entire day at Potts Hill observing the prominence with the small optical telescope; Davies did not want to miss the laboratory bus ride back to RPL.

So, in the end, Payne-Scott’s long term dream of a joint radio and optical solar observatory was finally fulfilled, partly as the fruit of her vision, her keen intellect, unflagging enthusiasm, drive and energy. She was not present, however, to enjoy the results. The interim Potts Hill Hα imager was installed in September 1952, but in July 1951 she had left RPL and ended her career as a solar physicist. Nevertheless Payne-Scott played an important role in establishing the importance of optical and radio comparisons for detailed studies of the sun.

[66]Earlier Potts Hill data from January 1950 to June 1951 were published by R.D. Davies, “An Analysis of Bursts of Solar Radio Emission and their Association with Solar and Terrestrial Phenomena” (1954), which includes an impressive data set based on observations at 62, 98, 200, 600, 1,200, 3,000, 9,400 MHz as well as data on flare incidence and solar (crochet) magnetic data. Fade outs at 18 MHz were obtained from Hornsby 18 MHz galactic background data. Davies took over the Mount Stromlo data when C.W. Allen left for London in October 1951. Davies (2009) has suggested that the last use of the Mount Stromlo spectrohelioscope was described by Giovanellli and Roberts (latter is J.A. Roberts, 1958), comparing radio and optical data on Type II events in 1956–1957; the optical instrument was then on the roof of the National Standards Laboratory on the University of Sydney campus. Marie McCabe (retired, Institute of Astronomy, Honolulu) participated in the observations and analysis (interview, mid-June 2008). John Jefferies (founding Director of the Institute of Astronomy, University of Hawaii from 1964 to 1983 and Founding Director of the National Optical Astronomy Observatories in Tucson, Arizona from 1983 to 1987, interview, 21 June 2008) has pointed out that the instrument used was, in fact, another instrument. Probably this was a spectroheliograph, consisting of birefringent filter to simultaneously image a large fraction of the solar disk. Ron Giovanelli had designed this instrument with support from W.H. Steel.

Appendix G

Payne-Scott and G.C. Southworth of Bell Telephone Laboratories[67]

In Chap. 5 we have described the impact the 1945 Southworth paper (detection of the thermal radiation of the sun at microwave frequencies) had on the early solar work at RPL. Surprisingly, the Pawsey et al. paper submitted to *Nature* in October 1945 made no reference to the Southworth results, though the Payne-Scott summary paper of December 1945 did refer to the microwave detections at Bell Laboratories. G.C. Southworth's paper was published in the April 1945 issue of the *Journal of the Franklin Institute*, with an erratum published in February 1946. This journal was not seen by many astronomers or radio engineers. Fortunately an unsigned *Nature* review was published on 1 September 1945 which pointed out the significance of the Southworth detections at the wavelengths of several cm. The extension of the solar spectrum from the infrared (near 1 μm) to wavelengths 10–100 thousand longer from a source 93 million miles distant would lead to "[incorporating] the best ideas of both the radio engineer and the designer of astronomical equipment with the hope that additional accuracy will bring out features so far overlooked". The Southworth observations were carried out at Bell Laboratories in New Jersey from June to October 1942 at X band (3.2 cm) and at S band (10 cm) and K band (1.25 cm) from June to August 1943. The antenna was a 1.5 m converted search light mirror. The angular resolution at these wavelengths was in the range 4, 1.3 and 0.53° (S, X and K band, respectively; the solar diameter is about 0.5°).

The published paper contained several serious errors. Charles Townes (a colleague at the Bell Labs) had pointed these out to Southworth, who published the erratum in February 1946 with an acknowledgement to Townes's input; at almost the same time Payne-Scott also found these errors. On 7 December 1945, Pawsey sent a long letter to Southworth with the Payne-Scott text (three pages). This text had no impact on the preparation of the erratum, since it was received on 29 January 1946 (the letter was addressed to "Bell Telephone Laboratories, New York" but had been sent first to the Australian Scientific Liaison Office in Washington), after the erratum had already been submitted to the scientific journal.

The major error was due to the fact that Southworth had considered the volume of the sun, instead of the *surface area.* In addition, his equations for the Planck function had the wrong units. In the end, the error was remarkably close to unity,

---

[67] Based on the Southworth reports and NAA: C3830: A1/1/1, Part 1 and NAA:C3830, F1/PAW/1, Part1. (Also footnote 71, Appendix G) The 7 December 1945 letter from Pawsey to Southworth was obtained from the Sullivan archive. Sullivan had found the letter in the Southworth archive at the Bell Telephone Laboratories in New Jersey.

1.33 R/c where R is the radius of the sun and c is the speed of light. In the usual units used by astronomers the error is a factor of only 3.1, with the units of seconds.[68]

The Pawsey letter to Southworth pointed out the early RPL solar work at 200 MHz (Chaps. 6 and 7). He wrote to Southworth, referring to his two War time memos (footnote 71 Appendix G) and the 1945 published article:

> We have naturally been re-reading these with some care. A colleague of mine, Miss Ruby Payne-Scott has pointed out what appears to be an error in your calculations, which, if correct seriously affects your outstanding conclusions that, at your wavelengths, the level of radiation approximates to that given by a black body at 6,000 K. I agree with ... [her] working and would like to ask your opinion on the subject.

Pawsey then pointed out that the derived equivalent temperatures should be much higher than 6,000 K, closer to 20,000 K. He went on to describe two calculations done by Payne-Scott, one based on the "radiation concept" (the Planck derivation) and the other "radiation noise concepts".[69] These elegant and simple calculations were included. The first was a derivation of the received power at an antenna from an object of unknown temperature (the sun) and known angular size, based on the noise power from a matched load in the receiving apparatus. The derivation is similar to the work of Burgess (1941); Payne-Scott claimed that the Burgess derivation was carried out "in a less direct fashion".[70] The second

---

[68] Sullivan (2009) has an extensive discussion of this fascinating story of the errors. If the error had been larger, then the coincidental rough agreement of his predicted intensity of the sun (based on an assumed temperature of the solar surface of 6,000 K) with the measured intensity would have attracted attention. By another coincidence, based on current knowledge of the temperature of the quiet sun at 3.2 cm (around 12,000 K), Southworth's previous error is almost cancelled. It is ironic that the erratum has another serious typographical mistake. In the erratum, Southworth corrected the units of the Planck constant (h) from erg/s to the correct erg s. However, the value of h printed was 6.5×10^{27} erg s instead of the correct value of 6.5×10^{-27} erg s. In the letter to Pawsey from 4 February 1946, Southworth was clearly embarrassed by his mistake: "To have made a mistake is, of course, quite inexcusable .... in the nature of a blunder rather than one of the many elusive errors that may... creep into an experiment." He was also upset that the journal delayed by 10 months in publication of the erratum. NAA: 3830, A1/1/5, Part 1. The error in the erratum was apparently due a typo introduced by the journal. The copy of the erratum send to Pawsey by Southworth with the letter had the correct value.

[69] In a February 1949 internal document ("Notes on the Detailed Balabcing [sic, should be Balancing] Derivation of the Laws of Thermal Equilibrium at Radio Frequency from Ionized Gas", RPL 32) Pawsey (1950c) reported his attempt to sort out possible discrepancies in methods of computing thermal radiation from an ionized gas. The two groups at RPL who seemed to disagree were: (1) Westfold and Smerd (Smerd 1950; Westfold 1950; Smerd and Westfold 1949) and (2) Payne-Scott and Pawsey from their 1944 report RP 209 (Chap. 6). Pawsey was able to show that the two groups did, in fact, agree. Detailed balancing was defined as the principle that every process of transformation in thermodynamic equilibrium is capable of direct reversal. In addition the transformation in the two directions occurs at an identical rate.

[70] Burgess wrote: "The mean energy density of the radiation ... is the Rayleigh-Jeans distribution for black-body radiation. Thus the radiation produced by fluctuation currents in an aerial, and the radiation which induces the 'Nyquist' noise e.m.f. [electromotive] in an aerial in a uniform temperature enclosure are to be identified with the usual temperature radiation ... Despite the relatively high temperature of the sun (6,000 K), its distance is so great that it cannot contribute

derivation was based on the Rayleigh-Jeans law and the known physical size and distance. The two derivations agreed; Payne-Scott then derived the expected signal from the sun at the three wavelengths based on an assumed temperature of 6,000 K. At 10 cm the predicted value was 3.3 times smaller than in the Southworth War time report and the 1945 publication. Many of these results regarding the calculation of the brightness of the microwave solar emission were also repeated in the December 1945 Payne-Scott summary paper (Chap. 6).

The arrival of the Southworth paper in 1945 and the War time reports in Sydney in late 1944 had a significant influence on the thinking of the RPL radio astronomy group in the course of 1945. The first War time report was MM 42-160-140, dated 20 November 1942, by G.C. Southworth, "Evidence of Microwave Components in the Sun's Spectrum", classified "SECRET".[71] This report contains the 1942 June 3.2 cm data. The date stamp indicates that this report was sent to the Scientific Research Liaison Office at the Australian Legation in Washington, D.C. on April 15, 1943. Additional date stamps appear on the cover from April 13, 1943 (US format for dates). A date stamp appears on the first page in the Australian format 3 June 1943, next to a similar stamp from the RPL library. As discussed in Chap. 6, it is likely that Pawsey and Payne-Scott did not know about the Bell Laboratory detection of the microwave sun in March 1944, when the first radio astronomy experiment was carried out at RPL at S band (11 cm).

The second report, MM-44-160-30, of 1 June 1944, was "Microwave Radiation from the Sun". This report, which was in essence a first draft of the 1945 publication, summarised the 1943 data at the three frequencies. This report was apparently received at the Australian Legation on 1 November 1944, with the explanation "United States Secret equals British Secret". A letter, dated 23 September 1944, is attached from M.J. Kelly, Director of Research of the Bell Telephone Laboratories. This letter may provide clues to possible faulty communication among the Australians. It is addressed to Maston Beard at the ASRLO office on Dupont Circle.[72] Kelly pointed out that Bell Laboratories had received a copy of Pawsey

---

appreciably to the noise field on earth. This is confirmed by Jansky (1935), who states that no solar radiations at radio frequencies can be detected." The quiet sun at 20 MHz is very faint compared to the galactic background.

[71] Copies of both reports were found in the RPL archives in Epping, NSW, in March 2007. The first report contains more information about the details of the apparatus used (size of mirror, exact frequency, etc.). The second report mentioned the size of the antenna but only to the "intermediate" (X band), "low" (S band) and "high" (K band) frequencies. The 1945 publication was censored by the US Patent Office with no mention of the detailed frequencies (only the "high", etc. designations) nor even of the antenna size. However, as Sullivan (2009) has pointed out, Southworth got around this restriction by relaying the information indirectly in the figures. In the letter referred to in footnote 68 Appendix G to Pawsey, Southworth gave the exact wavelengths used at Bell Labs (9.8, 3.2 and 1.25cm) as well as the dish size of 5 feet (1.5m). He also gave the noise figures. As an example the noise figure at 1.25cm was 18 to 20 db or noise temperature of over 20,000 K.

[72] Beard was a long term RPL scientist who later played a major role in the construction of the computer networks for the Culgoora Radioheliograph. He also participated in the early CSIRO RPL computer projects after the War (Beard and Pearcey 1984).

and Payne-Scott's report RP 209 of April 1944, "Measurement of the Noise Level Picked Up by an S-band Aerial" (Chap. 6). Fred White (former Chief of RPL) had suggested that this report be passed on to Southworth. Kelly wrote:

> We assume from the test [sic, must be text] of this memorandum that the authors may have seen a memorandum by Mr. Southworth, dated November 20, 1942, reporting X-band measurements of solar radiation, made during the summer of 1942, but that have not yet seen a similar memorandum covering measurements made in 1943 in the S and K-bands, as well as the in the X-band. Noise effects very similar to those reported by Drs. Pawsey and Payne-Scott are described in this second memorandum. We have made some comparisons of the noise inside a closed room with that coming from the open sky, but more often we have compared the noise received by a highly directive antenna from points below and above the sky. We are naturally very glad to know that Drs. Pawsey and Payne-Scott have independently noted these effects and that they have also confirmed our conclusions that there is no very considerable amount of microwave radiation coming from the Milky Way. We are, however, rather surprised that they have found no radiation coming from the sun.[73] We are attaching a copy of memo ... We shall be glad if you will forward it to Dr. Pawsey with best regards from Mr. Southworth. We remember very well [his] visit here 3 years ago. We are very interested in the work covered by the memorandum which you have sent us and shall be glad to receive any further information on the subject which you may publish.

Probably the 1944 Southworth report reached RPL in Sydney in late 1944; the arrival date of the earlier report from late 1942 remains a mystery. It is possible that it was in the RPL library in mid-1943 but not noticed in the rush of War time activities; another possibility is that it was posted in late 1944 with the newer Bell Laboratory report. Certainly by early 1945, Bowen at RPL knew about both reports; on 20 March 1945 he wrote to Richard Woolley (Director of the Commonwealth Solar Observatory at Mount Stromlo in Canberra) with copies of the two Bell Laboratory reports and the Pawsey and Payne-Scott internal report RP 209 (Chap. 6). Bowen wrote:

> This work may be of particular interest to you and you may have further ideas on experimental work which should be carried out. We ourselves have unfortunately ceased active work in this direction due largely to the stress of war-time requirements and not because we consider such investigations unprofitable.[74]

The poor communication (plus poor memories) continued in the 1956 retrospective article by Southworth, "Early History of Radio Astronomy". The article contained a number of personal recollections of his association with Karl Jansky at Bell Laboratories, with a short description of the solar work and a mention of his first associate, A.P. King, during the campaign of 1942. Southworth also mentioned

---

[73]The Australian team did not, however, observe the sun in March 1944 (Chap. 6). Southworth described negative 3.2 cm observations of the moon, planets and stars. Sullivan (2009) reports that the Milky Way between "Cygnus and Scorpio" was also observed by Southworth, with negative results.

[74]NAA: C3830, A1/1/1, Part 1. See also Chap. 5 (footnote 17) for a possible solution of the reason the Southworth results were only cited by RPL after the publication of the results in 1945 in *Journal of the Franklin Institute*.

the contacts with his Australian colleagues. His surprising statement about the solar work at Bell Laboratories read:

> In addition, many of the current visitors to our Holmdel laboratory saw the work while it was in progress. Included were several who have subsequently become very active in this field – for example A.G. [sic, should be E.G.] Bowen and J.W. [sic, should be J.L.] Pawsey of the now famous Radiophysics Laboratory of Sydney, Australia.

We know that this cannot be correct; Pawsey only visited Bell Laboratories in 1941,[75] a year before the solar research began in 1942. Bowen was certainly at Bell Laboratories for a visit during the War, while he was working at the Radiation Laboratory in Cambridge, MA, but he appears to have had no knowledge of the solar work at Bell Laboratories until late 1944 or early 1945.

For the next few years, Pawsey and Southworth continued an infrequent correspondence. On 14 March 1946, Pawsey wrote with some details of the RPL 200 MHz solar burst work of late 1945 and early 1946.[76] He pointed out that intense burst radiation could not be thermal emission from the sun at 10^8K. "This seems a bit too hot". Southworth had obviously answered Pawsey's letter of 7 December 1945 (above); this letter has not been found. Later in the year, on 2 August 1945, Pawsey sent Southworth a preprint of the 1947 *Proceedings of the Royal Society* paper (McCready, Pawsey and Payne-Scott) with a question about the Bell Laboratories work on the sun in mid-1946 (footnote 76 Appendix G). Southworth answered that occasional monitoring observations continued at X band (9,400 MHz) and K band (24,000 MHz): "We do not have enough personnel to do a very satisfactory job." (footnote 76 Appendix G) Few day-to-day variations had been detected. A new "Dicke radiometer" had been constructed following the precepts of Dicke (1946, a technique still widely used in the twenty-first century). The final letter in this era was a report from Pawsey to Bowen on 29 January 1948.[77] Pawsey had been staying with his brother-in-law in Princeton, New Jersey, and visited Bell Laboratories at Holmdel; he met with numerous colleagues including Karl Jansky: "Jansky has dropped cosmic noise but is still interested." And he could not find Southworth: "[He] is still away ill."

The interactions with Southworth certainly provided a stimulus to the RPL group; the thermodynamics of radio noise were understood by Payne-Scott influenced by Southworth's mistake in the 1945 publication. In addition, the recognition that at microwave frequencies the solar emission was thermal and arose close to the visible surface of the sun, contributed to an understanding of the thermal balance of the both the solar photosphere and corona. Pawsey and Yabsley (1949) in their influential publication, "Solar Radio-Frequency Radiation of Thermal Origin", began their paper with a detailed summary of Southworth's ground breaking observations of the sun.

---

[75]Pawsey was in the US and Canada from mid-July to mid-October, 1941. (NAA: A8520, PH/PAW/1B, Part 1)

[76]NAA; C3830, A1/1/1, Part 1.

[77]NAA: C3830, F1/4/PAW/1, Part 1.

Appendix H

Council for Scientific and Industrial Research (CSIR), Commonwealth Scientific and Industrial Research Organisation (CSIRO) Officers' Association: Contributions of Ruby Payne-Scott

Payne-Scott was an active member of the CSIRO Officers' Association (CSIRO OA). This organisation was a professional trade union. The name changed from the CSIR to the CSIRO Officers' Association in 1949. The ASIO file (Chap. 12) indicates that she was the branch correspondent of NSW (New South Wales) in April 1950; John (Jack) Warner of RPL was the National President at this time.

Two major issues were addressed by Payne-Scott in the publications of the CSIR (O) OA *Bulletin* from September 1948 to December 1949. Two letters were written individually by her, while a third letter was written with 39 colleagues.

The major issues were:

(1) The contentious issue of military research in the CSIR during 1948 was a factor in the creation of the new CSIRO in May 1949.[78] Much of this debate remains relevant for scientific endeavours in the twenty-first century. A heated discussion occurred in the OA *Bulletin* in September 1948, "CSIR and Defence Research". Two unsigned articles appeared: "One Point of View," an article with a pro-Rivett point of view [no secret research in the CSIR(O)], and "-And the Other", with the point of view of the "cold [War] warriors" (Wilde 1998).

The pro-Rivett article, which represented the majority of the 610 members of the OA, began with a stirring message: "Science has always flourished best in an atmosphere of freedom ... No important weapon or techniques used in the last War originated from work done in Service [military] laboratories." The author referred to the US example where secrecy started in military laboratories and then spread to other research institutes, adversely affecting original research. "An atmosphere of suspicion stunts initiative, breeds intimidation and makes recruiting of scientists increasingly difficult".[79] The author also quoted Edward U. Condon (1902–1974), a well known physicist at the National Bureau of

[78] See Schedvin (1987) for a history of CSIR and the creation of CSIRO (Chap. 8, "Reorganisation") and Collis (2002, Preface). The former has a short description of the "Kaiser affair," involving Payne-Scott's former colleague T.R. Kaiser, who had a CSIR research student scholarship (1947–1949) at Oxford to do a PhD in physics. On 28 July 1949 he was sighted protesting in front of Australia House in London against the goaling of trade unionists, who were protesting against the Chifley government's strike-breaking actions back in Australia. Deery (1999) has written a fascinating account of the repercussions of this event for CSIRO and Kaiser. Kaiser, a well known left-winger, was sacked by CSIRO later in the year. Also note footnote 17 Chap. 4 and footnote 7 Chap. 13.

[79] This statement is remarkably similar to the well known lecture on 25 March 1947 given by Sir David Rivett, then Chairman of CSIR) at the Canberra University College commencement: "Science and Responsibility" (Schedvin 1987). In another context, Rivett wrote: "As to all this business about classified information ... I just loathe it. Of course we shall be prepared to give whatever guarantees may be required if that is the only way we can engage in research work of any

Standards in the US; Condon was quite critical of the witch-hunts for communists going in the US in the post War period.

The opposing author adopted the contrary point of view using strong language:

> However, when warfare implies a total national effort, as it does today, embracing all aspects of industry and civilian activity, the suggestion that CSIR should hold itself aloof from defence matters on so-called ethical or other grounds is sheer humbug ... It is indeed flagrant impertinence as well as humbug for us to refuse a request for assistance from the Commonwealth, the parent upon whom the very existence of the CSIR depends.

A letter (29 July 1948) to the two prominent Sydney papers, *The Sydney Morning Herald* and *The Daily Telegraph*, was signed by 40 scientists,[80] mainly from RPL and the Division of Physics. The letter was strongly supportive of Rivett's position, with a strong response to the press criticism of both CSIR and Rivett. The letter contained a strong plea for the maintenance of non-secret research at the CSIR laboratories. This group suggested that classified military research should be carried out within separate military research institutes. (During World War II, RPL, as well as other CSIR divisions, had carried out War related research in strict secrecy.) The letter pointed out that the basic principles of two major tools of warfare from World War II (atomic weapons and radar) were not discovered during the War, but were based on fundamental research of the previous decades. "Such work was practically stopped during the War".

Sally Wilde[81] has pointed out the remarkable fact that the OA editors of the *Bulletin* were "out of step with the views of members". Readers objected that the "cold Warriors" were given equal time in the letters columns. The most serious charge of bias came from the editorial in the September 1948 *Bulletin*, in response to the letter to the newspapers of July 1948. The editors of the *Bulletin* were critical of the fact that the letter had been sent to the Sydney newspapers: " ... it is felt that the writing of this letter was unfortunate ... it was thought that newspaper propaganda might have an undesirable effect." The OA suggested that any controversial views should be expressed by the CSIR Executive or even the Minister for the CSIR.

Wilde has stated that Rivett was quite heartened by this letter; he sent a telegram to the authors: "... thanks for very fine statement of the principle for which we are fighting." Soon there was a red scare campaign directed against the CSIR during August and September 1948.[82]

---

value. I have however the utmost distrust of secrecy ... when they are influenced by military people." (Cited by Deery 2000)

[80]The signers included Payne-Scott and many of her colleagues at RPL. The list included future Chiefs of RPL (Minnett and Wild, also a future Chairman of CSIRO), Chris Christiansen (future Professor of Electrical Engineering at the University of Sydney), and the well known solar optical astronomer Ron Giovanelli. See also Chap. 13; ASIO was aware of this letter.

[81]In *Unions in CSIRO: Part of the Equation*, 1998, page 21.

[82]NAA: C3830, E2/3A. Bowen wrote Sir David Rivett on 1 October 1948: "... how sorry I am that your name and that of CSIR has been used so badly by the minority in the House [Australian parliament in Canberra] ... Need I say that we are wholeheartedly behind you and your point of view."

In the same issue of the OA *Bulletin*, Payne-Scott had her own letter to the editor, with a ringing defence of Condon,[83] who had been attacked in the US House of Representatives. It was this letter that concerned ASIO (Chap. 13). She was fearful of the restrictions that secrecy would bring to scientific research: "If we try to carry over such rules into the peace-time lives of our scientists, it spells death to any form of international co-operation, hence death to the benefits we get from contact with scientists abroad, and it spells death to our own activity."

The new CSIRO was not involved in secret research.[84] The new institution was formed on 19 May 1949; sadly the career of the CSIRO chairman, Sir David Rivett, was compromised. As Collis (2002, p. xvi) has written: "Rivett . . . was beaten . . . dragged down by politicians and editors more interested in denigration than discourse. . . . His one remaining consolation [was] that the . . . . investigation found that no secret information of any kind had ever been wrongfully disclosed by anyone connected with CSIR, and that the relevant authorities in the US and Britain were also satisfied there had been no leakage of confidential information from the CSIR." Rivett was replaced by Clunies Ross in May 1949 as Chairman, while the young Fred White (39 years) became the new CEO, replacing A.E.V. Richardson, who died shortly afterwards on 5 December 1949.

The other controversial topic in which Payne-Scott was involved in the pages of the OA *Bulletin* was the issue of women's pay. This issue did affect Payne-Scott personally (Chap. 4). In the course of 1949 there was a blow to women's equality as the WEB on 6 June rescinded the War time regulation that women professional staff were to be paid at roughly the male rate. Thus it was no longer obligatory that CSIRO follow the equal salary ruling of the WEB from September 1944.

[83] Edward Condon (1902–1974) was a famous US atomic physicist who had been active in the Manhattan Project to build the atomic bomb during World War II in the US. At one point he was the Director of the US National Bureau of Standards. Condon's loyalty was attacked in the post World War II period by J. Parnell Thomas and the young Congressman Richard M. Nixon. Condon was supported in the House by Chet Holifield, a prominent Democratic member of the US House of Representatives from 1943 to 1974. Holifield was a supporter of Truman against the anti-communist attacks in post World War II US and of Edward Condon.

[84] And fortunately for the future of radio astronomy, RPL was not transferred to the Department of Supply (for secret research) as happened to the CSIR Division of Aeronautitics (NAA: C3830, E2/3A). Rivett wrote to Bowen on 22 September 1948 fearing that both this Division and RPL would be transferred by the Department of Supply (secret defence research). Rivett initially suggested that he, Bowen and White would visit Dunk, the Chair of this Board who "... should be given a little necessary education . . . the sad fact is that CSIR has somehow or other become a pawn in the game". However, within a few days this scare passed. On 29 September 1948 Bowen wrote to Rivett with a sense of relief. "The politicians also seem to forget that it isn't enough to have a bag of money and a laboratory full of scientists. Just as important is the existence of a problem, the appropriate stimuli and the right kind of atmosphere." Finally Rivett wrote back from Melbourne on 4 October 1948: ". . . I am very glad that Radiophysics has escaped from the threat of being taken over by ... Supply under the PSB (Public Service Board) auspices . . . Tragedy will arise if other parts of our work [such as RPL] are similarly taken from us on any wrong assumption that they, too, belong to defence.. I cannot see us carrying on CSIR in the atmosphere that is proper and inevitable for studies in war technology."

Payne-Scott was the convener of a sub-committee of the NSW Branch Committee of the CSIRO OA, together with three other women colleagues, to look into the situation and make recommendations. The National Council of the OA had reaffirmed its adherence to the principle of equal pay. She then wrote a column in the OA *Bulletin* (October 1949), reporting on the current status. If the ruling of the WEB were to be accepted by CSIRO at this time, women would receive about £A100 per annum less than men. For many professional staff this would lead to a reduction in wages of 15–25%. For this period in 1949, CSIRO had agreed that the wage equality would be maintained only for women appointed before 6 June 1949; new employees and personnel transferred to other Divisions or to a different position within the same Division would be paid at the reduced rate. Payne-Scott wrote a tongue-in-cheek response: "The best advice to the women concerned is to stick like glue to their present fields of work till the situation is defined."[85]

Payne-Scott concluded her report with a detailed discussion of the complex negotiations going on at the time with the Public Service Board. An additional respondent (Shirley Andrews) from Victoria wrote:

> The necessity for prompt action in our case is all the more important because of the changed economic position. This is very different now from when the rates [during the War] were first granted us; it would be very foolish to ignore the obvious signs in other parts of the world that it will probably deteriorate very soon ... We are still doing exactly the same work, and presumably the same logical argument should hold ... Women members ... are very firmly convinced of the justice of the claim, and we urge our advocates to put our case very strongly.[86]

A few months later, in December 1949, Payne-Scott wrote again about the issue of equal wages. This article was written in an ironic tone, responding to a previous

[85]Ironically she also described a situation that she was to face a year later in April 1950 (Chap. 4), when she was forced to become a temporary employee after her marriage was announced. "The fact that the appointment of temporary officers (including a number of married women) is reconfirmed each year will not affect their rate of pay as long as they continue in the same position."

[86]A striking coincidence occurs in the OA *Bulletin* of October 1949, immediately following the article by Payne-Scott and Andrews on "Women's Rates of Pay." Helen Newton Turner (1908–1995 AO, OBE), the animal geneticist, contributed an article, "Report on Analysis of Questionnaires from Technical Officers Concerning Salary and Status." Technical Officers generally had lower educational qualifications than Research Officers; her survey was related in an indirect manner to the issue of women's rates of pay. Newton Turner did not differentiate in the survey between men and women. She concluded that there were real differences between various Divisions of CSIR and age groups. Newton Turner's fascinating career has been summarized by Nessy Allen (1992a,b); her life was also influenced by Sir Ian Clunies Ross, her mentor, starting in 1931 when she was his secretary at the McMaster Laborary, Division of Animal Health, CSIR, on the University of Sydney Campus. Later in her career she became a prominent geneticist at the CSIRO. Two prominent CSIRO awards were announced in 2008: the Newton Turner career award and the Payne-Scott career award. The Newton Turner award is a recognition for senior scientists at CSIRO. We do not know if Newton Turner and Payne-Scott knew each other; since both worked on the Sydney University campus during the same period in the 1930s and 1940s, there is a good chance that they were acquaintances.

unsigned article (October 1949) about anomalies of men's pay scales. This previous author had written about "Henry, Harry and Hal" who were in different job classifications; due to the vagaries of the salary scales, the man (Hal) with the highest academic qualifications would receive the lowest salary. Thus this author was urging technicians to avoid getting a technical college degree and moving to a higher level with a lower salary! Payne-Scott discussed the even more absurd situation awaiting their female counterparts "Joan and Jane". Here the young "Joan" studied for a degree and was promoted to the lowest level of Technical Officer (not covered by the WEB equality rates, the male rate minus £A100 per annum), from her previous position of Assistant Female (covered by WEB). Her salary would be reduced after she was promoted. Thus Payne-Scott suggested that "Joan" who has been happily acquiring learning and experience as a Laboratory Assistant with the hope of turning into a Technical Officer, had better hastily lose her ambitions, lest she suffer the fate that appears to be becoming common in CSIRO for those who pursue knowledge. For individuals remaining as Technical Officers, the wage parity would be maintained.

Immediately in the next paragraph of the letter to the *Bulletin*, it becomes clear that Payne-Scott had met the new Chairman of CSIRO (Sir Ian Clunies Ross) in Sydney to discuss this confusing and unfair situation. As pointed out in Chap. 4, at this meeting Clunies Ross heard about Payne-Scott's marriage of 1944, 5 years earlier. Clunies Ross had discussed the wage issue with the sub-committee of the OA on women's pay in Sydney, sometime in the period September–November 1949:

> He stated that the Executive favoured the principle of equal pay but might be in a difficult situation if it attempted to set a precedent in the Public Service. ... disturbed by the anomalies being created between Laboratory Assistants and Technical Officers ... Dr. Ross promised to investigate the position and to discuss the whole matter with [the Minister] ...

As Wilde (1998) has described, the outcome was discouraging. The OA joined the Professional Officers' Association with an appeal to the Public Service Arbitrator. In 1951, they lost the case and the WEB "concessions". The CSIRO had argued for equal pay for the same skill, but only for professional classifications. The Public Service Board (of the Australian Government civil service) had argued for the prevailing point of view:

> The various decisions of the Arbitration Court which dealt with female rates of pay ... had all accepted the principle that the basic wage was designed to cover the needs of a man, his wife and children, and as a female has not the responsibility of a married man, her wage should be lower.[87]

However, the energy that Payne-Scott brought to the issue of equal wages remains striking; in 1951 she resigned a few months before her son was born (Chap. 4).

[87] Cited by Wilde (1998, page 32).

The fight for equal wages for women was not over, however. The campaign continued in 1954 as the OA set up a new Committee on Equal Pay, led by Payne-Scott's friend and colleague, K. Rachel Makinson (Chap. 11). "She began to campaign on the grounds that equal pay was being adopted elsewhere in the world and Australia was behind the times . . . Rachel Makinson and her colleagues were to have a long fight."[88]

Only in 1969 did the Australian Conciliation and Arbitration Commission introduce the principle of "equal pay for equal work," affecting mainly teachers and nurses. The next major milestone was in 1972 when the Commission granted "equal pay for work of equal value"; this ruling affected only a small fraction of women since men and women worked under different industrial awards. The Commission stated: "the male wages takes account of family considerations and it will not apply to females." But there was continued improvement and by 1977 the gender gap was reduced to about 20%.[89]

As discussed in Chap. 4, the regulation that affected Payne-Scott most severely was the ban on permanent employment for married women in the federal public service; this was only rescinded in 1966, 15 years after her resignation.

Appendix I

Letter written by the Chairman of CSIRO Sir Ian Clunies Ross to Ruby Payne-Scott on 3 March 1950 about her marriage.

3 March 1950

PERSONAL

Miss R. Payne-Scott
Radiophysics Laboratory
University Grounds,
CHIPPENDALE, N.S.W.

Dear Miss Payne-Scott,

Thank you for your letter of 20th February. First of all, it might be helpful if I outlined the legal position regarding married women on our staff. It is true that neither in the Act constituting C.S.I.R. nor in the Act constituting C.S.I.R.O. does any mention of married women occur. In the former Act, however, there was a clause (Section 14A (2)) which stated that officers should be engaged subject to the conditions as were prescribed or as the Council, with the approval of the Minister, determined. A somewhat similar clause applied in the case of employees

[88] Cited by Wilde (1998, pages 32–33).

[89] Based on web sites from the Women's Electoral Lobby, Victoria, the Australian Government Office for Women, Workers Online of Labornet (Labor Council of NSW,) and "Women towards Equality" from the Department of Foreign Affairs and Trade (Australian Government).

(temporaries). Under the Act, Regulations were in due course prescribed and one such was to the effect that a female officer shall be deemed to have retired from the service of the Council upon her marriage, unless the Minister, upon a recommendation by the Council, certified that there were special circumstances which made her employment desirable.

The above Regulation was based on a similar Regulation promulgated under the Public Service Act.

In the present Act under which C.S.I.R.O. is operating, there is a clause which states that officers shall hold office on such terms and conditions as are, subject to the approval of the Public Service Board, determined by the Council. Up to date, the Executive has not completed its determination of all terms and conditions, and has not discussed them finally with the Board. At the moment it is carrying on under the old C.S.I.R. terms and conditions, under which a female officer on marriage shall be deemed to have retired from the service of the Council, unless the Minister, upon a recommendation by the Council, certifies that there are special circumstances which make her employment desirable.

There is the further point, and in this you, as an officer under superannuation, are directly concerned. This is that in the Superannuation there is a clause which reads as follows:

"A female officer who marries after the commencement of this Section shall for the purposes of this Act be deemed to have resigned from the date of her marriage."

(The above section appears in Act 15 of 1945 which was assented to on 3.8.1945.)

You will see that with respect to the Superannuation Act there is no ground for equivocation or varying interpretation.

You also discuss in your letter the question of the moral obligation people like yourself may feel towards letting us know about such change in circumstances as marriage, and you go on to say that personally you felt no such moral obligation. There can, of course, be two opinions on that point, but I will content myself with pointing out that if everyone thought as you do or acted as you apparently think proper, the administration of C.S.I.R.O. would be greatly complicated, and we would have to introduce a system of rigid scrutiny of the actions of officers instead of relying on their discretion and good sense. You may remember that it was only in December last that we received a notification from you that you had recently opened an account in a Sydney bank in the name of "Ruby Violet Payne-Scott".

The usual procedure in the case of our women officers is, of course, that they are perfectly frank and open about their marriages and in that way help us to administer the law as it affects them and us. I cannot think that your Chief or clerical officers of the Division know of your marriage since they would have felt bound to have acquainted us of this. In this office there was certainly no knowledge of your marriage, nor do we appear to have received the form to which you refer as one which you were asked to complete and return to Head Office, in which your married state is mentioned. I should be grateful if you would let me have details of that form and to whom and when it was submitted.

In conclusion, I think the simplest way of regularizing the whole affair would be for you to tell us the date of your marriage. We will then look into the matter and tell you what should be done in your own and our best interests.

Yours sincerely,

CHAIRMAN

Appendix J

Ruby Payne-Scott: Bushwalking Photos

From the Hall family and Betty Hurley (later Hall), we have collected numerous photos of bushwalking trips taken by Ruby Payne-Scott and her husband, Bill Hall, who met while bushwalking with the Sydney Bushwalkers, SBW.[90] This family's passion for bushwalking has been described in detail in Chap. 12;

Here we show nine photographs, capturing some of the spirit of the events.

During Easter 1941, during a bushwalking trip near Nowra (about 160 km south of Sydney), a dramatic event occurred. The leader, Allex Colley, having organised the bus transport from Sydney, became ill and could not participate. The group of about 18 people started from Drury's farm, east of Pigeon House Mountain; as they walked up the Nerriga Valley to Mt Wog Wog towards Nerriga,[91] they were spotted by a group of three farmers, including two old-timers, gathering eucalyptus leaves, which are the source of eucalyptus oil. As the SBW group came out of Yadboro Creek, the three farmers intercepted the hikers who, to the farmers, looked like a stranded group of Japanese airmen or even paratroopers who had begun an invasion of Australia! The farmers had seen them engaged in gymnastics, an activity probably instigated by Dorothy (*née* English) Butler; she later drew an imaginative cartoon of the event, which appeared in the SBW magazine (Fig. J.1). The farmers were ready to give battle with pitchforks, but the misunderstanding was quickly clarified when they met face to face.

A series of photos was then taken (see Fig. J.2 for one of them); note that Ruby is looking directly towards her new friend, Bill Hall. In another photo (not shown) she is looking straight ahead, with her hand shading her face. Based on input from a number of SBW members (see figure caption) we have attempted to identify the

[90] Joan Freeman has also written about Payne-Scott as a well known bushwalker (1991). Freeman wrote: "On Saturday mornings she often appeared with a huge back-pack, and would stride off purposefully after work for a weekend camping expedition in the [Blue Mountains] . . . she would return with a lurid tale of some hazardous experience she had had on the trip. She certainly added colour to RP . . ." Bernard Mills (letter, 14 September 1997) never did any walks with Payne-Scott, but got her expert advice about challenging hikes in Tasmania.

[91] Letter from Alex G. Colley, 18 March 1999.

Fig. J.1 A poem by Dot English Butler after the Easter 1941 adventure (near Pigeon House Moutain). This poem appeared in the SBW Magazine. Used by permission of Alex Colley

brave individuals in this adventure at Easter 1941. There is no consensus on the identity of a few of the bushwalkers; we give the most likely name with a query.

Figure J.3 shows two images from an unidentified bushwalking trip, perhaps a weekend excursion from Sydney to the Blue Mountains. As is often the case with Payne-Scott, she is eating. The tea is being prepared in a "billy" in Fig. J.3b; note the formal tie worn by Payne-Scott's companion, perhaps Grace Jolley.

For several years Bill Hall and Ruby Payne-Scott organised Christmas trips in the Kosciuszko National Park of the Snowy Mountains in New South Wales. At an elevation of 2,228 m, Kosciuszko is the highest Australian mountain; Mt Kosciuszko and Mt Twynam (at 2,196 m, the third highest mountain on the mainland of Australia) encompass much of the Australian Alpine region.

Fig. J.2 Photo of a meeting at Easter 1941 with the three eucalyptus leaf gatherers, including two old-timers. Alex Colley had organized this bush walk and then could not go since he was ill. A number of colleagues (Alex Colley, Bob and Christa Younger, the late Arthur Gilroy, the late Merle Waltman and Betty Hall) have attempted to identify the participants. Uncertainty in the spelling of the names is indicated by a "?". Back row from left to right: Alan Hardy(?, also called "Dormo"), Bill Hall, farmer, Doreen Helmich, Bill Cosgrove (hardly visible), Edna (Stretton) Gentle, farmer, Tim Coffey, Dot (English) Butler, farmer, Max Gentle. Sitting in the middle from left: Betty Pryde (or Doris Pryor?), Elsa (Issacs) McGregor, Grace Jolley (or Jolly? – also figure A11-4), Hilma (Gailliott) Colley, Ruby Payne-Scott (looking at Bill!), Norm Helyer (very uncertain, may be Hillyer of Hillier?), John Johnson (Johno Johnson?), Jessie Martin. The two men in the front from left: Frank Leyden and Bert Whillier (?). This is an impressive collection of bush walkers! Photo by Reg Alder. Bill Hall family collection, used by permission of Peter Hall

Figures J.4, J.5 and J.6 show three images from a trip at Christmas 1947 to this area. Betty Hurley, who had a camera, took all these photos, except for the one of herself and Ruby on top of Mt Twynam (Fig. J.6) which was taken by an unknown member of the group with Betty's camera. In 1947, the group of about 12 left Old Adaminaby in an open truck in a blizzard. They started walking at O'Keefes, then arrived at Pretty Plains (see Fig. 12.2), Grey Mare, Mt Tate (2,068 m, Figs. J.4 and J.5), Mt Twynam (Fig. J.6) and finally at Charlotte Pass. At this point the party split up after Betty Hurley hurt her ankle.[92] She remained at Red Hut (Charlotte Pass on the Summit Road), while the rest of the group went on to Geehi Flat, to Geehi Hut on Swampy Plain River, and then ascended to the southwest to Dead Horse Gap and Seaman's Hut.

The final photos (Figs. J.7 and J.8) were taken on a bushwalk on the Overland Track in Tasmania in 1949; the photographer was the late Arthur Gilroy. Figures 12.3 and J.8 are from the walk near Mt Pelion-West in the Cradle Mountain-Lake St. Claire National Park. Figure J.8 shows Ruby Payne-Scott in the historic

[92] She stayed at the Red Hut with a group of three men who, it turned out, were engaged in small-scale rustling. She advised them to hide a sheep's leg which was protruding from a rucksack before Bill, Ruby and the others returned. But she did enjoy the lamb chops! (email to Goss, 16 April 2009).

Fig. J.3 (**a**) and (**b**) The top photo may be a companion to Fig. 1.3; Ruby continues to eat. Both image here are likely from the same bush walk. The bottom photo shows Ruby and Grace Jolley (uncertain of the identification) are making tea after boiling the "billy." Bill Hall family collection, used by permission of Peter Hall

Fig. J.4 Bush walking in Kosciuszko National Park. A trip organized by Ruby and Bill Hall at Christmas 1947 to Mt Tate: Ruby Hall, Bill Hall and possibly Bill Caw. Photo by Betty Hall. Bill Hall family collection, used by permission of Betty Hall

Fig. J.5 The entire group of bush walkers on Mt Tate. From left and behind: Gladys Roberts, Edna Garrard, Bill Hall, Mavis Jeans, Geoff (or Bill?) Bradley, Merle Watman, New Zealand nurse (standing, name unknown) and Ruby Payne-Scott (front of the nurse). Three men in the foreground are Roy Braitwaite (slightly in front of Ruby), and then Bill Caw (lying down) and a cousin of Braithwaite (name unknown) at the extreme right. Identifications done in 1999 by Merle Watman. Bill Hall family collection, used by permission of Betty Hall

Fig. J.6 A marvellous photo of Betty Hall and Ruby Payne-Scott (standing) on Mt Twynam in Kosciuszko National Park. Photo with Betty Hall's camera by an unknown companion. Bill Hall family collection, used by permission of Betty Hall

Fig. J.7 Trip to Tasmania in 1949. Ruby and Bill Hall with Arthur and Val Gilroy, Photo by Arthur Gilroy. Mt Pelion West. From left Bill, Ruby and Val. (see also a photo from this trip Fig. 12.3). From Bill Hall family collection. Used by permission of Lindsey Baudinet

Du Cane hut (built in 1910) on the Overland Track[93] making damper.[94] Two additional photographs from this series (not shown) are from Mt Field National Park (closer to Hobart) and Mt Rufus (in the Lake St. Clair region) in Tasmania; they all show Bill Hall, Ruby (Payne-Scot) Hall and Val Gilroy.

[93]Typically a 6-day walk (Douglas Bock communication, March 2009).

[94]An iconic Australian dish – soda bread often made in the coals of a campfire.

Fig. J.8 Photo of Ruby making damper in the historic De Cane Hut on the Overland Track in Tasmania. Photo by Arthur Gilroy. See Fig. K.8 for source

Appendix K

Ruby Payne-Scott Lectures at Danebank School: 1999–2009[95]

The RPS lectures 1999–2007

1999 Professor Anne Green, radio astronomer

2000 Dr. Jessica Chapman, radio astronomer

[95] See Chap. 14 for a discussion of Payne-Scott's career as a science teacher at Danebank School from 1963 to 1974.

2001 Dr. Mary White, palaeobotanist
2002 Professor Maria Skyllas-Kazacos, chemical engineer
2003 Professor Marcela Bilek, physicist
2004 Professor Susan Clark, geneticist
2005 Dr. Kate Brooks, radio astronomer
2006 Professor Michelle Simmons, nanotechnologist
2007 Dr. Kate Jolliffe, chemist
2008 Dr. Samantha Ginn, research scientist in gene therapy
2009 Dr. Ilana Feain, radio astronomer

Dr. Anne Green, 1999, now Professor of Physics, graduated B.Sc. (Hons) from the University of Melbourne, then became the first woman Ph.D. student in the School of Physics at Sydney University. She studied with Professor B.Y. Mills and Associate Professor Alec Little, and is Australia's third woman radio astronomer. Professor Green is currently Head of the School of Physics at the University of Sydney, the first woman to hold the position. She is also the first female Director of the Science Foundation for Physics and was for 10 years the Director of the Molonglo Observatory, owned and operated by the University of Sydney. Her research is concerned with supernova remnants and astrophysical masers in the Milky Way Galaxy. Anne's address, titled "Microwaves in Space", described how masers worked, where they are found and what they tell us about their environment, as well as interesting details of her progress in a male-dominated field. She was a worthy choice as inaugural Ruby Payne-Scott lecturer.

Dr. Jessica Chapman, Dr. Jessica Chapman, 2000, is the Operations Research Program Leader at the Australia Telescope National Facility and is a leader for the operations of the ATNF radio telescopes at Parkes and Narrabri. She obtained her PhD in radio astronomy at the Jodrell Bank radio observatory in England in 1985. She has worked in Australia since 1988 with considerable experience at both radio and optical observatories. Her research interests include the evolution of stars, astrophysical masers and the structure of our Galaxy. In 2007 she was the Chair of an International Astronomical Union Symposium held in Alice Springs, Australia, on *Astrophysical masers and their environments*, In her talk on the "Birth, life and death of stars", Jessica described how stars are born and evolve and how radio and optical telescopes can be used to study their different stages of evolution.

Dr. Mary White, 2001, was particularly requested by Fiona Hall to whom she had become close while working on her Gondwana project at Government House Sydney. Fiona felt that Mary White embodied much of what she admired in her mother. Dr. White grew up in what was Southern Rhodesia and studied at the University of Cape Town where she obtained her Masters degree. She was granted a Doctor of Science degree by Macquarie University in Sydney in 1995 in recognition of her contributions to science through her books. An acclaimed and awarded palaeobotanist, a prolific author of works such as "The Greening of Gondwana", her chief interest is in writing and lecturing about the pre-historic world and the evolution of the Australian continent and its biota. Mary titled her talk "The Natural Sciences change your view of the World". In 2008, she had her property on the NSW north coast dedicated as a conservation area.

Professor Maria Skyllas-Kazacos, A.M., 2002, professor of chemical engineering at UNSW, is the inventor of the vanadium redox battery, which can be recharged an unlimited amount of times, and has been successfully marketed commercially. She sees a move towards a wider use of solar power in society as a result. Maria continues to work on refinements of the battery. In her talk, she described her challenges and successes in balancing the competing demands of work and family. She has in excess of 250 publications and patents and was invested as Grand Lady of the Byzantine Order of St Eugene of Trebizond on Australia Day 2009.

Professor Marcela Bilek, 2003, was appointed professor of applied physics at the University of Sydney at the age of 32. Her main research interest involves preparing plasma solutions for thin film and surface preparations. As well as having international recognition for her work, she is an accomplished sportswoman. Marcela described many of the applications of her work, and the reliance her team places on funding from business enterprises. She is the author and co-author of 140 international journal articles and was a delegate to the Prime Minister's 2020 Summit in 2008.

Associate Professor Susan Clark, 2004, associated with the Garvan Institute of Medical Research and UNSW Faculty of Medicine, is a specialist in epigenetics – the modification of DNA that changes gene expression without altering their sequence. She applies this to cancer. She is internationally recognised and awarded in her field. Susan's address greatly encouraged girls to persevere with their chosen careers in the sciences, again outlining her success in balancing work and family. She is currently Principal Fellow at both the Garvan Institute (Cancer Research) and the University of NSW Faculty of Medicine.

Dr. Kate Brooks, 2005, is a radio astronomer working at CSIRO Australia Telescope National Facility as a research scientist. Her research is focused on how massive stars form in our Galaxy. Kate and her research team use data from telescopes both in Australia and Chile to study sites in our Galaxy where massive stars are currently forming. Kate spent 4 years working as an astronomer in Chile for the European Southern Observatory and has two young boys. Her talk "Head in the Clouds" inspired all students, especially girls, to create career opportunities that maximise career, adventure and family.

Professor Michelle Simmons, 2006, is associated with the School of Physics, UNSW, and is director of the Atomic Fabrication Facility in the ARC Centre for Quantum Computer Technology. Her core research areas are Quantum State Computing and Quantum Electronic Devices. Another inspiring and awarded scientist, Michelle identified Australia as an ideal environment for young researchers to tackle difficult scientific problems. In 2005, she was awarded the Australian Academy of Science's *Pawsey Medal*. She has filed three patents and has served on several international committees including the C8 Commission for Semiconductors. In 2009 she was awarded her second Federation Fellowship.

Dr. (now Professor) Kate Jolliffe, 2007, deputy head and senior lecturer of the School of Chemistry at Sydney University. Her interests are Biological and Medicinal Chemistry, for which she received an RACI award in 2006. She researches the design and synthesis of organic molecules which act on other molecules such as cyclic peptides which in turn can act as antibacterial, antifungal and anticancer drugs. In 2009, Kate was promoted to full professorship.

Dr. Samantha Ginn, 2008, works at the Children's Medical Research Institute (Sydney) in gene therapy and was one of the first researchers in Australia to use specific lentiviral vectors for gene transfer protocols. She achieved recognition for her work in treating the immunodeficiency disorders of "the boy in the bubble". Her address on this subject inspired many students. In 2007 she won the inaugural Australian Gene Therapy Society Panos Ioannou Young Investigators Award (a reciprocal award with the British Society of Gene Therapy). Samantha describes herself as a Research Scientist in Gene Therapy.

Dr. Ilana Feain, 2009, is Project Scientist for the Australian Square Kilometre Array Pathfinder (ASKAP), attached to the CSIRO Australia Telescope National Facility. Her research interests are spread over a number of areas, including supermassive black holes and active galaxies, and formation and evolution of massive galaxies. Her involvement with the new ASKAP project is her current focus at both national and international levels. In 2007, she won the inaugural L'Oreal Australia for Women in Science Prize, as well as a British Council Young Researcher Exchange Award. In this international year of astronomy, her talk is aptly titled "Black Holes, Cosmic Corkscrews and Building state-of-the-art Astronomical Facilities".

Appendix L

Refraction Effects using the Sea-Cliff Interferometer: Bolton's Criticism

Optical refraction was well understood in 1946 (Smart 1962); however, radio refraction differed from optical refraction due to the increased importance of water vapour at radio frequencies (Thompson et al. 2001). As an example, at the horizon, optical refraction is about 35 arc min, while the radio refraction is about twice this amount (refraction displaces sources or stars towards the zenith). As an example, Fig. 7.7 shows that the radio interference pattern from the sun was observed *before* the optical sun had reached the eastern horizon at sunrise. The time difference was about 5 minutes. For dry atmospheres (no water vapour), the optical and radio refraction (due to resonances of oxygen and nitrogen in the ultraviolet) are quite similar. For the MPP (McCready, Pawsey and Payne-Scott, *Proceedings of the Royal Society*, 1947), T. Pearcey[96] derived an empirical equation that was used to account for the wet component of refraction, using an assumed

[96] Pearcey, a mathematician, was a new employee at the RPL. He joined RPL on 1 February 1946 in charge of a "Mathematical Physical Section, responsible to the Propagation Committee. He will have under him, Reed, Ferris and the computing staff." Minutes of PC, 15 January 1946. See also footnote 31, Chap. 6.

scale height and surface opacity derived from values based on daily data from the Weather Bureau in Sydney.

As discussed in Chap. 7, *refraction does have an impact on the fringe phase*[97] for the sea-cliff interferometer (in contrast to a "horizontal interferometer", as is the case in a modern instrument such as the Very Large Array). The reflected ray experiences an additional atmospheric path compared to the direct ray, since it reflects off the sea; the variation of refractivity over the height of the cliff, h, can be neglected. The extra path in the reflected ray, which must twice traverse the layers of air below the cliff top, will cause a shift (r) in fringe phase roughly equivalent to a change in elevation (in radians):

$r = (N-1)/\sin(\theta)/\cos(\theta)\,(\text{radians}),$

where θ is the elevation and N is the index of refraction with values of $N-1$ near 360×10^{-6} for typical values of temperature and pressure. This equation agrees to first order with the Pearcey equation used by MPP. The higher order terms in the Pearcey equation may not be correct. A comparison of the above equation with the Pearcey equation shows good agreement for values of θ greater than 2°; MPP stated that the corrections were not used at lower elevations due to increased uncertainty.

MPP attempted to check the validity of the Pearcey equation; the observed times of the minima and maxima from the sun required a correction for the total refraction. As shown above, the effective phase difference between the direct and reflected ray (from the sea) depended directly on this refraction correction. This in turn provided the estimate of the *location* of the enhanced solar emission; the location was, in fact, the centroid of the combined effect of the radio emission of the quiet sun and the sum of all the individual sources of enhanced radiation, associated with sunspots.

There are two misconceptions in the claim made by Bolton in his 1982 article: (1) the MPP equation contained a substantial term due to the ionosphere and (2) the assumption was made by MPP that radio enhanced radiation and optical sunspot positions were coincident (see Chap. 6).[98] Indeed, the MPP appendix is quite ambiguous, regarding the method used to check the Pearcey equation over the low elevation range from 0 to 14°. We think that it is unlikely that MPP were guilty of circular reasoning; they did not assume that radio and optical positions were coincident. The process was probably executed in the following manner: from the observed times of fringe minima, the predicted elevations of the centre of the sun were obtained using the known optical ephemeris of the sun and then compared with the observed radio position derived using the formulation derived in MPP.

---

[97] B.G. Clark, A.R. Thompson, J.M. Moran, W.T. Sullivan, III, and Goss have contributed to the analysis of the impact of refraction on the sea-cliff interferometer; these discussions occurred between November 2006 and January 2007. The equations for the phase shift due to refraction were derived by Clark and Thompson.

[98] The interview carried out by Sullivan with Bolton at Jodrell Bank in 1976 has this same claim. In a letter to Sullivan (30 July 1986; Sullivan archive), Paul Wild doubted that the MPP results were based on the assumption that the radio enhanced radiation agreed with the sunspots: "I do not believe that the results were 'fortuitous'."

Thus a comparison of the observed minus predicted elevation was carried out, the angle of refraction causing the observed sun to be higher in the sky. The calculated mean radio elevations (accurate at the level of a few arc min) referred to a mean value between that of the centre of the sun (the thermal radiation of the quiet sun at the level of 10^5 Jy at 200 MHz) and a typical region of enhanced emission, several times this value. Payne-Scott described this in RPL 9: "Analysis . . . shows that the lobe pattern produced by a single spot source superimposed on the thermal sun will . . . again give an apparent elevation which is the weighted mean between that of the centre of the sun and that of the spot group . . . for a group producing power equal to the thermal level [quiet sun], the apparent width will be about 0.3° [18 arc min] deg or about 0.6 of the sun's diameter." As stated in MPP: "For any day there should be a constant difference between the two sets of angles of elevation, equal to the displacement of the radiating source from the centre of the sun . . . Their mean value . . . was taken as giving the true displacement."

In the comparison shown in Fig. L.1, all values for the entire data set from late January 1946 to the end of February 1946 were used. For each value of θ, the refraction correction was calculated based on the assumption that the origin of the emission arose from the centre of the sun. This operation was repeated over many days, i.e., the refraction corrections shown in Fig. L.1 are the collection of all the data in early 1946, at a range of elevations ranging from close to the horizon (zero) to 14°. As we now know, the maximum possible error in the correction was about $\pm$ 20 arc min, the radius of the radio sun at 200 MHz (see below; determination of this value was made a few years later). The error of each observation was a few arc min.[99] The determined refraction for each observation – the difference between the radio measured elevation and the predicted (from the optical ephemeris) centre of the sun at this time – was then compared with the theoretical calculation to check its validity. As shown in Fig. L.1, the agreement was quite reasonable with typical errors of less than 0.1° (6 arc min), except for elevations close to the horizon (elevations less than 2°). These low elevations, where the corrections were larger than the model by 0.2°, were not used in deriving positions or sizes.[100] Thus the calibration of the radio refraction was determined by using the original sea-cliff

[99]The radio emission was *constrained* to arrive within the solar disk as observed at 200 MHz; probably in 1946, the RPL Solar Noise Group was not yet aware that the 200 MHz emission had to arise at radii of 1.03 to 1.25 solar optical radii, due to the index of refraction in the solar corona. (The exact value depended on the solar cycle.) This realisation became clear in the course of the swept-lobe interferometer data obtained by Payne-Scott and Little at Potts Hill (Chap. 9) a few years later.

[100]The minutes of the 19 March 1946 PC meeting stated that "an attempt is being made to determine the position on the sun of the source of the energy, from a study of the lobe patterns obtained . . ." The theoretical treatment of refraction right through the earth's atmosphere has been worked out with some approximations. A calculation in one case gave a result in which the source of energy lay somewhat on a horizontal band passing through the large sunspot. Little activity was noticed during the second appearance of the large spot in March [the giant spot of early February which had occurred one solar rotation (27 days) earlier].

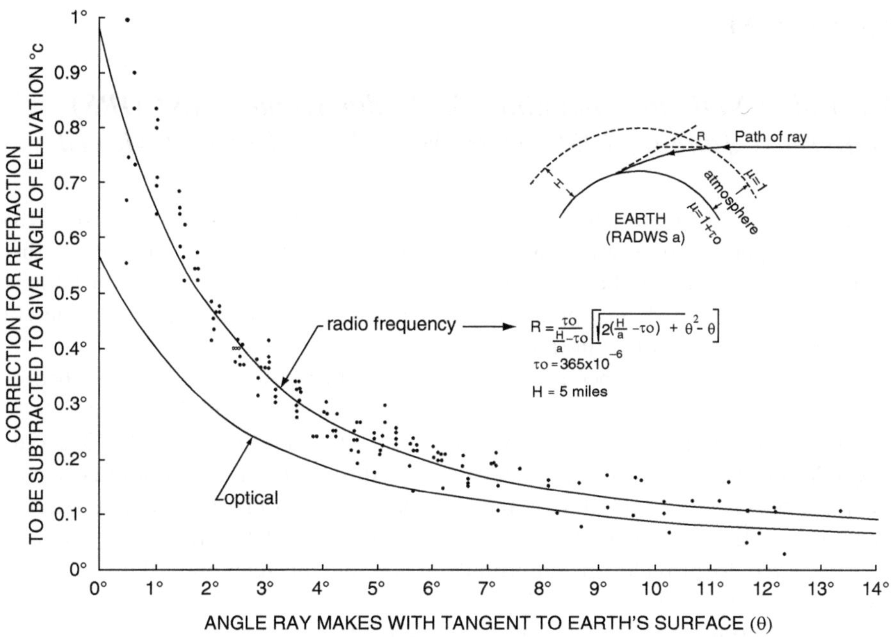

Fig. L.1 The calibration of radio refraction at 200 MHz using the sea-cliff interferometer at Dover Heights in 1947. The calibration of the radio refraction near the horizon discussed by McCready, Pawsey and Payne-Scott (MPP 1947). Both the optical refraction (stars and radio sources are displaced in elevation toward the zenith due to the lower atmosphere of the earth) and radio refraction are compared; the radio refraction is about one degree at the horizon compared to about 0.5° for optical radiation. Radio refraction theory is the solid line and the known optical refraction is the dashed curve. The experimental data are the dots, based on the assumption that the mean radio source position would be equal to the known centre of the optical sun. The increased radio effect is due to the greater refractive index of water vapour at radio frequencies compared to optical frequencies. The radio data in early 1946 was used to calibrate the validity of the adopted model for radio refraction. Figure 8 of MPP. *Proceedings of the Royal Society, Series A*, vol 190:357, "Solar Radiation at Radio Frequencies and its Relation to Sunspots." Used by permission of the Royal Society

interferometer data; since the radiation was restricted to arise from the radio sun at 200 MHz, an additional, plausible constraint could be applied.[101]

Finally, Bolton's claim that the Pearcey equation contained a substantial contribution from the ionosphere was not correct. This equation contained no frequency dependence terms; thus no relation to ionospheric effects was implied.

[101] Bolton has discussed this restraint in a letter to Sullivan dated 29 April 1986; he pointed out that Pawsey et al. were only interested in determining the *altitude* of the solar emission with respect to the horizon. An incomplete version of this letter is in the Bolton archive at the National Library of Australia; the complete version was located in the Sullivan archive.

Appendix M

Australian National Committee for Radio Science (ANCORS) Conference January 1950: Metre Wave Solar Bursts Reviewed

In mid-1949, the Australian National Committee for Radio Science (ANCORS) planned a special conference to be held in early 1950.[102] The details of this conference illustrate how Payne-Scott's thinking about the nature of Type III bursts evolved, and provided an opportunity to present early results from the newly commissioned swept-lobe interferometer observations.

ANCORS had recently been formed by Australian radio scientists. The subcommittee consisted of George H. Munro of the Radio Research Board, Joseph L. Pawsey of RPL and Richard d.v.R. Woolley of the Commonwealth Solar Observatory at Mount Stromlo. The special occasion for this Australia wide conference was the planned visit to Australia of the well known ionospheric scientist, Sidney Chapman (University of Oxford), in late 1949–early 1950; the conference was also expected to provide an opportunity to try out presentations that might be given at the 1950 Zurich General Assembly of URSI. All commissions of URSI were to be represented. The meeting was to be held from 16 to 20 January 1950 and Chapman was to present a public lecture on Tuesday night (17 January), sponsored by the New South Wales Division of the Institute of Physics.

The meeting was held at the Department of Electrical Engineering at the University of Sydney. The major sessions of the conference were under the auspices of Commission III (Ionosphere and Wave Propagation) and Commission V (Extra-terrestrial Radio Noise), both to last a day and a half. In the Commission III sessions Chapman presented two talks, one of 30 min ("The Treatment of Ionospheric Data for Geophysical Investigations") and the other of 90 minutes ("Some Theoretical Implications of Magnetic Storms"). The radio astronomy sessions (Commission V) consisted of 12 presentations, and included three papers by Bolton and a 30 min presentation by Payne-Scott, "Some Characteristics of Non-Thermal Solar Radiation at Metre Wave-Lengths". Paul Wild presented a paper on the new Penrith observations, "The Spectrum Analysis of Solar Bursts at Metre Wave-Lengths". A fascinating paper given by Pawsey, "Proper Fields for Radio Astronomy", concluded Commission V.

After the conference, the papers were quickly published in an international journal. Within a few months, Munro had organised with Merle A. Tuve[103] that short papers would be published in the June 1950 issue of the *Journal of*

[102] Most of this material comes from NAA: C3830, C6/3/2 and *Journal of Geophysical Research*, 1950, vol. 55, pages 191–210.

[103] Tuve (1901–1982) was the Editor of the *Journal of Geophysical Research*, and Director of the Department of Terrestrial Magnetism of the Carnegie Institute of Washington.

Geophysical Research (JGR). Tuve was impressed with the quality of the extended summaries from the ANCORS conference; he wrote to D. F. Martyn (Honorary Secretary and Convenor of ANCORS) on 4 April 1950:

> I suspect that the referees will agree with me that material of this kind, which is so much more vigorous and alive than a study of dead abstracts for a meeting, is just what we have been wanting to make the JGR a lively journal.

The summaries of the 33 presentations and the discussion sessions are an invaluable snap-shot of the state of radio science in Australia in 1950; the radio astronomers had come a long way since late 1945. The radio astronomy presentations show the evolution of the understanding in Australia about both "solar noise" and galactic radiation; no distinction was made at this time between galactic and extragalactic noise. In particular, the conference participants were all aware of the fundamental problem of identifying the observed non-thermal emission from both the galaxy and the sun with a known physical mechanism.

There were even some whimsical remarks after Bolton's presentations:

> Dr. Woolley remarked that he was glad to see this work [on galactic radio radiation] going on and happy that the fluctuations are in the ionosphere [at this time intrinsic variations had been ruled out using spaced receiver data with two radio telescopes spaced by 10 s to 100s km and even New Zealand to Australia at a distance of about 2000 km]. He hoped that it would not be necessary to invoke extragalactic nebulae which would be a headache for theorists, and he wished the laboratory [RPL] every success in the prosecution of its experiments. Dr. Bowen thanked Woolley for his remarks . . .

The Payne-Scott (1950) article on metre wave solar radiation provided a valuable summary of all solar work at RPL in 1950. Not only did she provide a summary of the achievements of 1945–1950, but she also presented an introduction to the following paper by Wild. She pointed out that the metre wave intensity of the sun was usually at a low level, associated with thermal radiation from the corona with a brightness temperature of about a million degrees. However, on occasions the intensity rose by a factor of a million, due to abrupt changes arising from non-thermal radiation. The time scales for the increased emission were 15 min to weeks; there was an association of these events with sunspots. The association of the noise storms (Type I) with circular polarisations suggested an influence of the sunspot magnetic field. A summary of the non-polarised bursts (Type III) was then presented; the work at Hornsby had been completed at this time (Chap. 8).

The most fascinating aspect of her publication from the January 1950 meeting was a progress report about the new observations occurring at Potts Hill, using the swept-lobe interferometer with the collaboration of Alec Little. She gave a short description of this new novel interferometer (Chap. 9), which allowed the observer to image the motions of the 100 MHz radio emission in position and polarisation at a time rate of 25 times per second. The more precise identification of Type I events with sunspots was now possible. The deductions about the origins of Type I events, higher in the corona (Chap. 8), had already been made in 1950. The deductions about the motions of the Type II and IVM events, however, were still preliminary at this time. Payne-Scott clearly had much work to do to sort out the motions of these moving events (Chap. 8).

Paul Wild (1950c) then described the groundbreaking work from the Penrith spectrum analyser observations (swept-frequency from 70 to 130 MHz); the data had been obtained from February to June the previous year, 1949. His understanding of solar bursts had developed quickly in the intervening months. The ANCORS report provided the first opportunity to report these results to a wider audience. In the publication in JGR, Wild described three types of bursts: α, β and γ. Surprisingly, Wild did not use the terms Type I, II and III at this time; however, only 6 or 7 weeks afterwards (in March 1950 when the publication by Wild and McCready was submitted to the *Australian Journal of Scientific Research*), he had changed to the new nomenclature of Type I, II and III.[104] The α bursts were Type III, β were Type II and γ were Type I. The basic properties of these bursts were described, e.g., the narrow band of the Type Is and the frequency drift rates (MHz per sec drift from the high to low frequency) of the Type II and III bursts. He compared his new results with Payne-Scott's data, pointing out that she had valuable data about Type I and Type III events. He ended his short presentation with the remarkable sentence: "... the rates of drift of the α bursts would be consistent with the existence in the solar atmosphere of corpuscular streams with radial velocities of the order of 3×10^4 km/s [10% of the velocity of light]."

A long discussion followed after the presentations by Payne-Scott and Wild; the two presentations were discussed simultaneously. Remarkably, none of the questions (from Father Hagemann, Jaeger, Pawsey, Bailey, Martyn, Christiansen, Bracewell, Bowen or Munro) cast doubt on the surprising claim of the relativistic velocities of the α or Type III bursts.[105] Jaeger was quite concerned about the possible agreement of the Payne-Scott and Wild data; he was assured by Payne-Scott that her Easterline Angus (type of pen recorder) records were in agreement with the spectrum photographs obtained by Wild on events where joint data existed. Many questions concerned the perplexing physical origin of the solar bursts; some years were to elapse before improved understanding of the physics of solar bursts developed.

The last presentation in the radio astronomy session was a remarkable talk given by Pawsey (1950c), "Proper Fields for Radio Astronomy". Pawsey was clearly worried about the place the new field of radio astronomy could occupy in the world of optical astronomy (or even just astronomy), which had been the only window on the universe for centuries:

> The use of radio waves in astronomy is a recent addition to astronomical techniques. Its value will depend on differences in generation and propagation between radio waves and

---

[104] Sullivan (2009) has also commented on this surprising use of Greek letters to describe the types of solar bursts in early 1950. We have no idea what caused Wild to change terminology within such a short period.

[105] In his publication later in 1950 about the detailed nature of the Type III bursts, Wild (1950b) was worried about this issue: "Corpuscular streams with these velocities have not been observed in the solar atmosphere, but in any case such streams would be likely to be highly ionised and may consequently escape optical detection."

> light which permit the observations of phenomena by radio waves which were not detectable optically. These differences are discussed and related to current research from the point of view of inferring lines of investigation likely to yield new knowledge. Such lines are termed "proper fields" for radio astronomy.

He then discussed the obvious differences in physical mechanisms for the origin of radio waves and light waves, e.g., in order to produce radio radiation, plasma oscillations and even possible non-coherent radiation from accelerated electrons were plausible mechanisms in the ionised solar corona. Within a few years, both mechanisms were to become accepted by the astronomical world as probable physical mechanisms for the generation of radio radiation. He was also impressed by "the most fundamental study of radio astronomy to date", the investigation of the spiral structure of the galaxy. This research was, of course, only possible because of the transparency of interstellar space at radio wavelengths, as compared to the highly opaque view of the Milky Way at optical wavelength, caused by interstellar dust particles. Clearly the pioneering work of Bolton and Wesfold, also presented at this conference, on the radio continuum structure (100 MHz from Dover Heights) of the Milky Way, impressed Pawsey.

Pawsey was becoming aware at the time that radio astronomy was the tool to be used to investigate the diffuse universe, including the solar corona and even the interstellar medium. Radio astronomy was also the tool to be used to study the cold universe, e.g., the surfaces of planets and in particular the moon (which had already begun at RPL by Piddington and Minnett at wavelengths near 1.3 cm). Finally, Pawsey was seeking approval from the astronomical (non-radio astronomy) world:

> The outstanding deficiency in such solar-noise studies to date is that no one has yet "seen" the phenomena producing solar noise, so that the conflict between such theories cannot yet be resolved by appeal to observation.

Some years were to elapse before radio astronomers could accept that their data also constituted a type of "seeing", producing a new reality which could lead to an improved understanding of the nature of the universe.

Appendix N

Henry Payne-Scott, Estranged Brother (1913–1970)

Ruby Payne-Scott was estranged from her only sibling, Henry, from the 1930s to his death in 1970. The estrangement may have been a result of the fact that while Ruby was still a student, she came home from a vacation and found that he had sold her books.[106] Neither Peter nor Fiona ever met their uncle.

[106] Interview with Peter and Fiona Hall, February 1999.

Fig. N.1 The Royal Australian Air Force identification photo of Henry Payne-Scott, estranged brother of Ruby Payne-Scott. NAA: A9301, 20769. From the collection of the National Archives of Australia

Since Henry was in the RAAF during World War II, we can follow many aspects of his life from 1940 to 1946 based on his extensive service record of 127 pages.[107] This record began with his initial enlistment on 28 August 1940 in the RAAF Reserve, then on 16 January 1941 in the RAAF Permanent Forces, and finally to demobilisation in Sydney on 12 March 1946

Henry was a troubled airman. He spent time in Port Moresby in New Guinea, and was there during the Japanese attacks in early to mid-1942. During most of the remaining period of World War II, he was stationed in Sydney as a General Clerk. His photograph at the time of enlistment is shown in Fig. N.1.

Henry's ill luck began in December 1940. He tried to qualify as a Wireless Trainee (radio operator trainee), but failed the Morse aptitude test because of numerous errors. Based on his 8 years of accounting and auditing experience, he became an apprenticed clerk and after qualifying, worked for the Permanent Trustee Company of New South Wales up to the time of his enlistment. In late March 1941 he was posted to New Guinea, where he served for 14 months during the period of the intense Japanese air attacks in early 1942; this was shortly after the Pearl Harbour attacks of December 1941 which led to a widening of World War II.

[107] NAA: A9301, 20769.

After he arrived in Port Moresby, capital of the Territory of New Guinea (administered by Australia in this period), his troubles began.[108] He developed a speech impediment, and suffered extreme anxiety. His Commanding Officer (Station Accounting) complained about Payne-Scott's general inefficiency, while acknowledging that he had "a heavy responsible job". At some point in late 1941 or early 1942, it was apparently decided that the best solution for him was an eight week recovery period back in Sydney. But in early 1942, his bad luck continued. The Japanese air raids on Port Moresby started on 2 February 1942 and continued to 12 April 1943. After only 3 days, Henry was sent back to Port Moresby, a continuation of his posting at RAAF Headquarters.

Within a short period the problems intensified. In late February 1942, a serious knee injury was reported by his squadron leader to Air Force Headquarters in Melbourne. The knee injury was a recurring problem related to a football accident which had led to a cartilage injury in his fourth year at high school. In March 1942, Payne-Scott was flown to Sydney to the 113 Australian General Hospital (AGH) (later the Concord Repatriation Hospital) for a meniscus operation on his left knee by Wing Cdr. Dr. Fischer. Apparently he then returned to Port Moresby for further duty. After a few months, it appears that the RAAF decided that Payne-Scott should not be near the front lines and he was transferred back to the RAAF installation at Bradfield Park in North Sydney in early June 1942. He was in the 113 AGH again within a few weeks for "fainting spells [indicative of] severe anxiety condition with depression..." In December, the psychiatrist, Dr. Fraser, recommended a complete evaluation by a medical board.

The Medical Board of three doctors interviewed Payne-Scott in mid December 1942 and issued their report on 11 January 1943. The board was quite sympathetic in trying to find a solution for Henry Payne-Scott's future in the air force, deciding that he would be given Class 2 duties, what today would be called "light duties." He was to do only "desk work" with a minimum of standing; of course he was also ineligible for air crew duty. The decision was apparently a wise one as he seems to have performed satisfactorily in his accounting position at various RAAF bases in the Sydney area from early 1943 to his discharge in March 1946. In addition, Payne-Scott was able to restart his private accounting practice while continuing in the air force.

The question of "next of kin" in the Henry Payne-Scott RAAF record is quite relevant in inferring the level of estrangement between him and his sister. At the

[108] Sgt. Henry Payne-Scott's War record (NAA: A9301, 20769) is filled with many pages about his medical problems, starting with a medical accident form on 23 February 1942. In late 1942 and early 1943, a comprehensive Medical Board evaluated him in Sydney, and included a detailed psychiatric examination by Flight Lt. W.H. Fraser a few days earlier. The Fraser report has many relevant details about his family, childhood, and medical and psychiatric condition. For example, the issue of home schooling in the Payne-Scott family was discussed (see Chap. 2) as well as his pride in his sister's (Ruby's) accomplishments. In the end the Medical Board assessment by Dr. Fraser was, "... [it] is doubtful if he is fit for further service in any capacity ... he lacks strong motivation in any direction except in avenues of escape. He should be medically boarded."

beginning of the War, Henry had listed his mother, Amy. His father died in 1942 and his mother in October 1943; his only living relative at this point was his sister, Ruby. Probably because of the estrangement, he chose as his "next of kin" a next door neighbour, Mrs. Elsie Leuty, at 12 Warrane Road, Chatswood; Henry claimed that she was an "aunt". (The Payne-Scott address in Chatswood had been 10 Warrane Road as discussed in Chap. 2.)[109] Henry even used her address as his postal address, until he found new housing for himself.

Finally, his contentious correspondence to secure early discharge from the RAAF, starting in June 1945, shows that he was certainly tenacious. He had to get the approval of both the RAAF and the Deputy Director-General of Manpower in New South Wales (DDGMP). This latter organisation was used in order to regulate the flow of military personnel back into civilian employment in Australia. Payne-Scott was turned down twice, once by the RAAF and once by DDGMP. At this point Payne-Scott showed some initiative by asking for help from his member of the Australian House of Representatives, W. M. ("Billy") Hughes (1862–1952), one of the more formidable figures of Australian politics in the twentieth century.[110] In the final years of Hughes's participation in Australian government, he was the member in the Lower House for the North Sydney area (Bradfield), and so Payne-Scott's member. Hughes wrote a long letter (12 November 1945) to the Hon. A.S. Drakeford, the Minister of Air in the Labor government from 1941–1949, with details of Payne-Scott's service, including the injury of 1942 from Port Moresby. The main purpose of the letter was to secure Payne-Scott's discharge from the air force so that he could begin his accounting career again, as well as to marry a "Nursing Sister in the RAAF Nursing Service, whom he would like to marry but is unable to do so until he gets his discharge." Henry Payne-Scott did not marry and the source of this statement is puzzling.[111] Given the prestige of Hughes, it is not surprising that Drakeford replied within a few weeks, promising to look into the question of Henry Payne-Scott's demobilisation. The result was that the air force announced that Payne-Scott was likely to be discharged during April 1946; in fact he was demobilised on 11 March 1946. A medical board on 4 March 1946 found no disability, which is surprising given the negative assessment of the previous medical board in early 1943.

[109]The only plausible candidate would have been his aunt Elvira Neale Davey. The NAA spelling for the "next of kin" is Elsie LENTY, inconsistent with our reading of the handwritten notes at two places in the RAAF file (NAA: A9301, 20769). In correspondence with the RAAF after his demobilisation in March 1946, Henry listed his address as 12 Warrane Road, Chatswood. Possibly he used his neighbour's address while he resettled into his old job as a private accountant and as an employee of the Permanent Trustee Company in Sydney.

[110]Hughes ("Little Digger") was the Prime Minister during World War I (1915–1923). He was in the Australian parliament for 51 years. He was a member of six political parties and was expelled from two. He was a member of the Liberal Party from 1944 to his death in 1952. A letter from Hughes to the Minister of Air about Henry Payne-Scott was certain to get attention.

[111]Payne-Scott made the same claim in a letter to the Commanding Officer of RAAF Personnel at Bradfield Park: "I intend to marry in the near future and it is very important to me to be in a position to avail myself of the prospects now offering."

Henry Payne-Scott spent the remainder of his life as an accountant; he probably had a private practice as well as continuing his employment at the Permanent Trustee Company of New South Wales. At the end of his life he lived in Bondi Junction. He died on 12 January 1973, probably while on holiday, in Athens, Greece. There was no contact with his sister and her family. Peter Hall remembers that when he was a young student, he was sent by his mother Ruby to collect his uncle's belongings in Sydney.[112] Henry Payne-Scott is referred to as "Harry" Payne-Scott in the short announcement of his death on 17 January 1973 in *The Sydney Morning Herald.*

[112]Peter and Fiona Hall (interview, February 1999) have reported that Ruby and Bill visited Henry's neighbours after his death; the neighbors told them that Henry had been quite proud of the achievements of his sister, even though the two were not on speaking terms.

Bibliography: S82–also found in the historical collection of radio astronomy papers by Sullivan (1982)

Adam-Smith P (1984) Australian women at war. Thomas Nelson Australia, Melbourne

Alexander E (1945) Report on the investigation of the Norfolk Island Effect. Radio Development Laboratory, Dept. Scientific and Industrial Research, Wellington, New Zealand RD.1/518

Alexander E (1946) The Sun's Radio Energy. Radio Electronics (New Zealand) 1:16

Alfvén H, Herlofson N (1950) Cosmic radiation and radio stars. Physical Review 78:616 S82

Allen CW (1947) Solar radio noise of 200 Mc./s. and its relation to solar observations. Monthly Notices of the Royal Astronomical Society 107:386

Allen CW (1957) The Variation of Decimetre-Wave radiation with solar activity. Monthly notices of the royal astronomical society 117:174

Allen N (1990) Australian women in science – a comparative study of two physicists. Metascience 8:75

Allen N (1992a) Australian women in science. Two unorthodox careers. Women's Stud Int Forum 15:551

Allen N (1992b) Helen Newton Turner and the wool industry. J. Aust. Stud. 33:56

Allen N (1993) Textile physics and the wool industry: an Australian woman scientist's contribution. Agri Hist 67:67

Allen CW, Gum CS (1950) Survey of galactic radio-noise at 200 Mc/s. Aust J Sci Res (A) 3:224

Appleton EV (1945) Departure of long-wave solar radiation from black-body intensity. Nature 156:534

Appleton EV (1950) Solar and galactic radio noise. URSI special report 1:1

Appleton EV, Hey JS (1946a) Solar radio noise – I. Phil Mag 37:73

Appleton EV, Hey JS (1946b) Circular Polarization of Solar Radio Noise. Nature 158:339

Barnard M (1946) One single weapon unpublished manuscript

Baade W, Minkowski R (1954) Identification of the radio sources in Cassiopeia, Cygnus a, and Puppis a. Astrophys J. 119:206 S82

Bailey DK (1948) On a new method for exploring the upper ionosphere. Terrestrial Magnet Atmos Elect 53:41

Basnett TN, Payne-Scott R (1942) oscillator frequency required for shore defense and gun laying calibrator. CSIR Radiophysics Laboratory Report RP138/1

Bayne M (ed) (1943) Australian women at war. Research Group of the Left Book Club of Victoria, Melbourne

Beard M, Pearcey T (1984) The Genesis of an early stored-program computer: CSIRAC. Ann Hist Comput 6:106

Bird TS (1993) Going strong after 50 years – antenna R&D at the CSIRO Division of Radiophysics. IEEE Antennas Propag Mag 35:39

Boishot A, Denisse JF (1957) Les Émssions de Type IV et l'origine des Rayons Cosmiques Associés aux Éruptions Chromosphériques. Comptes Rendus d'Académi 245:2194

Bolton JG (1953) Radio Astronomy at U.R.S.I. The Observatory 73:23

Bolton JG (1955) Australian work on radio stars. Vistas in astronomy 1:568

Bolton JG (1982) Radio astronomy at Dover Heights. Astronomical Society of Australia, Proceedings 4:349

Bolton JG, Slee OB (1953) Galactic radiation at radio frequencies. V. The sea interferometer. Aust J Sci Res (A) 6:420

Bolton JG, Stanley GJ (1948a) Variable source of radio frequency radiation in the constellation of cygnus. Nature 161:312 S82

Bolton, JG, Stanley GJ (1948b) Observations of the variable source of cosmic radio frequency radiation in the constellation of cygnus. Australian Journal of Scientific Research (A) 1:58

Bowen EG (ed) (1947) A textbook of radar. Sydney: Angus and Robertson, 2nd ed, 1954. Cambridge University Press, Cambridge

Bolton JG, Stanley GJ (1949) The position and probably identification of the source of galactic radio-frequency radiation Taurus-A. Aust J Sci Res (A) 2:139

Bolton JG, Westfold KC (1950a) Galactic radiation at radio frequencies. I. 100 Mc/s. survey. Aust J Sci Res (A) 3:19

Bolton JG, Westfold KC (1950b) Galactic radiation at radio frequencies. III. Galactic structure. Aust J Sci Res (A) 3:251

Bolton JG, Westfold KC (1951) Galactic radiation at radio frequencies. IV. The distribution of radio stars in the galaxy. Aust J Sci Res (A) 4:476

Bolton JG, Stanley GJ, Slee OB (1949) Positions of three discrete sources of galactic radio-frequency radiation. Nature 164:101

Bowen EG (1984) The origins of radio astronomy in Australia. In: Sullivan (ed) In the early years of radio astronomy: reflections fifty Years after Jansky's discovery. Cambridge University Press, Cambridge

Bowen EG (1987) Radar days. Adam Hilger, Bristol

Bowen EG (1988) From wartime radar to postwar radio astronomy in Australia. J Elect Electron Eng, Australia – IE Australia and IREE Australia 8:1

Bowen EG, Pulley OO, Gooden JS (1946) Application of pulse technique to the acceleration of elementary particles. Nature 157:840

Bracewell RN (1950) An instrumental development in radio astronomy observatory 70:185

Bracewell RN (1952) An honour to Australian science. Aust J Sci 14:173

Bracewell RN (1984) Early work on imaging theory in radio astronomy. In: Sullivan WT III (ed) In the early years of radio astronomy: reflections fifty years after Jansky's discovery. Cambridge University Press, Cambridge, p 167

Briton JN (1947) Lightweight air warning and G.C.I. radar in Australia. J Inst Eng Aust 18:121

Brown L (1970) Marjorie Barnard: A checklist: 1920–1969. Biblionews (Book Collector's Society of Australia), p 5

Brown RH (1991) Boffin: A personal story of the early days of radar, radio astronomy and quantum optics. Adam Hilger, Bristol

Brown L (1999) Technical and military imperatives: a radar history of World War II. Taylor & Francis, New York

Brown L (2001) Fire and water. Collection of poems. Privately published by Lyn Brown, Sydney

Brown R Hanbury, Minnett HC, White FWG (1992) Edward George Bowen 1911–1991. Hist Rec Aust Sci 9:151

Buckley-Moran J (1986) australian scientists and the cold war. In: Martin, Baker, Manwell and Pugh (eds). Intellectual suppression: Australian case histories, analysis and responses. Angus and Robertson, Sydney

Budden, KG (1988) John Ashworth Ratcliffe- 1902–1987. Biographical Memoirs of the Royal Society 34:671

Bullock EM (1999) History of radar manufacture by the NSWGR during World War II. The Bulletin of the Australian Railway Historical Society 50:287

Burgess RE (1941) Noise in receiving aerial systems. Proc R Soc 53:293

Burgess RE (1946) Fluctuation noise in a receiving aerial. Proc R Soc 58:313
Butler D (1991) The barefoot bush walker: the remarkable story of adventure, courage and romance. Australian Broadcasting Corporation, Sydney
Carey J (2002) Departing from their sphere? Australian Women in Science, 1880–1960. In: Xavier Pons (ed) Departures: how Australia reinvents itself. Melbourne University Press, Melbourne
Carlslow HS, Jaeger JC (1947) Conduction of heat in solids, 1st edn. Oxford University Press, Oxford
Christiansen WN, Yabsley DE, Mills BY (1949) Measurements of solar radiation at a wavelength of 50 centimetres during the eclipse of November 1, 1948. Aust J Sci Res (A) 2:506
Christiansen WN, Hindman JV, Little AG, Payne-Scott R, Yabsley DE, Allen CW (1951) Radio observations of two large solar disturbances. Aust J Sci Res (A) 4:51
Clark RW (1971) Sir Edward Appleton. G.B.E., K.C.B., F.R.S. Pergamon Press, Oxford
Collis B (2002) Fields of discovery. Allen and Unwin, Sydney
Conklin ND (2006) Two paths to heaven's gate. National Radio Astronomy Observatory, Charlottesville, Virginia
Davies RD (1953) Radio observations at the time of an ascending solar prominence. Nature 172:447
Davies RD (1954) An analysis of bursts of solar radio emission and their association with solar and terrestrial phenomena. Monthly Notices of the Royal Astronomical Society, vol 114, p. 74
Davies RD (2005) A history of the Potts Hill radio astronomy field station. J Astronom Hist Heritage 8:87
Davies RD (2009) Recollections of two and a half years with Chris Christiansen. J Astronom Hist Heritage 12:4
Deery P (1999) Science, security and the cold war: an Australian dimension. War and Security 17:81
Deery P (2000) scientific freedom and post-war politics: Australia, 1945–1955. Hist Rec Aust Sci 13:1
Dicke RH (1946) The measurement of thermal radiation at microwave frequencies. The Review of Scientific Instruments 17:268 S82
Domb C (2003) Fred Hoyle and naval radar 1941–1845. Astrophys Space Sci 285:293
Eddington AE (1926) The internal constitution of stars. University of Cambridge Press, Cambridge
Evans WF (1970) History of the Radiophysics Advisory Board 1939–1945. Melbourne: Commonwealth Scientific and Industrial Research Organization, Australia
Evans WF (1973) History of the Radio Research Board. 1926–1945. Melbourne: Commonwealth Scientific and Industrial Research Organization, Australia
Ewen HI, Purcell EM (1951) Observation of a line in the galactic radio spectrum: radiation from galactic hydrogen at 1,420 Mc./sec.' Nature 168:356
Ewington J (2005) Fiona Hall. Piper Press, Sydney
Fielder-Gill W, Bennett J, Davidson JA, Pollard A, Porter FH, Slayter RT (1999) The 'Bailey Boys': The University of Sydney and the Training of Radar Officers. In: MacLeod (ed) The Boffins of Botany Bay: Radar at the University of Sydney, 1939–1945. Hist Rec Aust Sci 12:419
Fränz K (1942) Messung der Emfängerempfindlichkeit bei kurzen elektrischen Wellen. Hochfrequenztechnik und Electroakustik 59:143
Freeman J (1991) A passion for physics: the story of a woman physicist. The Institute of Physics Publishing, Bristol
Friis HT (1945) Noise figures of radio receivers. Proceedings of the Institute of Radio Engineers 32:419
Geophysical Research (Journal of Geophysical Research-JGR). 1950. Summary of Proceedings of Australian National Committee of Radio Science, URSI, Sydney, January 16–20, 1950. JGR, vol. 55, p 191
Gillmor CS. (1991) Ionospheric and radio physics in Australian science since the early days. In: Home and Kohlstedt (eds) International science and national scientific identity. Kluwer, Dordrecht

Ginzburg VL (1951) Cosmic Rays as the Source of Galactic Radio Emission Doklady Akademii Nauk SSSR 76:377 S82

Giovanelli RG, Roberts JA (1958) Optical observations of the solar disturbances causing type ii radio bursts. Aust J Phys 11:353

Glen W (1988) Why scientists should do history. EOS 69:1572

Goss AM (2009) Decent colonialism? Pure science and colonial ideology in the Netherlands East Indies, 1910–1929. J Southeast Asian Studies 40:187

Goss WM, McGee RX (1996) The discovery of the radio source Sagittarius A (Sgr A).' In: Gredel (ed) The Galactic Center. 4th ESO/CTIO Workshop. San Francisco: Astronomical Society of the Pacific

Green AL (1941) Superheterodyne tracking charts- I. AWA Technical Journal 5:77

Greenstein JL, Henyey LG, Keenan PC (1946) Interstellar origin of cosmic radiation at radio-frequencies. Nature 157:805

Hall F (2008) Force Field (exhibition guide with contributions from Fiona Hall, Elizabeth Ann Macgregor, Vivienne Webb, Paula Savage and Gregory O'Obrien). Sydney and Wellington: Museum of Contemporary Art (Sydney) and City Gallery (Wellington)

Hamersley H (1988) Cancer, physics and society: Interactions between the Wars. In: Home (ed) Australian Science in the Making. Cambridge University Press, Cambridge

Hayward R (2007) A few comments on Ruby Payne-Scott's 1944–1945 CSIRO Radiation Laboratory Reports, unpublished NRAO

Haynes R, Haynes R, Malin D, McGee RX (1996) Explorers of the southern sky: a history of Australian Astronomy. University of Cambridge Press, Cambridge

Hey JS (1946) Solar radiation in the 4–6 metre radio wave-length band. Nature 157:47

Hey JS (1949) Reports on the progress of astronomy: radio astronomy. Monthly Notices of the Royal Astronomical Society 109:179

Hey JS (1955) Solar radio eclipse observations. Vistas in Astronomy 1:521

Hey JS (1973) The evolution of radio astronomy. Science History Publications, New York

Hey JS, Parsons SJ, Phillips JW (1946) Fluctuations in cosmic radiation at radio frequencies. Nature 158:23 S82

Hey JS, Parsons SJ, Phillips JW (1947) Solar and terrestrial radio disturbances. Nature 160:371

Hey JS, Parsons SJ, Phillips JW (1948) Some characteristics of solar radio emissions. Monthly Notices of the Royal Astronomical Society 108:354

Higgins C, Shain CA (1954) Observations of Cosmic Noise at 9.15 Mc/s. Australian Journal of Physics 7:460

Hooker C (2004) Irresistible forces: Australian Women in science. University of Melbourne Press, Melbourne

Hoyle F (1994) Home is where the wind blows: chapters from a cosmologist's life. University Science Books, Mill Valley, CA

Isaacson W (2007) Einstein: his life and universe. Simon and Schuster, New York

Jaeger JC, Westfold KC (1949) Transients in an ionized medium with applications to bursts of solar noise. Aust J Sci Res (A) 2:322

Jansky KG (1932) Directional studies of atmospherics at high frequencies. Proc Inst Radio Eng 20:1920 S82

Jansky KG (1933) Electrical disturbances apparently of extraterrestrial origin. Proc Inst Radio Eng 21:1387 S82

Jansky KG (1935) A note of the source of interstellar interference. Proc Inst Radio Eng 23:1158 S82

Jansky KG (1937) Minimum noise levels obtained on short-wave radio receiving systems. Proceedings of the Institute of Radio Engineers 25:1517

Jones RV (1978) Most secret war: British scientific intelligence 1939–1945. Hamish Hamilton Limited, London

Kai K, Melrose DB, Suzuki S (1985) Storms. In: McLean and Labrum (eds) Solar radiophysics. Cambridge University Press, Cambridge

Kellermann KI, Moran JM (2001) The development of high-resolution imaging in radio astronomy. Annu Rev Astron Astrophys 39:457

Kellermann KI, Orchiston W, Slee OB (2005) Gordon James Stanley and early development of radio astronomy in Australia and the United States. Publications Astronom Soc Australia 22:13

Kerr FJ (1953) Radio astronomy at the URSI Assembly. Sky Telescope 12:59

Kerr FJ (1953) Radio astronomy at Potts Hill. Sydney Water Board J Jan 1953:123

Kerr FJ, Shain CA, Higgins CS (1949) Moon echoes and penetration of the ionosphere. Nature 163:310

Kerr FJ, Hindman JF, Robinson BJ (1954) Observations of the 21 cm line from the Magellanic clouds. Aust J Phys 7:297

Kipenheurer KO (1950) Cosmic rays as the source of general galactic radio emission. Physical Review 79:738 S82

Kundu MR (1965) Solar radio astronomy. Interscience, New York

Lehany FJ, Yabsley DE (1948) A solar noise outburst at 600 Mc./s and 1,200 Mc./s. Nature 161:645

Little CG, Lovell ACB (1950) Origin of the fluctuations in the intensity of radio waves from galactic sources: Jodrell Bank Observations. Nature 165:423

Little AG, Payne-Scott R (1951) The position and movement on the solar disk of sources of radiation at a frequency of 97 Mc/s. I. equipment. Aust J Sci Res (A) 4:489

Lovell B (1983) Impact of World War II on radio astronomy. In: Kellermann and Sheets (eds) Serendipitous discoveries in radio astronomy. Green Bank West Virginia (National Radio Astronomy Observatory)

Lovell B (1990) Astronomer by chance. Basic Books, New York

Lovell B (1991) Echoes of war: the story of H_2S radar. Adam Hilger, Bristol

Lovell ACB, Banwell CJ (1946) Abnormal solar radiation on 72 megacycles. Nature 158:517

MacLeod R (1999) Introduction: revisiting Australia's wartime radar programme. In: MacLeod (ed) The Boffins of botany Bay: radar at the University of Sydney, 1939–1945. Hist Rec Aust Sci 12:419

Makinson KR (1979) Shrinkproofing of wool. Marcel Dekker, New York

Martyn D (1946a) Polarization of solar radio-frequency emissions. Nature 158:308

Martyn DF (1946b) Temperature radiation from the quiet sun in the radio spectrum. Nature 158:632 S82

McCready LL, Pawsey JL, Payne-Scott R (1947) Solar radiation at radio frequencies and its relation to sunspots. Proc R Soc (A), Math Phys Sci 190:357

Mellor DP (1958) The role of science and industry. In: The series Australia in the war of 1939–1945. Series 4 – civil. Australian War Memorial, Canberra

Melrose DB, Minnett HC (1998) Jack Hobart Piddington 1910–1997. Hist Rec Aust Sci 12:229

Menzel DH, Salisbury WM (1948) Audio –Frequency waves from the sun. Nature 161:91

Mills BY (1945) Scanning considerations in LW/AWH Mk. II. CSIR Radiophysics Laboratory Report RP TI 137/4

Mills BY (1952) The positions of six discrete sources of cosmic radiation. Aust J Phys 5:456

Mills BY (1984) Radio sources and the log N–log S controversy. In: Sullivan (ed) The early years of radio astronomy: reflections fifty years after Jansky's discovery. Cambridge University Press, Cambridge

Mills BY, Little AG (1953) A high-resolution aerial system of a new type. Aust J Phys 6:272

Minnett HC (1943) SH.D aerial array and feeder system. CSIR Radiophysics Laboratory Report PD 5/1

Minnett HC (1999) The radiophysics laboratory at the University of Sydney. In: MacLeod (ed) The Boffins of botany Bay: radar at the University of Sydney, 1939–1945. Hist Rec Aust Sci 12:419

Minnett HC (1999) Light-weight air warning radar. In: MacLeod (ed) The Boffins of botany bay: radar at the University of Sydney, 1939–1945. Hist Rec Aust Sci 12:419 (with Bruce Alexander, Ernest Bullock, George Day, Walter Fielder-Gill, Bernard Mills and Bob Richardson)

Minnett HC, Robertson R (1996) Frederick William George White 1905–1994. Hist Rec Aust Sci 11:239

Mitton S (2005) Conflict in the cosmos: Fred Hoyle's life in science. Joseph Henry Press, Washington, D.C

Nature (1945) Unsigned article. Radio waves from the sun. 156:273

Newton HW (1955) The lineage of the great sunspots. Vistas Astron 1:666

Norman L (1983) The brown and the yellow: Sydney Girls' High School 1883–1983. Oxford University Press, Melbourne

North JD (1965) The measure of the universe. Clarendon Press, Oxford

Orchiston W (2004) From the solar corona to clusters of galaxies: the radio astronomy of Bruce Slee. Publications Astron Soc Aust 21:23

Orchiston W (2005) Dr. Elizabeth Alexander: first female radio astronomer. In: Orchiston (ed) The new astronomy: opening the electromagnetic window and expanding our view of planet earth. Springer, Dordrecht

Orchiston W, Slee OB (2005) The radiophysics field stations and the early development of radio astronomy. In: Orchiston (ed) The new astronomy: opening the electromagnetic window and expanding our view of planet earth. Springer, Dordrecht

Orchiston W, Slee OB, Burman R (2006) The genesis of solar radio astronomy in Australia. J Astronom Hist Heritage 9:35

Paterson MS (1982) John Conrad Jaeger 1907–1979. Hist Rec Aust Sci 5:64

Pawsey JL (1946) Observations of million degree thermal radiation from the sun at a wavelength of 1.5 metres. Nature 158:633 S82

Pawsey JL (1950a) Solar radio-frequency radiation. Proc Inst Electr Eng Part III 97:290

Pawsey JL (1950b) Possible use of solar radiation to check sensitivity of microwave radio receiving equipment. CSIRO Radiophysics Laboratory Report RP111

Pawsey JL (1950c) Notes on the 'detailed balancing' derivation of the laws of thermal emission at radio frequency from ionized gas. CSIRO Radiophysics Laboratory Report RP 32

Pawsey JL (1950d) Proper fields for radio astronomy. Journal of Geophysical Research 55:206 (In collection of papers Summary of Proceedings of Australian National Committee of Radio Science, URSI, Sydney, 16-20 January, 1950)

Pawsey JL (1957) Preliminary statistics of discrete sources obtained with the mills cross. Proceedings from 4th IAU Symposium. Editor H.C. van de Hulst. Cambridge University Press, Cambridge

Pawsey JL, Payne-Scott R (1944) Measurements of the noise level picked up by an S-band aerial. CSIR Radiophysics Laboratory Report RP 209

Pawsey JL, Smerd SF (1953) Solar radio emission. In: Kuiper and Middlehurst (eds) The sun. University of Chicago Press, Chicago

Pawsey JL, Yabsley DE (1949) Solar radio-frequency radiation of thermal origin. Aust J Sci Res (A) 2:198

Pawsey JL, Minnett HC, Dobbie LG (1940) Aerial arrays, support of elements. CSIR Radiophysics Laboratory Report, RP 6/1

Pawsey JL, Payne-Scott R, McCready LL (1946) Radio-frequency energy from the sun. Nature 157:158

Payne-Scott R (1933) Relative intensity of spectral lines in indium and gallium. Nature 131:365

Payne-Scott R (1936) Notes of the use of photographic films as a means of measuring gamma ray dosage. J Cancer Res Committee Univ Sydney 7:170

Payne-Scott R (1937) The wavelength distribution of the scattered radiation in a medium traversed by a beam of X or gamma rays. Br J Radiol 10:850

Payne-Scott R (1943) A note on the design of iron-cored coils at audio frequencies. AWA Tech J 6:91

Payne-Scott R (1944a) A thermal noise generator for absolute measurement of receiver noise factor at 10 cms. CSIR Radiophysics Laboratory Report RP 211

Payne-Scott R (1944b) The present position of low-power S-Band measurements in the radiophysics laboratory. CSIR Radiophysics Laboratory Report TI 121/1

Payne-Scott R (1945a) Present position of fundamental R.F. [Radio Frequency] measurements in the radiophysics laboratory. CSIR Radiophysics Laboratory TI 191/1

Payne-Scott R (1945b) Solar and cosmic radio frequency radiation; survey of knowledge available and measurements taken at radiophysics laboratory to Dec. 1, 1945. CSIR Radiophysics Laboratory Report SRP 501/27

Payne-Scott R (1947) A study of solar radio frequency radiation on several frequencies during the sunspot of July–August, 1946. CSIR Radiophyscis Laboratory Report, RPL 9

Payne-Scott R (1948a) The visibility of small echoes on radar RRP displays. Proc Inst Radio Eng 36:180

Payne-Scott R (1948b) Solar noise records taken during 1947 and 1948. CSIR Radiophysics Laboratory Report. RPL 30

Payne-Scott R (1949a) Bursts of solar radiation at metre wavelengths. Aust J Sci Res (A) 2:214

Payne-Scott R (1949b) The noise-like character of solar radiation at metre wavelengths. Aust J Sci Res (A) 2:228

Payne-Scott R (1950) Some characteristics of non-thermal solar radiation at metre wave-lengths. J Geophys Res 55:223. (In collection of papers 'Summary of proceedings of Australian National Committee of radio science, URSI, Sydney, 16–20 January, 1950.')

Payne-Scott R, Green AL (1941) Superheterodyne tracking Charts – II. AWA Tech Rev 5:251

Payne-Scott R, Little AG (1951) The positions and movement on the solar disk of sources of radiation at a frequency of 97 Mc/s II. noise storms. Aust J Sci Res (A) 4:508

Payne-Scott R, Little AG (1952) The positions and movement on the solar disk of sources of radiation at a frequency of 97 Mc/s III. outbursts. Aust J Sci Res (A) 5:32

Payne-Scott R, Love WH (1936) Tissue cultures exposed to the influence of a magnetic field. Nature 137:277

Payne-Scott R, McCready LL (1948) Ionospheric effects noted during dawn observations on solar Noise. J Geophys Res 53:429

Payne-Scott, Ruby (1944b) The present position of low-power S-Band measurements in the radiophysics laboratory. CSIR Radiophysics Laboratory Report TI 121/1

Payne-Scott, Ruby (1948b) Solar noise records taken during 1947and 1948. CSIR Radiophysics Laboratory Report. RPL 30

Payne-Scott R, Yabsley DE, Bolton JG (1947) Relative times of arrival of bursts of solar noise on different radio frequencies. Nature 160:256

Penzias AA, Wilson RW (1965) A measurement of excess antenna temperature at 4080 Mc/s. Astrophys J 142:419

Pfeiffer J (1956) The changing universe. Random House, New York

Pfisterer S (2002) Brave red witches: communist women playwrights and the Sydney New Theatre. In: Pons (ed) Departures: how Australia reinvents itself. Melbourne University Press, Melbourne

Pick M, Vilmer N (2008) Sixty-five years of solar radio astronomy: flares, coronal mass ejection, sun-earth connection. Astronomy and Astrophysics Review 16:1

Piddington JH (1951) The origin of galactic radio-frequency radiation. Monthly Notices R Astron Soc 111:45

Piddington JH, Minnett HC (1949) Solar radiation of wavelength 1.25 centimetres. Aust J Sci Res (A) 2:539

Piddington JH, Minnett HC (1951) Observations of galactic radiation at frequencies of 1200 and 3000 Mc/s. Aust J Sci Res (A) 4:459

Reber G (1940a) Cosmic static. Proc Inst Radio Eng 28:68 S82

Reber G (1940b) Notes: cosmic static. Astrophys J 91:621 S82

Reber G (1944) Cosmic static. Astrophys J 100:279 S82

Reber G (1946) Solar radiation at 480 Mc/sec. Nature 158:945

Reber G (1948) Solar intensity at 480 Mc. Proc Inst Radio Eng 36:88

Reber G, Greenstein JL (1947) Radio-frequency investigations of astronomical interest. The Observatory 67:15

Rice SO (1945) The mathematical analysis of random noise. Bell Systems Journal 24:46

Robertson P (1992) *Beyond southern skies: radio astronomy and the parkes telescope*. Cambridge University Press, Cambridge

Robinson RD (1985) Flare continuum. In: McLean and Labrum (eds) Solar radiophysics. Cambridge University Press, Cambridge

Robinson BJ (2002) Recollections of the URSI Tenth General Assembly, Sydney Australia, 1952. Radio Sci Bullet 300:22

Rossi B (1957) *Optics* Adison-Wesley, Reading Massachusetts

Rowe AP (1948) One story of radar. Cambridge University Press, Cambridge

Ryle M (1948) The generation of radio-frequency radiation in the sun. Proc R Soc (A) 195:82

Ryle M (1950) Radio astronomy. Rep Prog Phys 13:184

Ryle M (1952) A New radio interferomter and its application to the observation of weak radio stars. Proceedings of the Royal Society A211: 351 S82

Ryle M, Vonberg DD (1946) Solar radiation on 175 Mc./s. Nature 158:339 S82

Schedvin CB (1987) Shaping science and industry: a history of australia's council for scientific and industrial research, 1926–1949. Allen and Unwin, Sydney

Shain CA, Higgins CS (1954) Observations of the general background and discrete sources of 18.3 Mc/s Cosmic Noise. Australian Journal of Physics 7:130

Sheridan KV, McLean DJ (1985) The quiet sun at metre wavelengths. In: McLean and Labrum (eds) Solar radiophysics. Cambridge University Press, Cambridge

Slee OB (1994) Some memories of the Dover Heights Field Station, 1946–1954. Aust J Phys 47:517

Smart WM (1962) Text-book on spherical astronomy, 5th edn. Cambridge University Press, Cambridge

Smerd SF (1950) A radio-frequency representation of the solar atmosphere. Proc Inst Elect Eng 97:447

Smerd SF, Dulk GA (1971) The solar type II burst of October 13, 1969. Aust J Phys 24:185

Smerd S, Westfold KC (1949) The characteristics of radio-frequency radiation in an ionized gas, with application to the transfer of radiation in the solar atmosphere. Phil Mag (Series 7) 50:831

Smith FG (1950) Origin of the fluctuations in the intensity of radio waves from galactic sources: Cambridge observations. Nature 165:422

Smith FG (1951) An accurate determination of the positions of four radio stars. Nature 168:555 S82

Smith FG (1952) The determination of the position of a radio star. Monthly Notices R Astron Soc 112:497

Southworth GC (1945) Microwave radiation from the sun. J Franklin Inst 239:285

Southworth GC (1956) Early history of radio astronomy. Sci Month 82:55

Stanley GJ, Slee OB (1950) Galactic radiation at radio frequencies. II. The discrete sources. Aust J Sci Res (A) 3:234

Steel WH (1987) Optics in Australia. The Australian Physicist 24:242

Steel WH, Smartt RN, Giovanelli RG (1961) A 1/8Å Birefringent filter for solar research. Aust J Phys 14:201

Stewart RT (1985) Moving type IV bursts. In: McLean and Labrum (eds) Solar radiophysics. Cambridge University Press, Cambridge

Sullivan WT III (1982) Classics in radio astronomy. D. Reidel, Dordrecht

Sullivan WT III (1984) The early years of radio astronomy: reflections fifty years after Jansky's discovery. Cambridge University Press, Cambridge

Sullivan WT III (1988) Early years of Australian radio astronomy. In: Home (ed) Australian science in the making. Cambridge University Press, Cambridge

Sullivan WT III (1991) Some highlights of interferometry in early radio astronomy. In: Cornwell, Perley (eds) Radio interferometry: theory, techniques and applications, IAU Colloquium No 131. Astronomical Society of the Pacific, San Francisco

Sullivan WT III (2009) Cosmic noise: a history of early radio astronomy. Cambridge University Press, Cambridge

Suzuki S, Dulk GA (1985) Bursts of type III and type V. In: McLean and Labrum (eds) Solar radiophysics. Cambridge University Press, Cambridge

Thackeray AD (1947) A short-lived solar phenomenon in high latitude. Nature 160:439

Thomas BM, Robinson BJ (2005) Harry Clive Minnett 1917–2003. Hist Rec Aust Sci 16:199

Thompson AR, Moran JM, Swenson GW Jr (2001) Interferometry and synthesis in radio astronomy. Wiley, New York

Townes CH (1947) Interpretation of radio radiation from the Milky Way. Astrophys J 105:235

Vonwiller OU (1938) Cancer Research in the University of Sydney: a review. J Cancer Res Comm Univ Sydney, October, 1938:69

Wendt H (2009) The contribution of the division of radiophysics Potts Hill and Murraybank field stations to international radio astronomy. PhD thesis, James Cook University, 2009

Westfold KC (1949) The wave equations for electromagnetic radiation in an ionized medium in a magnetic field. Australian Journal of Scientific Research (A) 2:169

Westfold KC (1950) The refractive index and classical radiative processes in an ionized gas. Phil Mag (Series 7) 41:509

Wild JP (1949) The radio-frequency line spectrum of atomic hydrogen. 2. The calculation of transition probabilities. CSIR Radiophysics Laboratory Report RPL 34

Wild JP (1950a) Observations of the spectrum of high-intensity solar radiation at metre wavelengths. II. outbursts. Aust J Sci Res (A) 3:399

Wild JP (1950b) Observations of the spectrum of high-intensity solar radiation at metre wavelengths. III. Isolated bursts. Aust J Sci Res (A) 3:541

Wild JP (1950c) The spectrum analysis of solar bursts at metre wavelengths. Journal of Geophysical Research 55:205 (In collection of papers Summary of Proceedings of Australian National Committee of Radio Science, URSI, Sydney, 16-20 January, 1950.)

Wild JP (1951) Observations of the spectrum of high-intensity solar radiation at metre wavelengths. IV. Enhanced radiation. Aust J Sci Res (A) 4:36

Wild JP (1952) The radio-frequency line spectrum of atomic hydrogen and its application in astronomy. Astrophys J 115:206

Wild JP (1953) Techniques for observation of radio-frequency radiation from the sun. In: Kuiper, Middlehurst (eds) The sun. University of Chicago Press, Chicago

Wild JP (1955) Observational radio astronomy. Advances in Electronics and Electron Physics 7:292

Wild JP (1968) The exploration of the sun by radio. Aust Physicist, August 1968:117 (Fourth Pawsey Memorial Lecture from 30 April 1968, Brisbane.)

Wild JP (1974) A new look at the sun. Highlights Astron 3:3

Wild JP (1985) The beginnings. In: McLean, Labrum (eds) Solar radiophysics. Cambridge University Press, Cambridge

Wild JP (1987) The beginnings of radio astronomy in Australia. Proc Astron Soc Aust 7:95

Wild JP, McCready LL (1950) Observations of the spectrum of high-intensity solar radiation at metre wavelengths. I. The apparatus and spectral types of solar burst observed. Aust J Sci Res (A) 3:387

Wild JP, Murray JD, Rowe WC (1953) Evidence of harmonics in the spectrum of a solar radio outburst. Nature 172:533

Wild JP, Murray JD, Rowe WC (1954) Harmonics in the spectra of solar radio disturbances. Aust J Phys 7:439

Wild JP, Roberts JA, Murray JD (1954) Radio Evidence of the ejection of very fast particles from the sun. Nature 173:532

Wild JP, Sheridan KV, Neylan AA (1959) An investigation of the speed of the solar disturbances responsible for type III RADIO bursts. Aust J Phys 12:369
Wild JP, Sheridan KV, Trent GH (1959) The transverse motions of the sources of solar radio bursts. In: Bracewell (ed) Paris symposium on radio astronomy: IAU/URSI Symposium. Stanford University Press, Stanford
Wilde S (1998) Unions in CSIRO: part of the equation. Hyland House, Melbourne
Woodlands School (1973) Woodlands 1923–1975. The first half century of Woodlands Glenelg Church of England Girls' Grammar School, Inc., South Australia. Rigby Limited, Adelaide
Woodlands School (1999) Woodlands reflections. Memories of Woodlands School aver 75 Years as recalled by some of old scholars. Alpha Visuals, Adelaide

Index

C

J

K

L

M

T

U

MIX
Papier aus verantwortungsvollen Quellen
Paper from responsible sources
FSC® C105338

If you have any concerns about our products,
you can contact us on
ProductSafety@springernature.com

In case Publisher is established outside the EU,
the EU authorized representative is:
Springer Nature Customer Service Center GmbH
Europaplatz 3, 69115 Heidelberg, Germany

Printed by Libri Plureos GmbH
in Hamburg, Germany